MW01633979

SOCIAL CARTOGRAPHY
MAPPING WAYS OF SEEING SOCIAL AND EDUCATIONAL CHANGE

EDITED BY
ROLLAND G. PAULSTON

GARLAND PUBLISHING, INC.
NEW YORK AND LONDON
1996

Library of Congress Cataloging-in-Publication Data

Social cartography : mapping ways of seeing social and educational change /
    edited by Rolland G. Paulston.
            p.   cm. — (Garland reference library of social science ; v. 1024.
    Reference books in international education ; v. 36)
        Includes bibliographical references and indexes.
        ISBN 0-8153-1994-0 (alk. paper)
        1. Social change—Maps.   2. Educational change—Maps.   3. Cartogra-
    phy.   4. Postmodernism.   I. Paulston, Rolland G.   II. Series: Garland
    reference library of social science ; v. 1024.   III. Series: Garland reference
    library of social science. Reference books in international education ; v. 36.
    HM101.S6913   1996
    303.4—dc20   96–22198
                                                                              CIP

Cover illustration: "Jila Japingka," a social, mythic and cultural map/text, painted
by Peter Skipper (Walmadjarri). From *Dreamings: The Art of Aboriginal Australia*,
edited by P. Sutton et al. New York: George Braziller, 1989, page 100. © The Asia
Foundation. Reprinted with permission.

Printed on acid-free, 250-year-life paper
Manufactured in the United States of America

Contents

Orbis Terrarum by Jeroen van Westen. Copyright 1994 by ORIENTARIUM.

ILLUSTRATIONS

Series Editor's Foreword

This series of scholarly works in comparative and international education has grown well beyond the initial conception of a collection of reference books. Although retaining its original purpose of providing a resource to scholars, students, and a variety of other professionals who need to understand the role played by education in various societies or world regions, it also strives to provide accurate, relevant, and up-to-date information on a wide variety of selected educational issues, problems, and experiments within an international context.

Contributors to this series are well-known scholars who have devoted their professional lives to the study of their specializations. Without exception these men and women possess an intimate understanding of the subject of their research and writing. Without exception they have studied their subject not only in dusty archives, but have lived and traveled widely in their quest for knowledge. In short, they are "experts" in the best sense of that often overused word.

In our increasingly interdependent world, it is now widely understood that it is a matter of military, economic, and environmental survival that we not only understand better what makes other societies tick, but that we make a serious effort to understand how others, be they Japanese, Hungarian, South African, or Chilean, attempt to solve the same kinds of educational problems that we face in North America. As the late George Z.F. Bereday wrote more than three decades ago: "[E]ducation is a mirror held against the face of a people. Nations may put on blustering shows of strength to conceal public weakness, erect grand façades to conceal shabby backyards, and profess peace while secretly arming for conquest, but how they take care of their children tells unerringly who they are" (*Comparative Methods in Education*, New York: Holt, Rinehart and Winston, 1964, p. 5).

Perhaps equally important, however, is the valuable perspective that studying another education system (or its problems) provides us in understanding our own system (or its problems). When we step beyond our own limited experience and our commonly held assumptions about schools and learning in order to look back at our system in contrast to another, we see it in a very different light. To learn, for example, how China or Belgium handles the education of a multilingual society; how the French provide for the funding of public education; or how the Japanese control access to their universities enables us to better understand that there are reasonable alternatives to our own familiar way of doing things. Not that we can *borrow* directly from other societies. Indeed, educational arrangements are inevitably a reflection of deeply embedded political, economic, and cultural factors that are unique to a particular society. But a conscious recognition that there are other ways of doing things can serve to open our minds and provoke our imaginations in ways that can result in new experiments or approaches that we may not have otherwise considered.

Since this series is intended to be a useful research tool, the editor and contributors welcome suggestions for future volumes, as well as ways in which this series can be improved.

Edward R. Beauchamp
University of Hawaii

PREFACE: FOUR PRINCIPLES FOR A NON-INNOCENT SOCIAL CARTOGRAPHY

Regimes of discourse and representation can be analyzed as places of encounter where identities are constructed . . . where violence is originated, symbolized, and managed. Charting regimes of representation . . . attempts to draw the 'cartographies' or maps of knowledge and power . . . and of struggle.
   Arturo Escobar, *Encountering Development: The Making and Unmaking of the Third World.*

We believe the major task . . . is to invent concepts that allow us to describe knowledge as an existing, but undiscovered, and hence unmapped realm. Conceptual inventions open new avenues for further exploration, allow new phenomena to be discovered, . . . and stimulate interesting conversations. Knowledge can be revealed and represented in so-called cognitive maps, using a cartography of logic. Ways of representing knowledge cartographically involve semantics and argument mapping.
   George von Krogh and Johan Roos, *Organizational Epistemology*

By complementing the findings of one partial perspective with the findings of other approaches, we can hope to create an overall picture [map] which will be fairly comprehensive and reasonably accurate, and will maintain a sense of proportion.
   Norman Davies, "The Misunderstood Victory in Europe"

Is social cartography—the art and science of mapping ways of seeing—a valuable new idea for comparative studies? While this is the thesis of our book, the general reader may be interested, but in need of illustration. My initial plan to introduce the argument was to show here how the diverse chapter texts that constitute this multidisciplinary work can be variously mapped as an intertextual field. Instead, these rather complex constructs have been placed in the concluding Envoi section, where the reader—after becoming better acquainted with our arguments—may wish to compare his or her mental map of the chapter texts with mine and engage in a bit of re-mapping, or counter-mapping.

So instead of maps I will first present a brief narrative—or personal mapping—of how I came to understand possibilities for a spatial turn in comparative education during what seemed to be a near-magic time of movement and possibility, of fragmentation and aporia:

> And he wrote in his field book after the date, after the hour of that sighting, not of why he had so far found nothing of what he believed he would find: he wrote, both deceiving and not deceiving himself: we are sailing off the map. (Kroetsch, 1975, p. 39)

BORDER CROSSINGS

I went to the University of British Columbia (UBC) in Vancouver as a visiting professor in the summer of 1991 with the hope that a trip to the "frontier" might provoke some new ideas about representing knowledge and visualizing difference. Given the collapse of the cold war with its totalizing stories, and the emergence of provocative new ways of seeing in poststructuralist, postmodern feminist and postcolonial studies, the time seemed alive with opportunities to rethink our world, to sail off our brutal old maps. UBC is situated in a setting of vast panoramas of sea, forest, city and sky. I had ample time to converse, to read and discover. "Phenomenography" (Marton, 1988), the postmodern geographers (Harley, 1988; Huggan, 1989; Soja, 1989), a related study by Bourdieu (1989), the French poststructuralists and some illuminating feminist cartographers (Anzaldúa, 1987; Mohanty, 1991; Rich, 1986)—all helped me to understand better possibilities to remap my mind and my field. I also reflected on the failure of my conference paper of the year before, "Comparing Ways of Knowing across Inquiry Communities," to specify exactly how contradictory ideas and views of reality might be represented and compared in a more open or "free-form" manner (Lawson, 1975).

On returning to the University of Pittsburgh that fall, I had begun to understand how a spatial turn in comparative studies would focus less on formal theory and competing truth claims and more on how contingent knowledge may be seen as embodied, locally constructed, and re-presented as oppositional yet complementary positionings in shifting fields. As Bateson (1979) points out, maps not only emphasize spatial relations, they also recognize and help to pattern difference. By naming and classifying, maps help us "know" something so we can "see" something different. The problem with getting comparativists to think more globally, for example, may be "that this task is difficult to map because there is nothing but difference. What a confused global thinker needs is patterns interspersed among the differences" (Weick, 1990, p. 2). This view would help me both to reconceptualize comparative studies as comparative mapping and to see it as situated, provisional and contested, that is, as a non-innocent practice. With the opening up of our maps to multiple perspectives, we might also better move beyond the two great modernisms of positivism and Marxism with their rigid categorical thinking and abhorrence of the Other. My efforts then turned to the crafting of a ground-level social cartography project with critical potential, that would build upon and extend earlier postmodern mapping contributions in geography and also in feminist, literary and postcolonialist studies. Work in this new genre uses spatial tropes to map discursive fields. It favors the rejection of essentialism and scientism found in most feminist theory. It views the "ground" of our era as akin to a space of shifting sites and boundaries definable only in relational terms (Mohanty, 1991). Where texts of modern geographers usually represent space as an innocent place of situated objects with fixed boundaries, coordinates and essences, texts of the postmodern cartographers mostly present an agonistic or contested space of continually shifting sites and boundaries perhaps best portrayed using "the transitory, temporal process of language," (Kirby, 1996, p. 21). Soja and Hooper (1993) explain this fascination with spatial analysis: "We suggest that this spatialized discourse on simultaneously real and imagined geographies is an important part of a provocative and distinctly postmodern reconceptualization of spatiality that connects the social production of space to the cultural politics of difference in new and imaginative ways" (p. 184).

At about this time, Don Adams invited me to write an encyclopedia entry (Paulston, 1994) titled "Comparative Education: Paradigms and Theories." I accepted, but with the proviso that the entry would in fact be post-paradigmatic; that is, it would use a perspectivist approach to "map" my view of increasingly complex conceptual relationships between the major discourse communities that compose the field. I presented this study, viewing comparison as a juxtaposition of difference, in July 1992 at the VIII World Congress of Comparative Education Societies at Charles University in Prague with a title more to my liking: "Comparative Education Seen as an Intellectual Field: Mapping the Theoretical Landscape." The paper sought to demonstrate how comparative education "after objectivity" can now make good sense "in perspective" (Moser, 1993, p. 227) by portraying a ludic play of different theoretical perspectives within the art form of social cartography. This cartography avoids the rigidities of modernist social models and master narratives, and shifts the research focus to current efforts by individuals and cultural groups seeking to be more self-defining in their sociospatial relations and in how they are represented. In this regard, Liebman (1994) has argued persuasively that while social mapping is open to all texts, it is a project of and for the postmodern era; it is a new method to identify changing perceptions of values, ideologies and spatial relations. In social cartography he sees an alliance of education and geography to develop a methodology consistent with the visualization of narratives in a time when people now realize their potential and place in the world quite differently than they did a few decades ago. In education, especially, he suggests that social mapping can assist students who desire to resolve personal questions of self in a world offering a multiplicity of truths and values. As in this volume, social maps are proposed as "a method of illustrating our vigorous social milieu composed of a profusion of narratives" (p. 236). This is done with an emphasis on layered fields of perception and intertextual space, an approach which draws in part upon the technique of chorography, that is, the mapping of domains or regions (Helgerson, 1986).

By the following summer members of our mapping venture had produced and made available three key studies addressing social mapping rationales, perspectives and methodology. In April 1994, Martin Liebman successfully defended his doctoral thesis (Liebman, 1994), the first book-length study of social mapping as a new method and resource for education and the human sciences. At the Comparative and International Education Society annual meeting held in Boston, during March 1995, seven authors of chapters in this work, E.E. Gottlieb, M. Liebman, C. Mausolff, T.W. Mouat, V.D. Rust, J.R. Seppi, and C.A. Torres, presented preliminary versions of their studies at a featured double session carrying the not-so-subtle title, "Social Cartography: Comparative Education as Mapping Ways of Seeing?"

LAYOUT

Now, with the project of social cartography or free-form mapping well underway, it is fitting perhaps to recognize Joseph Seppi's admonition in his chapter that "an attempt at formalizing the technique must follow." The nineteen multidisciplinary chapters that this book comprises all, in various ways and from diverse perspectives, address this need to sketch in some "first principles" for a social cartography oriented toward charting the variable topography of social space and spatial practices. In the opening section, Mapping Imagination, creative ideas from cultural geography, social history and comparative education, among others, are used to suggest how the human sciences might benefit from the use of a perspectivist approach. This section examines challenges facing all knowledge fields today as postmodernist sensitivity, with its rejection of universals and attention to difference, permeates the academy, the media and individual consciousness. The four chapters in this section use both modernist and postmodernist perspectives to query how mapping imagination can help comparativists to better identify and compare difference.

Imagination can also be seen to work through spatial representation at the individual level. Said, for example, suggests that space may acquire emotional and even rational sense through a poetic process where empty reaches of space and distance are converted into meaning in the here and now: "There is no doubt that imaginative geography or history help the mind to intensify its own sense of itself by dramatizing the distance and difference between what is close . . . and what is far away" (1978, p. 55). The concept of spatial imagination seen as an ability to reveal multiple intersections (Hayhoe, 1996; Lenski, 1994), to resist disciplinary enclosures (Price-Chalita, 1994) and to cross borders and come into critical dialogue with other imaginations (Gregory, 1994; Epstein, 1995) is a guiding principle of our project, our book, and the perspectivist-style "webbed" map found in Figure 2 in the concluding Envoi.

The second section, Mapping Perspectives, demonstrates how ways of seeing portray relationships—in this case from the viewpoints of the positivist, humanist, cognitive and literary traditions. These four chapters examine how the application of spatial ideas and techniques has elaborated mapping in specialized areas such as scientific geographical information systems (GIS) and land use planning, humanistic and environmental studies, management and business studies, and comparative literature, where maps are increasingly seen as rhetorical strategies that variously facilitate processes of learning and unlearning, of resistance and transformation or, perhaps, serve as agendas for coercion and containment (Huggan, 1994). The principle illustrated in this section is that disciplinary theory and practice continually interact in a process of mutual referral. Theory is not detached from the realities of everyday life. It is a construct with semantic commitments and spatial origins and, as Liggett and Perry (1995, p. 12) contend, "it is the responsibility of analysis [and mapping] to return it there."

Mapping Pragmatics, the third section, provides our invitation to social cartography with case study reports of mapping in practice and mapping as practice—i.e., studies that facilitate a spatial understanding of power relations and transitions. Here, contributors variously map ways of seeing the organizational space of third world educational interventions, a textual utopia-building effort, local perceptions of a rural development project, the expanding representational space of international corporations, intercultural communication problems in educational consultancies, the intertextual field of environmental education, and innovative social mapping techniques. While these reports on mapping practice evidence something of the indeterminate and incomplete aspects of provisional cartographic representation, they also suggest how maps can open space for present difference, represent conflicting visions of the future (as with Escobar's "maps . . . of struggle"), and enhance our ability "to ironize our own claims to truth" vis-à-vis competing claims (Foreman, 1995, p. 566).

In Mapping Debates, the authors use critical perspectives to engage and question a good deal of what is argued in the preceding three sections. Here we find the project's critical reflexive principle that interrogates all knowledge, and especially my contention that a ludic mapping practice can help to subvert mapping's colonizing role under modernity and open a site of resistance in postmodernity, all the while (as in this book) seeking to undermine its own authority as a new discourse of power.

Site-ings

These chapters strongly suggest that comparative education, as with the related fields of comparative literature, comparative politics and the like, now share a common interdisciplinary pursuit of cultural theory and situated knowledge-generation processes, as well as the more traditional cross-cultural comparison of national practices. Huggan (1994) argues that this new agenda moves alterity, or awareness of the Other, to the center of comparative studies:

> Comparativists are not syncretists: that they choose to outline similarities among works deriving from different cultures or disciplines, or written in different languages does not imply the erasure or compromise of their differences. . . . Comparativists are best seen as mediators moving among texts without seeking to "reconcile" or "unify" them. What is needed . . . is a flexible cross-cultural model [i.e., a map] that allows the nature of each country's [or actor's] schizoid vision of itself to be redefined as a source of creative power. (p. xi)

From this postmodern view, objectivity is no longer about unproblematic objects, but about always partial translations and how to portray and compare imbricated local knowledges (Smith and Katz, 1993).

Because social cartography allows the comparison of multiple views and contested codes in heuristic constructs, it will also have potential to serve as a metaphorical device for the provisional representation and iconographic unification of warring cultures and disputatious communities. Every social map is the product of its makers and open to continuous revision and interrogation. In the process of mapping meaning, the subject is seen to be mobile and constituted in the shifting space where multiple and competing discourses intersect. This view advances neither the self-sufficient Cartesian subject of modern Western humanism nor the radically de-centered Baudrillardian subject seen by extreme poststructuralism. Instead, the mapper is articulated around a core self that, as Flax (1990) argues, is nonetheless differentiated locally and historically. Social mapping, in this view, makes possible a way of understanding how sliding identities are created, and how the multiple connections between spatiality and subjectivity are seen to be grounded in the contested terrain between discourse communities (Kirby, 1996).

Feminist writers especially have used social cartography and spatial metaphors, in this manner, to expose and challenge what they see as patriarchal representations and to chart new social relations grounded in feminist knowledge and experience (Ardener, 1981). Kolodny (1975), for example, explains the strategic role of spatial metaphors in the engineering of social change in American history. The land-as-woman metaphor was central, while the map served both as a metaphor of male control and domestication of the continent (i.e., the virgin land) and for the continuing domesticity of women. Feminist metaphors and use of an empowering spatial language invert and counter this story. Feminist cartographers—and especially those using postcolonial perspectives—have effectively subverted the complicity of maps in attempts to maintain what they see as an oppressive status quo (Price-Chalita, 1994), and have much to offer our social cartography project. Vandana Shiva for example, contends that ways of thinking and seeing are not biologically determined but rather are culturally patterned, as in her example of the "masculinization of science" by the European "fathers" of modern science. She notes that Francis Bacon, for example, saw the birth of modern science as "the masculine birth of time." Bacon exhorted "the true sons of knowledge" to find a way into nature's "inner chambers," to turn their "united forces against the nature of things . . . to storm and occupy her castles and strongholds" (1995, p. 15). Donna Haraway (1992) has elaborated a useful counterepistemology informed by the feminist cultural community. She provides an agenda similar to ours in her advocacy of a shift from map-as-taxonomy to map-as-guide and record of movement, not stasis: "There is no safe place here, there are, however, many maps of possibility" (p. 326).

Ethnic, ecological and regional groups have also been active in creating alternative maps that disrupt or reject the truth claims of central authority (Aberley, 1993). Such "resistance" maps—both on the left and the right—seek to avoid capture in established power grids, to create counter mapping that presents alternative world views, to open new rhetorical spaces and to articulate postcolonial ambitions (Mohanty, 1991).

It would seem that the time is propitious for comparative educators to consider how a cartography of relations might help us move beyond our present Cartesian anxiety to a more open play of perspectives (V. Masemann, 1990). I believe that social cartography, with its deconstructive view of all modes of representation—including its own—and with its ludic tolerance of new ideas and diverse ways of seeing, can help us make this intellectual journey. In addition to its critical and demystification utility to make visible ideas and relations that otherwise might remain hidden, social cartography will also be useful to convert increasing volumes of data into usable information (D.R.F. Taylor, 1994). This will help comparativists recognize patterns and relationships in spatial contexts from the local to the global. In conceptual terms, cartographic visualization can also provide a common space for what were once viewed as incommensurable epistemological paradigms or perspectives (E.H. Epstein, 1988). But as counseled by Nelson Goodman, a readiness to recognize alternative worlds may be liberating, yet it builds none. Mere acknowledgment of many frames of reference charts no map. Awareness of varied ways of seeing paints no pictures, and "a broad mind is no substitute for hard work" (1978, p. 21).

I hope our efforts in this volume will better enable the reader to see utility in the practice of social mapping as it opens traditional cartographic representation to multiple perspectives and the play of difference. While mapping does not resolve the conflict of interpretations and sense of disorientation that would seem to be the defining characteristics of our era, this study in large measure argues—and seeks to illustrate—that social mapping will nevertheless be useful to construct, as Davies advocates, more "comprehensive and reasonably accurate" re-presentations of social and cultural phenomena. With the new conceptual tools of social cartography, comparative educators and other knowledge workers will be better able to visualize and pattern all the social "scapes" that are seen to constitute our challenging new world.

RGP

REFERENCES

Aberley, D. (1993) *Boundaries of home: Mapping for local empowerment*. Gabriola Island BC: New Society.

Anzaldúa, G. (1987). *Borderlands/La frontera: The new mestiza*. San Francisco: Spinsters/Aunt Lute.

Ardener, S. (ed.) (1981). *Women and space: Ground rules and social maps*. London: Croom Helm.

Bourdieu, P. (1989). Social and symbolic power. *Sociological Theory*, 7, 14–25.

Davies, N. (1995). The misunderstood victory in Europe. *The New York Review of Books*, 43 (9), pp. 11–14.

Epstein, E.H. (1988). The problematic meaning of "comparison" in comparative education. In J. Schriewer and B. Holmes (eds.), *Theory and methods in comparative education* (3–23). Frankfurt am Main: Peter Lang.

Epstein, I. (1995). Comparative education in North America: The search for other through the escape from self? *Compare* 25(1), 5–16.

Escobar, A. (1995). *Encountering development: The making and unmaking of the Third World*. Princeton, NJ: Princeton University Press.

Flax, J. (1990). *Thinking fragments: Psychoanalysis, feminism and postmodernism in the contemporary West*. Berkeley: University of California Press.

Foreman, P. (1995). Truth of objectivity, part 1: Irony. *Science*, 269, 565–567.

Goodman, N. (1978). *Ways of worldmaking*. Indianapolis: Hackett.

Gregory, D. (1994). *Geographical imaginations*. Oxford: Blackwell.

Haraway, D. (1992). The promises of monsters: A regenerative politics for inappropriated others. In L. Grossberg et al. (eds.), *Cultural studies*. New York: Routledge.

Harley, J.B. (1988). Maps, knowledge and power. In D. Cosgrove and S. Daniels (eds.), *The iconography of landscape* (277–312). Cambridge: Cambridge University Press.

Hayhoe, R. (1996). *China's universities, 1895–1995: A century of cultural conflict*. New York: Garland.

Helgerson, R. (1986). The land appears: Cartography, chorography and subversion in Renaissance England. *Representations*, 15, 51–85.

Huggan, G. (1989). Decolonizing the map: Post-colonialism, post-structuralism and the cartographic connection. *Ariel*, 20(4), 115–131.

Huggan, G. (1994). *Disputed terrain: Maps and mapping strategies in contemporary Canadian and Australian fiction*. Toronto: University of Toronto Press.

Kirby, K.M. (1996). *Indifferent boundaries: Spatial concepts of human subjectivity*. New York: Guilford.

Kolodny, A. (1975). *The lay of the land: Metaphor as experience in the history of American life and letters*. Chapel Hill: University of North Carolina Press.

Krogh, G., von and J. Roos (1995). *Organizational Epistemology*. New York: St. Martins Press.

Kroetsch, R. (1975). *Badlands*. Toronto: New Press.

Lawson, R. (1977). Free-form comparative education. *Comparative Education Review*, 19, 345–353.

Lenski, G. (1994). Societal taxonomies: Mapping the social universe. *Annual Review of Sociology*, 20, 1–26.

Liebman, M. (1994). *The social mapping rationale: A method and resource to acknowledge postmodern narrative expression*. Ph.D. dissertation, University of Pittsburgh.

Liebman, M., and R. Paulston (1993). *Social cartography: A new methodology for comparative studies*. Research Report #3, Occasional Paper Series. Pittsburgh: University of Pittsburgh Department of Administrative and Policy Studies.

Liggett, H., and D.C. Perry, eds. (1995). *Spatial practices: Critical explorations in so-*

cial/spatial theory. Thousand Oaks, CA: Sage.

Marton, F. (1988). Phenomenography: Exploring different conceptions of reality. In D.M. Fetterman (ed.), *Qualitative approaches to evaluation in education* (176–205). New York: Praeger.

Masemann, V.L. (1990). Ways of knowing: Implications for comparative education. *Comparative Education Review, 34*(4), 465–473.

Mohanty, C.T. (1991). Cartographies of struggle. In C.T. Mohanty et al. (eds.), *Third world women and the politics of feminism* (1–49). Bloomington: Indiana University Press.

Moser, P.K. (1993). *Philosophy after objectivity: Making sense in perspective*. Oxford: Oxford University Press.

Paulston, R. (1993). *Mapping knowledge perspectives in studies of social and educational change*. Research Report no. 1, Occasional Paper Series. Pittsburgh: University of Pittsburgh Department of Administrative and Policy Studies.

Paulston, R. (1994). Comparative education: Paradigms and theories. In T. Husén and N. Postlethwaite (eds.), *International Encyclopedia of Education* (923–933). Oxford: Pergamon.

Paulston, R., and M. Liebman (1993). *The promise of a critical postmodern cartography*. Research Report no. 2, Occasional Paper Series. Pittsburgh: University of Pittsburgh Department of Administrative and Policy Studies.

Price-Chalita, P. (1994). Spatial metaphor and the politics of empowerment: Mapping a place for feminism and postmodernism in geography? *Antipode, 26*(3), 236–254.

Rich, A. (1986). Notes toward a politics of location. In *Blood, bread and poetry: Selected prose, 1979–1985*. New York: Norton.

Said, E. (1978). *Orientalism*. London: Routledge.

Shiva, V. (1995). Monocultures, monopolies and the masculinization of knowledge. *IDRC Reports, 23*(2), 15–17.

Smith N. and C. Katz (1993). Grounding metaphor: Towards a spatialized politics. In M. Keith (ed.), *Place and the politics of identity*. New York: Routledge.

Soja, E. (1989). *Postmodern geographies: The reassertion of space in critical social theory*. London: Verso.

Soja, E. and B. Hooper (1993). The spaces that difference makes: Some notes on the geographical margins of the New Cultural Polities. In M. Keith and S. Pile (eds.), *Place and the politics of identity*. New York: Routledge.

Taylor, D.R.F. (1994). Perspectives on visualization and modern cartography (pp. 333–341). In A.M. MacEachren & D. R.F. Taylor (eds.), *Visualization in modern cartography* (333–341). Oxford: Pergamon.

Weick, K.E. (1990). Cartographic myths in organizations. In A.S. Huff (ed.), *Mapping strategic thought* (1–10). Chichester, UK: John Wiley and Sons.

ACKNOWLEDGMENTS

In appreciation and acknowledgment of the hard work that has gone into this social cartography project over the past several years, I would like to thank the following:

• The chapter authors who wrote with dispatch and great craft. Your essays beckon all to a promising new opening in comparative studies. Your creative intelligence and cheerful willingness to rework chapter drafts has helped to produce a more rigorous and integrated "rough guide" to social cartography.

• The staff and administration at the University of Pittsburgh who provided outstanding support services, especially Yvonne Jones and Mary Jane Alm.

• To the students of my social theory seminars who challenged, elaborated and counterargued the conceptual schemes that emerged over the years. You know who you are. I thank you one and all.

• To colleagues who shared their views—and alarms—and assisted in that special collegial way that helps to make it all worthwhile. These include Don Adams, Nicholas Beattie, Diana Brandi, Roger Boshier, Robert Chambers, Michel Dear, Erwin Epstein, Irving Epstein, Mark Ginsburg, Noreen Garman, Ruth Hayhoe, Sean Hughes, Torsten Husén, Dell Hymes, Robert Lawson, Ference Marton, Maureen McClure, Noel McGinn, Zbyszko Melosik, Gilbert Merkx, David Plank, David Post, Eugenie Potter, Fritz Ringer, Roland Robertson, Jurgen Schriewer, John Singleton, D.R.F. Taylor, John Weidman, John Yeager and many more.

• Christina Bratt Paulston for her patience and support.

• The editorial and production teams at Garland Publishing, including Marie Ellen Larcada, Adrienne Makowski, Susan Papa, Helga McCue, Sylvia Ploss, Jason Goldfarb, and Edward Beauchamp.

I also acknowledge permission granted to adapt the following previously published work and thank the publishers for their cooperation:

A. Buttimer, Geography, humanism and global concern. *Annals of the American Association of Geographer, 80* (1), pp. 1–33; Blackwell, 1990.

A. Huff, Mapping strategic thought. In A. Huff (ed.), *Mapping strategic thought*, pp. 11–52; John Wiley and Sons Ltd., 1990.

R. Paulston and M. Liebman, An invitation to postmodern social cartography. *Comparative Education Review, 38*(2), pp. 215–232; The University of Chicago Press, 1994.

Special Rider Music: *My Back Pages* by Bob Dylan. Copyright © 1964 by Warner Bros. Music. Copyright renewed 1992 by Special Rider Music. All rights reserved. International copyright secured. Reprinted by permission.

RGP

I Mapping Imagination

Painted by Peter Skipper (Walmadjarri), Jila Japingka is a social, mythic and cultural map/ text. See P. Sutton et al., eds. (1989), *Dreamings: The art of aboriginal Australia* (p. 100). New York: George Braziller © The Asia Foundation

"Aboriginal texts, written, and unwritten, deal directly with the fundamental issues facing Aboriginal people, torn as they are between alienation and a sense of belonging. The strategy they use is . . . constructing maps that are designed to represent broad stretches of space and time, to give meaning and perspective, direction and hope." From B. Hodge and V. Mishra (1991), *The dark side of the dream: Australian literature and the postcolonial mind* (p. 117). Sydney: Allen and Unwin.

INTRODUCTION

In their chapter, "Social Cartography: A New Metaphor/Tool for Comparative Studies," Rolland G. Paulston and Martin Liebman ask how comparative researchers might enhance the presentation of their findings, particularly when their findings focus on the postmodern diffusion of heterogeneous orientations. The authors are concerned with developing in comparative discourse a visual dialogue as a way of communicating how we see the social changes developing in the world around us. Visual images—depicting on the two-dimensional surface of paper or screen the researcher's perceived application, allocation or appropriation of social space by social groups at a given time and in a given place—offer such an opportunity. Applied to comparative education, social maps may help to present and decode immediate and practical answers to the perceived locations and relationships of persons, objects and perceptions in the social milieu. The interpretation and comprehension of both theoretical constructs and social events then can be facilitated and enhanced by mapped images.

Paulston and Liebman's chapter presents the concern of three academic practitioners—one in comparative education and two in geographic cartography—who have called on colleagues in these areas to open their respective academic fields to postmodernist ideas, to become more reflexive, comparative and aware of heterogeneous orientations in their academic discourse. Postmodernism is not promoted here, but rather the possibilities for comparative fields to expand their knowledge bases through an appropriate, thoughtful and skillful development and application of social maps. The postmodern turn opens the way to social mapping exercises.

Accordingly, the authors propose, first, that the structures of multiple education knowledge systems can be recreated in one or more maps, in a social cartography where the space of the social map reflects the effect of social changes in real space; and second, that comparative education researchers consider representing that space through the creation of maps. Their rationale for this proposal is that the map provides the comparative educator a better understanding of the social milieu and gives all persons the opportunity to enter a dialogue to show where they believe they are in society. The map reveals both acknowledged and perceived social inclusions while leaving space for further inclusions of social groups and ideas. Whether the map is considered a metaphorical curiosity or accepted as a more literal representation, it offers comparative researchers an opportunity to situate the world of ideas in a postmodern panorama, disallowing the promotion of an orthodoxy.

In his chapter "From Modern to Postmodern Ways of Seeing Social and Educational Change," Val D. Rust explores how social cartography represents a novel way of seeing the intellectual landscape of comparative and international education. Visual mapping of the diversity of the field not only allows for the beginning of a more open social dialogue that includes those who have been marginalized and excluded from social discourse, but it represents a different way of seeing. Our concepts are typically formulated and transmitted, at least in North American academia, in a language that is linear, scientific and logical. It is difficult to provide meaningful concepts that may not be compatible with such language unless we turn to new means of conceptual representation, including visual representation, that allow for multidimensionality, nonlinearity and turbulence to be more fully represented.

Rust notes that the process of social cartography, at least as initiated by Paulston, is based on a particular point of view that is reflected in his mapping process. He locates Paulston's texts on social cartography on the map itself. In addition, he suggests a limited number of theoretical orientations, which resonate with this orientation, but which have not yet become part of comparative and international education discourse. These include chaos theory, transformation theory, and self-organization theory all of which are part of what Rust considers to reflect postmodern ways of seeing.

Rust's opinion is that Paulston has made a distinct contribution to the field with his initial attempts to map the intellectual terrain of comparative and international education. Rust's contribution to this volume includes some provision for possible difference in interpretation as well as attempts to further elaborate Paulston's mapping activity. Paulston has provided a major two-dimensional framework for discourse mapping, and Rust wishes to engage in some elaboration of that process by creating a third dimension, related to the relative dominance of the various theoretical orientations in the field.

In his chapter, "Constructing Knowledge Spaces and Locating Sites of Resistance in the Modern Cartographic Transformation," David Turnbull observes that maps are everywhere in contemporary Western life. The cartographic trope is all pervasive. We talk of cognitive maps, computer maps, mental maps, genetic maps, and of mapping the intertextual, the cosmos, the social, the mind, the human genome. Minds, languages and cultures are described as maps. The anthropologist Edmund Leach suggests that "our whole social environment is map-like" and the sociologist Pierre Bourdieu notes that "culture is often referred to as a map." Postmodernists constantly resort to the cartographic figure in their explorations of the contemporary social landscape. While the mapping mode of Northern technoscience may currently be dominant, there are many ways of mapping. Different cultures, different periods and different groups in a culture have differing ways of assembling and representing knowledge.

Right now the mapping mode dominant for several centuries is undergoing what might be called a cartographic transformation. It is likely that the maps of the twenty-first century will look very different as the development of an optimal-conformal projection and fractal computer analysis will enable the incorporation of previously unutilized satellite data in topographical maps. The development of geographical information systems (GIS) is set to provide a radically new technology for obtaining "knowledge and reality." Over the past twenty years, there has also been vastly increased range of spatially portrayed domains. Scientists are currently mapping everything from the cosmos to the atom, from the brain to society, from concepts to theories. At the same time, there are moves to expand our linear, individual and causal style explanations to include spatial, relational and metaphorical styles. It is even claimed that it may become possible to map any or all knowledge as a "fuzzy cognitive map." This transition to spatial explanation has occurred not only in the physical sciences, but in the social sciences and humanities as well.

In his chapter, Turnbull contrasts longstanding modern/mimetic mapping practice with emergent postmodern social cartography that sees mapping as the creation of heterogeneous assemblages offering spaces for new knowledge, relations and antihegemonic construction.

Thomas W. Mouat's paper, "The Timely Emergence of Social Cartography," demonstrates that there is a threefold reason for the emergence of social cartography during the postmodern era, an era that began with a dawning awareness that "reality" is composed of disconnected fragments. Mouat argues that as early postmoderns sought reconnection, they discovered that the concrete presentation of interrelationships between and among fragments often eludes expression. As the struggle to discover and express interrelationships intensifies, it becomes apparent that the abstract representation of interrelationships is often possible when their concrete representation is not. Therefore, he contends, social cartography as mapping abstractions arises initially as a vehicle through which to express in highly condensed, abstract form, the interrelations between and among elements of systems that are not amenable to concrete description. However, once invented, social cartography also allows the mapping of abstraction upon abstraction. In this view, social cartography arises to meet the need for reintegration of the fragmented universe bequeathed by modernity. As it happens, full reintegration is impossible: Not only has the universe fragmented, the conceptual framework for knowledge has disintegrated. A new framework of knowledge is required and social cartography's techniques for mapping abstractions will eventually produce such a neomodern synthesis that will occur at the level of mapping abstractions. Mouat uses a framework from Piaget to create abstract maps of the cycle of individual thought and the pattern of cognitive development. These abstractions are overlain upon a map of Western social history. Through this process, the cyclic pattern of individual cognitive development and the cyclic pattern of social development are interrelated; education, the link between individual and social development is investigated; and the necessary purpose of social cartography is explicated.

Social Cartography

A New Metaphor/Tool for Comparative Studies[1]

Rolland G. Paulston and Martin Liebman

I view maps as a kind of language . . . as reciprocal value-laden images used to mediate different views of the world.

J.B. Harley, "Maps, Knowledge, and Power"

The ludic practices of postmodernism [i.e. mapping] should not be dismissed as mere frivolity since they function as a means of challenging the power of representation and totalizing discourses (discourses that present themselves as the final 'Truth' which explains everything) without falling into another and equally oppressive power discourse.

Robin Usher and Richard Edwards, *Postmodernism and Education*

This chapter demonstrates how social cartography—the writing and reading of maps addressing questions of location in the social milieu—may enhance social research in its struggles to distance itself from the restraints of modernism. Social cartography as a space of juxtapositions suggests an opening of dialogue among diverse social players, including those individuals and cultural clusters who want their "mininarratives" included in the social discourse. We propose that the social cartography discourse style has the potential to demonstrate the attributes, capacities, development and perceptions of people and cultures operating within the social milieu. It offers comparative educators a new method for visually demonstrating the sensitivity of postmodern influences in order to open social dialogue, especially to those who have experienced disenfranchisement by modernism and the totalizing metanarratives of progress.

Introduction

How might comparative researchers enhance the presentation of their findings, particularly when their findings focus on the postmodern diffusion of heterogeneous orientations? We are concerned with unfolding in our comparative discourse a visual dialogue as a way of communicating how we see the social changes developing in the world around us. Visual images, depicting on the two-dimensional surface of paper or screen the researcher's perceived application, allocation, or appropriation of social space by social groups at a given time and in a given place, offer such an opportunity. Mapping social geography is similar to both cognitive mapping and geographic cartography. Social cartography is created through "a process composed of a series of psychological transformations by which an individual acquires, codes, stores, recalls, and decodes information about the relative locations and attributes of phenomena in . . . [the] everyday geographical environment" (Downs and Stea, 1973, p. 9). This process consists of "aggregate information . . . acquisition, amalgamation, and storage" (Downs and Stea, 9), producing a product depicting space peculiar to a point in time. Applied to comparative education, social maps as a distinct mode of visual representation may help to present and decode immediate and practical answers to the perceived locations and relationships of persons, objects and perceptions in the social milieu. The interpretation and comprehension of both theoretical constructs and social events then can be facilitated and enhanced by mapped images.

It was Peter Hackett (1988) who concluded "that without metaphor, allegory and a thick description of the world around us there is no basis for comparative study or analysis" (p. 389). We concur in this observation, for it counsels anyone in comparative studies to be aware of the possibilities of exclusions. In what we as researchers seek, as well as what we report, should be found the "basic source of unity in our experience" (p. 391). We believe scholarly descriptions of society can be enhanced in this postmodern era of emerging mininarratives by the inclusion of a social cartography as a secondary discourse style, a style that by its nature is both metaphor and allegory, and also thick with descriptive characteristics.

Contemporary change advocacies have the attention of many who also express their ideas about how change should proceed. What rationales for new discourse methods have comparative researchers recently introduced? How does the map when seen as a frame of spatial reference fulfill the needs addressed by these rationales? How does the social map allow new ideas and knowledge communities to emerge and enter the field of discourse and dialogical community?

Invitations to a Postmodern Reflection
Presented in this essay are the concerns of three academic practitioners, one in comparative education and two in geographic cartography, who have

called on colleagues in these areas to move their respective academic fields toward a postmodernist integration, to become more explicative, comparative and open to heterogeneous orientations in their academic discourse. Postmodernism is not promoted here, but rather the possibilities for comparative fields to expand their knowledge bases through an appropriate, thoughtful and skillful development and application of social maps. The postmodern turn opens the way to social mapping exercises.

Offering the argument that postmodernism "should be a central concept in our comparative education discourse," Val Rust's 1991 presidential address to the Comparative and International Education Society, and his chapter in this volume, call for the application of postmodernist theories to strengthen emerging representations of reality (Rust, 1991, p. 610). Rust notes that Foucault believes there is a need to move beyond determinism and universals while Lyotard discerns in the postmodern a distrust of modernist metanarratives. Noted also by Rust is Richard Rorty's observation that metanarratives are "the theoretical crust of convention that we all carry and tend to universalize" (p. 616). Postmodernism calls for deconstructing those universal metanarratives of social valuation common to the modernist era, metanarratives seen as totalizing, standardizing and predominating.

Rust finds postmodern discussions and criticisms address the history of modernist society and culture as it was ingrained and justified by a worldview obsessed with focusing on time and history. These two measures of the modernist world were not always separate cognitive structures, but links holding each at least parallel to the other, if not often viewed as the same entity. Rust contends, moreover, that postmodernism's liberating influences transcend not only combined time and history, but combined space and geography as well. Space becomes more important than time in our postmodern mapping discourse.

Rust entreats comparative educators to relocate into this space, to extract from modernity the metanarratives to be dismantled, metanarratives containing the multiple small narratives previously hidden in the invisible space of modernist society. The small narratives that Rust suggests we draw our attention to can be the focus of comparative mapping efforts in postmodern social science.

Social cartography might also be seen to advance Heidegger's (1971) argument that "truth" is best understood not as correspondence or correctness of assertion or representation, but as the absence of concealment, i.e., what the Greeks called *aletheia*. When literary space is revealed in visual space, the map as a cultural text becomes a kind of language, the mode or *dichtung* (literally, a saying) in which what we see as truth happens.[2]

Dichtung is prior to the technical instrumental understanding of language. Like Cartesian metaphysics in general, fields such as linguistics, cybernetics, and artificial intelligence are seen by Heidegger to be impossible without the more primordial, prereflective realm with which *dichtung* proper is associated. This language realm inaugurates a "world" and gives to things their appearance and significance. It is perhaps best uncovered in poetry using literary theory. Here the essence of language is not propositional form, but openness to a resonance or nexus of relations out of which the "real" and the "human" may emerge.[3]

Suggesting, as does Rust, that the search for the silent blueprint to life means looking in areas of darkness, a searching for new growth in an old-growth forest, Star focuses our attention on these small and previously hidden narratives, on making the invisible visible.[4] Her five rules help us track omissions and understand the mechanisms of power tied to the deletion of certain kinds of practical and intellectual work. They also provide a powerful rationale for opening up mapping opportunities to all participants and communities in an intellectual field. Ringer (1990) advocates the study of intellectual fields as textual clusters composing a whole, "as a set of relationships, rather than as a sum of individual statements" (p. 275). He has however, yet to undertake the mapping of such spatial relationships, or intertextual fields.

We consider it possible for the comparative studies of social narratives to develop similarly to those of the studies of land formations and the cartographic mapping of land mass changes. As social cartographers, we look for the small and large erosions and eruptions of the social masses for the opportunity to map changes, to analyze and interpret events. We take events and make them consumable commodities for our readers by filtering, fragmenting and reelaborating them "by a whole series of industrial procedures . . . into a finished product, into the material of finished and combined signs" (Baudrillard, 1990, p. 22).

Mapping social space is an effective method for addressing Rust's thoughtful arguments calling for a postmodernist application to strengthen emerging representations of reality. There is, however, much we must learn and understand to become effective social mappers. This requires an association with an academic field experienced in representing geographic space on a map. For this reason we introduce in our invitation to a reflective social cartography two cartographers who have observed in their field several of the same concerns and needs addressed by Rust.

A leading advocate of the postmodern enterprise in geography and its practice of cartography, J.B. Harley, suggests that cartographers both in academia and in the field might consider postmodernity's potential for revitalizing their cartographic efforts. Harley (1992) contends that the premise of cartography has long been foundational, that map makers were compelled to create knowledge limited by scientific or objective standards. Earlier than Harley, however, Robert McNee (1981, p. 12) observed that the tenacity of the cartographic process and its practitioners in the retention of positivist traditions could be attributed to their attraction to both the label and the role-playing associated with being objective scientists. However, McNee and Harley differ in their explanations for the reasons cartography remained steadfastly grounded in positivism.

McNee argues that during the long history of cartography, this tenacious holding to the positivist ideal of the objective scientist resulted in the continued essentialist construction of textual metanarratives, both in the maps and in the semiotic and iconic representations used by the mapper. Harley, however, considers a more potent influence, arguing that after the last three decades, when much of academia moved toward or into the postmodern enterprise, cartographers adhered to a modernist style of application of knowledge, not only out of a concern for their reputations as objective scientists, but because of the influence modernist power structures had on the creation of maps. Harley states the field might better be served now if the power structures gave way to the new ideas postmodernism makes applicable for a social cartography, a cartography permitting the interpretation of the map as well as opening the map to the intent and need of those who use it and those who assume the responsibility for its creation.

Harley makes an important distinction between the external and internal powers regulating the creation and reading of maps, or, by extension, any texts. External power, emanating from patrons, monarchs and elite institutions, controlled what went into the map. Internal power was "embedded in the map text," determined by the inclusions and exclusions of information written into the map at the will of the external power. Internal power limited all map readers to only the knowledge included by the external power.

Figure 1 develops the relationships in Harley's suggested top-down power influences as they controlled what little knowledge the reader could gather from a modernist map.[5] These relationships are examples of what Baudrillard finds to be "bogus to the extent that they present themselves as authentic in a system whose rationale is not at all authenticity, but the calculated relations and abstractions of the sign" (Baudrillard, 1990, pp. 35–36). It is, as Baudrillard suggests, a finished product of combined signs, available to consumers who are expected to use it without altering its design or questioning its origins or purpose.

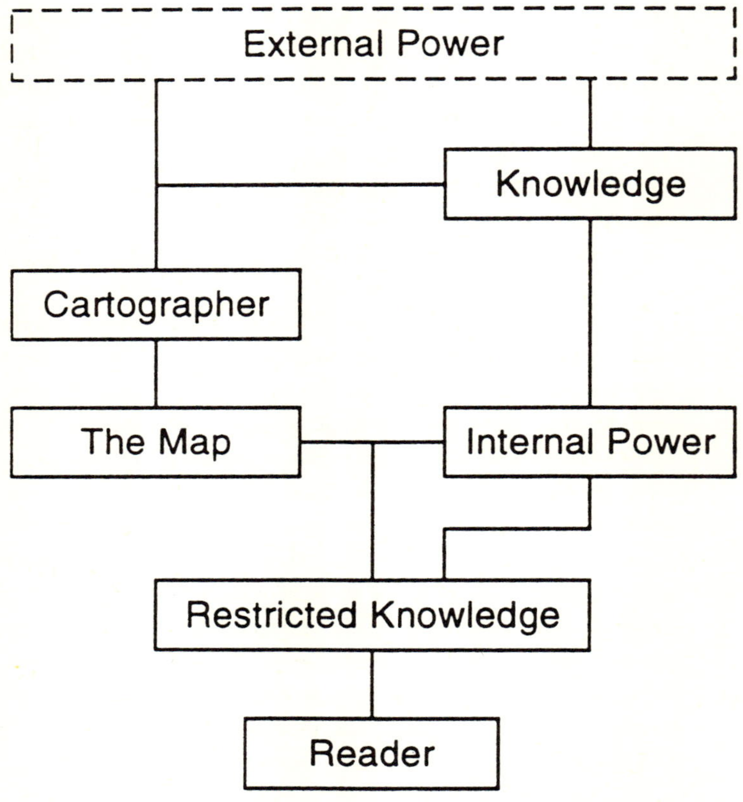

Figure 1. A figure we deliberately design in the modernist fashion for its heuristic value, illustrating what Harley considers the foundational constraints limiting both the scope and the function of cartography, as well as the reader's access to knowledge from maps created under modernism's authority. Adapted from J. B. Harley (1988), p. 278.

Note that in Figure 1 there are no "markedly different proposals also seeking to improve the rigor and relevance of research in education [or social cartography] by encouraging tolerance, reflection, and the utility of multiple approaches in knowledge production and use" (Paulston, 1990, p. 396). Foucault made a similar observation of modernist social science, finding it to be a contemplation of space and time that treated space as "the dead, the fixed, the undialectical, the immobile" while time was "richness, fecundity, life, dialectic."[6] These perspectives of positivist restrictions on knowledge and space are represented in the style presented in Figure 1.

Concerned as are Rust and Harley with overcoming the problems of modernism's positivist treatment of space is urban cartographer Edward Soja

(1989), who contends that in the past "space more than time, geography more than history, [hid] consequences from us" (p. 71). Arguing as we do for the use of space to represent space as it is claimed by cultural clusters, Soja advocates making space and geography the primary focus and framework for the critical study of social phenomena; situating the whereness of cultures and the events driving their realities is a better framing choice for the questions we ask and the answers we receive as we pursue meaning in the mininarrative-rich environment of the postmodern world.[7]

Soja (1989) portrays modernism's purpose and influence during its extended epoch as a deliberate obfuscation of the spatiality of the map, "blurring [the reader's] capacity to envision the social dynamics of spatialization" (p. 122). Postmodernism encourages us to detail the map, particularly where multiple mininarratives are revealed to occupy geographical and ideological space where only a metanarrative of progress served before (Thrower, 1996). Advocating space as the primary starting point for research diminishes the importance of time and creates the opportunity for researchers to apply to their craft the critical cartography advocated by Soja. Postmodern space is the research domain containing the multiple social ideologies and convictions arising from modernism. The postmodern researcher in comparative education, who may also become a postmodern cartographer, prizes both the space within the social milieu and the possibilities for a more open and inclusive mapping of that space, motivating the creation of multiple, inclusive and, therefore, antifoundational maps.

Recall how Figure 1 shows the external power's relationship to the creating and reading of knowledge from the map text, and consider whether this map represents a construction appropriate to Rust's argument for "the critical task of disassembling these narratives [while increasing] our attention to small narratives" (Rust, 1991, pp. 625–26). Clearly, Figure 1 is not an appropriate model for Rust's argument. Rather, this figure authenticates Charles Hampden-Turner's comment that the "visual-spatial imagery of the human is a style of representation largely missing from the dominant schools of psychology and philosophy, [so] there can be no pretence of impartially cataloguing the status quo. The image-breakers are still in charge" (1981, p. 8). Our advocacy of social cartography has as its purpose the breaking of the image-breakers,[8] the encouraging of comparative analysts to become image-makers and, in doing so, including a visual-spatial imagery of the human in comparative discourse (Mitchell, 1938).

Rust's and Harley's challenges to their respective fields of comparative education and cartography encourage illustrating the global vision reflecting the spatial as advocated by Soja. We suggest that the prospect of a critical cartography offers comparative education possibilities for examining educational problems "in the light of culturally determined needs, objectives, and conditions" (Raivola, 1985, p. 392). What is this social cartography we advocate? What is the benefit of social cartography to the practice of comparative education? How might we begin to see maps, like art, as mechanisms "for defining social relationships, sustaining social rules, and strengthening social values" (Geertz, 1983, p. 99)? How might social mapping help us see the intertexuality of all discourses?

As is true of any written discourse, a map begins as the property of its creator. It contains some part of that person's knowledge and understanding of the social system. As a mental construction representing either the physical world or the ideologies of cultures, maps can be characterized as what Baudrillard's translators describe as "art and life." They note that Baudrillard finds art and life shape the system of objects, that a purely descriptive system "carve[s] out a truth" (Foss and Pefanis, 1990, p. 13). While we find maps can shape the system of objects, we suggest that rather than carve out a truth, they portray the mapper's perceptions of the social world, locating in it multiple and diverse intellectual communities, leaving to the reader not a truth, but a cognitive art, the artist's scholarship resulting in a cultural portrait.

Viewed from this perspective, then, what Baudrillard (1990) calls the artistic enterprise includes the map in the sense that the map is a descriptive system consisting of a collection of knowledge objects around a "point where forms connect themselves according to an internal rule of play" (p. 27). The map reveals information about space by showing that information scaled within the boundaries of another space. Mapping the elements of comparison models can contribute to our comprehension of the social, and provide a point of departure for new research as well as for new maps.

An example of this type of antifoundational map is the macro-mapping of "paradigms and theories" uncovered using semiotic analysis in some sixty exemplary comparative education texts presented in Figure 2 (Paulston, 1994). This map embodies Soja's concern for "a social ontology in which space matters from the very beginning" (1989, p. 7). It is a study of society establishing "a primal material framework [of] the real substratum of social life" (Soja, p. 119). This heuristic map identifies intellectual communities and relationships, illustrates domains, suggests a field of interactive ideas and opens space to all propositions and ways of seeing in the social milieu. What appears as open space within the global representation is space that

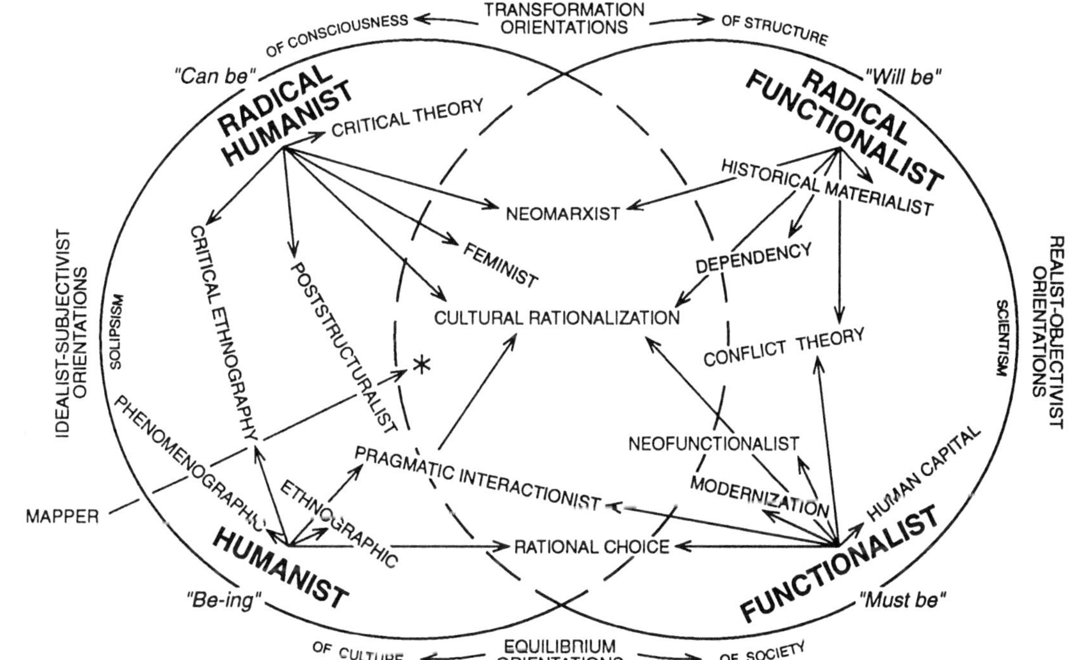

Figure 2. Paulston's (1994, p. 931) "global mapping" of paradigmatic discourse and territorial disputes in sixty recent comparative education texts. This assemblage opens to all claimants space for inclusion in the intellectual field (Ringer, 1990) and social milieu. Situating the mapper in this representation suggests that "... by the act of attributing spirit to everything, giving every element of the landscape its own point of view, shows [the mapper] to be alive to the fact that there are other powers in the world, [that social cartography] is not a fantasy of omnipotence. It is a matter of doing your best in a difficult, hostile world ... in which the spectator is alive to forces of a complexity we can barely grasp" (Fenton, 1996, p. 40).

can be claimed by intellectual communities whose discourse is not yet represented on the map. It is conceivable that the part of the world the mapper draws our attention to does look like this, but it is his perception of the world derived from textual exegesis (Ricoeur, 1971); however, it is probably not what Baudrillard would consider a map carving out the truth. If not truth, but only one possible way of rationally seeing some identifiable parts of the world, how should or can it be considered as a relevant contribution by those who read the map?

By creating on the spatial surface of paper an image depicting a social framework, Figure 2 represents one answer to Rust's recommendation that comparative education focus on mininarratives rather than metanarratives. This map situates both paradigms and theories on the spatial surface of paper, granting to each mininarrative the mapper's recognition of its space in the real world. Readers may question whether the depiction is accurate, whether the allocation of space is appropriate and whether the genealogy and relationships suggested by the arrows have developed or are developing in the directions the mapper indicates. Readers of this map who have answers to these questions need only redefine the space—as Rust does in his chapter in this volume. There is, however, one extremely important caveat to any reader of the map who may wish to redefine its space. This is that the map is not a voodoo doll available for social incantations and cultural pin-sticking.

The map illustrated in Figure 2 resulted from intensive research of multiple published scholarly articles, each framed in one or more of the theories located on the map. The map's creator found specific textual orientations and then created an agonistic intertextual field. The mapper's article accompanying the map both documents and defends the decisions made. Any attempt to redefine the space of this map or of any mapping of the mininarratives of the social milieu should be given equally demanding and scholarly attention. This is one reason why Figure 2 can be viewed as a "holistic, context dependent, and integrative" treatment of paradigmatic knowledge, not as "isolated facts, but as integrated wholes" (Masemann, 1990, p. 465). Spatial mapping of how paradigms and theories are represented in texts also moves comparative education away from a modernist "system for classifying societal data" (Holmes, 1984, p. 591), away from structuring knowledge as illustrated in Figure 1, so that knowledge is no longer viewed only as positivist data but as integrated forms of culture "open to the play of difference in meaning" (Usher and Edwards, 1994, p. 139). This postmodern view acknowledges the interrelatedness or intertextuality of cultural texts and cultural space, and suggests that for every text there must be a space and that every text is a space. As feminist and postcolonial cartographers have taught us, modernity does not offer space to every text.

Burbules and Rice's analysis of the postmodern notes Derrida's insistence "that the relations that bind and the spaces that distinguish cultural elements are themselves in constant interaction" (1991, p. 400), a consideration highly adaptable to the relations Figure 2 shows between the numerous paradigms and theories illustrated on the space of the map. Burbules and Rice find in Derrida the premise that any "particular formalization is . . . nothing more than the momentary crystallization and institutionalization of one particular set of rules and norms—others are always possible" (p. 400). The sense of institutionalization as a concept to be understood or read into postmodern maps, such as Figure 2, is located in the formalization of scholarly ideas. The map is not putty in the hands of map readers. The map cannot seek to authenticate an orthodoxy and remain a scholarly contribution. Thus, Figure 2 is a Derridean "momentary crystallization" of the space claimed by social and ideological ways of seeing only because it represents mutable space available to be either transferred or captured in an ongoing competition of meaning between interpretive communities.[9]

Figure 2 does not conform to the model for modernist maps shown in Figure 1. In Figure 2, there are no powers controlling the disbursement of knowledge. Rather, Figure 2 develops as proposed in Figure 3, where the power to read and map the world is so equally shared that it is not even a category in the developmental model. In contrast, Figure 3 supports Star's suggestion that "the silent blueprint to life means looking in areas of darkness" (1991, p. 265). Her first rule for the study of invisible things is the rule of continuity: Phenomena are continuous; objects are created by overleafing stratified networks originating from radically different points. These areas are represented in this figure by the layered mininarratives interwoven in the social milieu. With this figure, we attempt to incorporate into our postmodern studies what Rust identifies as Derrida's, Foucault's and Lyotard's "emphasis [on] the contingency of meaning and the slipperiness of language" (1991, p. 266). This slipperiness seems to be overcome in the ease of movement between levels enjoyed by scholars and the knowledge they uncover. Modernism's "deep structures of language . . . which allow us to attach ultimate meanings to words" (p. 611), as well as, we are arguing, the deep structure of a metaculture and the imposition of an ultimate meaning predicated on that culture, are overcome because of the ready access of the scholar to all levels, and the knowledge transferred and readily available through and across the levels previously hidden under the shadows of modernism's metanarratives.[10] The scholar's reading and mapping of the

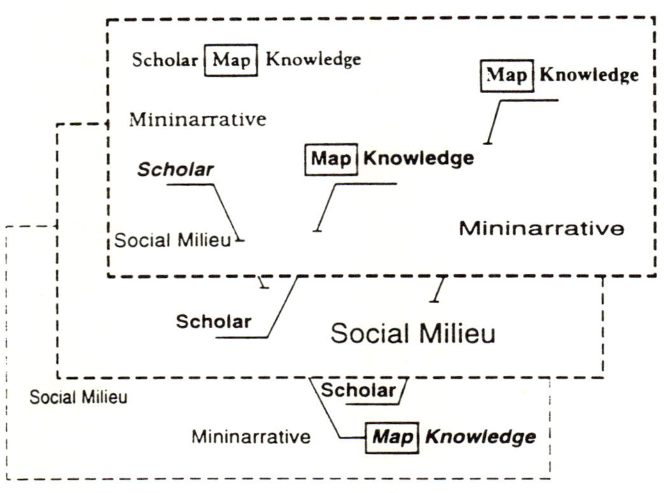

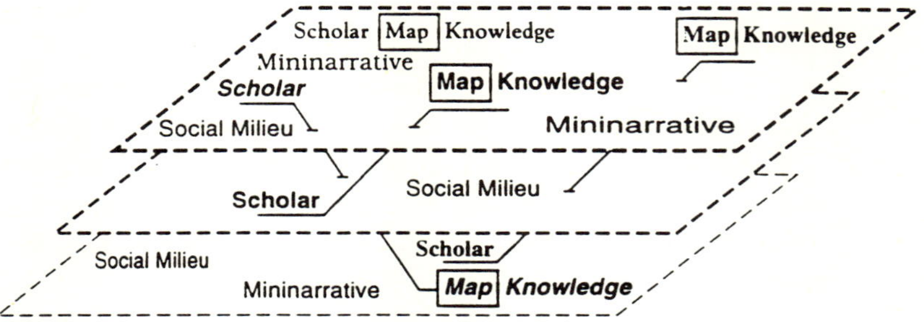

Figure 3. A heuristic mapping by Liebman and Paulston of Star's vision concerning the interwovenness of previously hidden narratives which postmodern inquiry has revealed to be overleafed networks originating from radically different points. Adapted from S. Star (1991), p. 277.

cultural clusters found in the social milieu provide a secondary format for the presentation of scholarship. The new knowledge has the potential for changing the social milieu by creating a changed understanding as well as providing new opportunities for research and critique.

The overleafed space in this map is much different from the space in Figure 1, where external power controlled and conditioned space, where the external power of the map restricted and contrived knowledge distribution. Figure 3 suggests the potential and the need for open, global mapping. Figure 3 suggests there is a continuity of space and time as Star suggests, but neither is dependent on the other. Time certainly is continuous and experienced; while it is biologically limited, it is socially flowing or rupturing. While space is in a time it is given to a social context and is subject to change just as the space of the land is made conditional by natural erosions and eruptions. For example, a calm sea or a dormant volcano slows geological time, leaving little trace of time's passing. A violent sea or an erupting volcano quickens the geological pace, creating changes for the cartographer to map. So it is with the measure and mapping of society. The more ebbing and flowing, the more movement and upheaval there are in a society, the more changes readers and mappers may perceive. However, social mappers need not await the abatement of the societal seas or the cooling of the societal land to begin their project. They may map immediately—as in Figure 2 and Figure 4 (below) and in the numerous social maps offered in this volume—as the erosions and eruptions affect the social milieu. The potential for comparative immediacies and simultaneities inspires the mapping project.

Rust has further opened comparative education to its postmodern potential, observing that "ours is a world, no longer of reality, but of simulation, where it is no longer possible to separate the real from the image" (1991, p. 622). Now maps offer comparative education a tool for expanding conceptual presentations and interpretations. In the hands of the comparative educator, maps can be a part of research directed, as Sack (1980, p. 16) suggests, "at reconstructing . . . chains of influence so that we will know what parts of the society are interrelated." Figure 3 provides a model for study and interpretation of that chain of influence, linking the components of the social milieu through social mappers and the social map. Revealing knowledge of the locations and interrelations in the milieu of diverse societies, cultures and ideas, and then mapping them in relation to one another, is the essence both of a social cartography and of comparative studies.

One study showing considerable potential for a critical social cartography is Apter's phenomenographic representation of the history of the Sanrizuka movement and its extensive use of nonformal education (Paulston,

1980) to oppose the construction of the Narita Airport outside Tokyo (Figure 4). Apter (1987) isolates a series of five distinctive episodes, each identified with a metaphor (i.e., transference) and a metonymy (i.e., substitute naming) "derived from interviews and written descriptions of events provided by those deeply involved in the movement" (p. 250). Apter describes the spatial bounds of his study as they were set by the participants of the revolt, "defining a larger cosmological space, underground to a sacred soil, above ground to the sky itself" (p. 248). In this way, Apter provides a readily visual three-dimensional physical cartography. The questions raised and considered at Sanrizuka not only addressed whether the land would be retained by traditional farmers or converted to use for a modern airport, but because the land was to be used for an airport, the questions involved the symbolic and real use of the air above the land.

In addition to the physical cartography of Sanrizuka is a moral cartography Apter identifies through the participants' metaphors and metonymies. This aspect of Apter's study coincides with our purpose noted above, that accuracy and inclusion in a postmodern social cartography consider not only the space being mapped, but the perceptions offered by the claimants of that space.[11]

The ordering of information in Apter's figure of the events at Sanrizuka offers opportunities to create multiple maps. Our single concern with the information provided is that the metaphors and metonymies Apter identifies with the five episodes of Sanrizuka would seem to be appropriate only from the perspectives of the farmers and militants. It is doubtful the other five participants he identifies on the map would use these terms to describe the events. So when Apter writes in his caption that the metaphors and metonymies "form a narrative of moral outrage and a radical text," it seems doubtful he is referring to the airport authorities or government officials, for example. We argue that not only would Apter's figure tell quite a different story when metaphors and metonymies from other participants became visible, but that the mapping of the Sanrizuka struggle begun by Apter would require multiple overleafings, such as those shown in Figure 3, to represent accurately the perceptual semiotics of the multiple participants. We also recognize that the metaphors and metonymies offered by Apter are not qualified by the metanarrative of persons empowered by modernism, but by the mininarratives of those persons involved in the educational, cultural and political struggle at Sanrizuka.

The comparative educator might consider the Sanrizuka event in terms of the knowledge communities and the systems of formal, nonformal and informal education touching the lives of all persons in Sanrizuka. We sug-

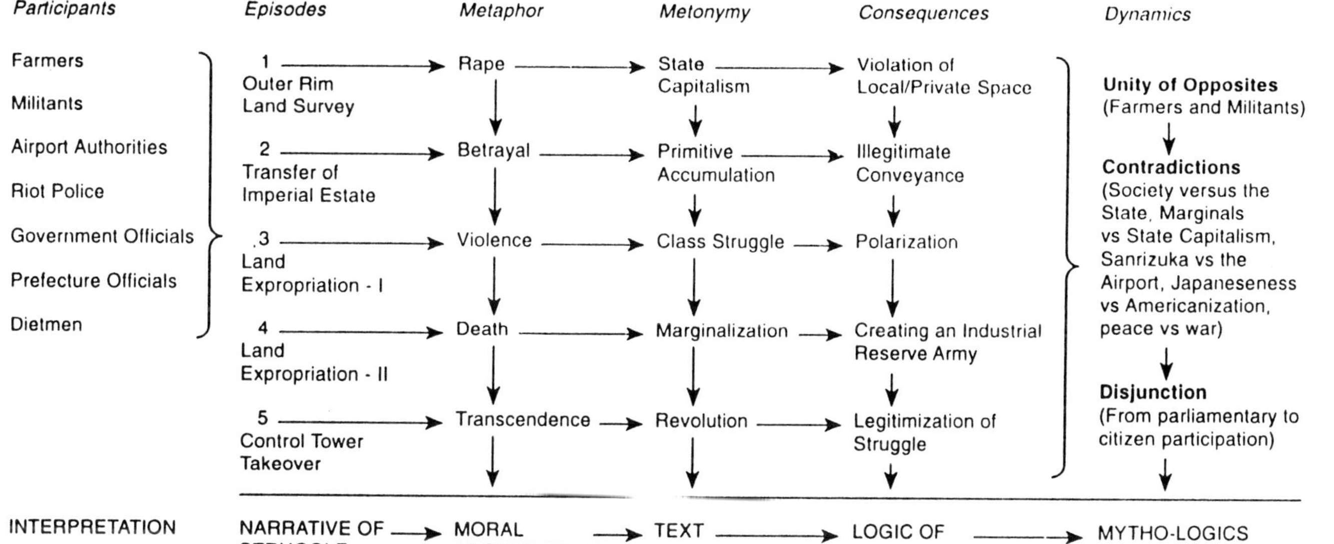

Figure 4. Captioned by Apter as "two crossroads and a terrain as a semiotic space," this figure derives its phenomenology (episodes and consequences) from phenomenography (narrative-dependent content). Apter's caption continues, "This diagram, derived in part from the work of Roland Barthes, Claude Lévi-Strauss, Edmund Leach, Paul Ricoeu, and Pierre Bourdieu, describes episodes of violence as they are interpreted by the participants involved in the movement. Together they form a narrative of moral outrage and a radical text. They constitute both the moral force and logical integrity of the movement and make convincing, at least to followers, the idea that such a small group of participants can win such a big victory. The ingredients of the ideology represent what Lévi-Strauss has called a mytho-logics. Evidence is provided by the actual episodes. A complete and total system, the mytho-logics serves as an interior discipline of language and an ordering of signs. By the same token, what orders within is disordering without. It captures certain critical ambiguities of modern life in Japan, ambiguities which are widely felt but rarely articulated, the shock value of the incidents attracting outside clienteles." Reprinted from Apter, D.E. Rethinking development: Modernization, dependency, and postmodern politics, p. 250. Copyright 1987 by D.E. Apter. Reprinted by permission of Sage Publications.

gest the systems of knowledge and education created to meet the needs of the participants at Sanrizuka and all other sites of educational resistance could provide an interesting source of new questions of postmodern educational implementation and analysis with the intent of opening comparative education discourse to a postmodern sensibility.[12]

It is our thesis that when scholars address the cultural values and differences revealed by different and often competing knowledge claims, they can enhance their research by developing and including in their findings a cognitive map showing their perceptions of how these multiple knowledge claims interrelate. Social cartography rejects no narrative, whether it is a metanarrative or that of a localized culture. Although metanarratives are accepted and mapped, they are neither privileged nor accepted in their previous role of dominating other narratives. Rather than legitimizing metanarratives—and their ideologies—in their modernist form, our mapping project introduces the concept of the mininarrativization of the metanarrative. Thus the breadth of research possibilities and understanding that social cartography envisions recognizes—in contrast to Apter—all points-of-view. Their general validity opens opportunities for comparison because mapping does not "deny integration of cultures and harmonizing values" (Rust, 1991). Social cartography arises from what Rust notes are the possible "metanarratives . . . [that] open the world to individuals and societies, providing forms of analysis that express and articulate differences and that encourage critical thinking without closing off thought and avenues for constructive action" (p. 616).

Our social cartography approach is similar to any geographic mapping (although Cartesian coordinates are not mandated) that reduces a "total" space to a much smaller space. While the purpose and goal of positivist geographic cartography is to create an empirically perfect model, our purpose is more aletheistic. Creating mapped social models cannot finalize with any exactitude a true representative. Maps created by social cartographers are not to be replicable uncritically by other social mappers. Social cartographs may be added to or amended, and they are certainly open to debate, change and even personalization. Thus, while Turnbull's geographic cartographer of empirical space can win an argument that a map should be altered because it does not replicate the physical world's measurements, social cartographers do not argue validity because they understand that others are encouraged to question the spatial relationships of mapped social realities: social maps are not empirical, mathematically correct representations. The social world cannot be measured, but it can be experienced, reported and compared. Because of this, we, like Mouat in his chapter, see social cartography as postparadigmatic: it will not create new paradigms, nor will it initiate a revolution of paradigms, as suggested by Thomas Kuhn. Rather, it provides a perspectivist orientation for which there are no eternal truths or universal facts, but only interpretations, only the competing constructs, as here, of various individuals and groups.[13] Social cartography, in short, helps comparative educators, along with all participants in the educational enterprise, to provisionally order and interpret today's multiple views of education and society.

CONCLUSION

We propose, first, that both the structures and relations of multiple education and knowledge systems can be recreated in one or more maps, in a social cartography where the space of the social map reflects the perceived effects of social changes in real space; and, second, that comparative education researchers consider representing that space through the creation of multiple social maps.

Our rationale for this proposal is that the map provides the comparative educator a better understanding of the social milieu and gives all persons the opportunity to enter a dialogue to show where they believe they are in society. The map as a particularly human way of looking at the world reveals both acknowledged and perceived social inclusions, while leaving space for further inclusions of social groups and ideas. Whether the map is considered a metaphorical curiosity or accepted as a more literal representation, it offers comparative researchers an opportunity to situate the world of ideas in a postmodern panorama, disallowing the promotion of an orthodoxy.

In this introductory essay we have sought to demonstrate how through the employment of a "social cartography"—the creation of maps addressing questions of location in the social milieu—social research may move one step further as it struggles to distance itself from the totalizing restraints of modernism. Social cartography suggests the further opening of dialogue among diverse social players, including those individuals and cultural clusters who want their mininarratives included in the social discourse. We propose that social cartography has the potential to be a useful discourse style for demonstrating the attributes and capacities, as well as the development and perceptions, of people and cultures operating within the social milieu. It offers comparative educators a new and effective method for opening social dialogue by visually demonstrating the sensitivity of postmodern influences to juxtapose seeming incommensurables that under modernity were presented, if at all, as worlds apart (Marcus, 1994, p. 566).

We should, perhaps, also note that our social mapping project is neither a total rebuttal of modernism nor a headlong plunge over the postmodern cusp. We do not advocate Derridean-Dionysian freewheeling ludic play or see reality as only "playing with the pieces" in a Baudrillardian house of mirrors (Baudrillard, 1984, p. 20). Rather, we agree with Rust's observation that metanarratives continue to have an important—if not predominant—place and societal influence, as well as with Habermas that modernity is not a project past renewal. What we envision is not the arbitrary rejection of any particular way of knowing, but an eclectic move toward encompassing the perspectives and methods we can find that serve to advance both knowledge and understanding. To replace one totalizing perspective with another would not improve social and comparative research but create a new focus for argument, misunderstanding and exclusions. Social cartography's method, however, decidedly favors the postmodern and a perspectivism open to the study of cultural clusters' narratives and ways of seeing. By using social maps as a part of our comparative studies, we may provide an inside view, a visual dialogue of cultural flow and changing influences, appropriate for future work in comparative education, particularly in those instances where cultural values and differences are revealed by competing knowledge claims. Social cartography offers a new space in which to arrange competing knowledge claims to reveal their interrelatedness and, subsequently, develop both spheres of understanding and grounds for resistance.

NOTES

1. An earlier version of this chapter appeared in 1994 in the *Comparative Education Review* 38(2), pp. 215–232. Permission to use from the University of Chicago Press is acknowledged with thanks.

2. The possibility for uncovering language in visual space also would advance Rust's postmodernist project, one dedicated to opening up the world to reveal its interwovenness of being and humanity. "Proponents of postmodernism suggest that a mass-oriented society is obsolete. Our decision-making apparatus must be altered to allow for a system based on multiple, rather than majority, rule" (Rust, 1991, p. 618). Following the advice of Burbules and Rice (1991, p. 412), who write "critical self-awareness is a step toward changing our practice," we gently suggest Rust's textual fashioning of an advocacy for a postmodern turn occasionally exhibits the orthodoxy of modernist rule-making. We have become aware of just how difficult is the task of breaking the modernist mold. In the present article we have struggled with modernist language through every draft. Postmodern thought, we have found—like its architecture and art—still prefers the readily identifiable building materials of modernity, although the array of the materials in postmodernity prefers subtle variations from the modernist style. Often perceptibly deplored by the postmodern turn is the "must" imperative (see Rust, pp. 625–626). While we do not insist or encourage others to join us in vacating that modernist imperative and its kith in their postmodern discourse, we are excited by the prospect of being met by others who choose to build and style their discourse as we have attempted to do.

3. In poetry, language can be seen as a mode of bringing a world to disclosure where the world and things are carried over and appropriated to each other in the

moment of disclosure. In his accessible analysis of Heidegger's ideas, Timothy Clark (1992, pp. 20–63) illustrates the opening-out and decentering possibilities of *dichtung* as a mode of appropriation with a poem by Charles Tomlinson (1972, p. 31), which begins:

> Poem
> space
> window
> that looks into itself
> a facing
> both and
> every way

Like poetry, social cartography may be seen to constitute something new. It does not attempt to merely copy or objectively describe what it appropriates. Rather, it creates new meanings by its spatial juxtaposition of images and signs. Exemplifying *dichtung*, mapping names "the open clearing whereby any object can emerge for any subject, (and) could not be reduced to the status of that which it renders possible" (Clark, 41).

4. See S.L. Star (1991). Star's useful rules to study (and map) invisible things include: 1) The rule of continuity: Phenomena are continuous. There is no dualism. Objects are created not by reacting to something, but by overleafing stratified networks originating from radically different points. Power is the imposition of a position in such space; 2) The rule of omniscience: Everybody has several viewpoints and every view is only part of some picture, but not the whole picture. The revealing and articulating of viewpoints is the way we can understand something about truth, a fundamentally interactional, social phenomenon; 3) The rule of analytical hygiene: i.e., concepts are verbs, not nouns; 4) The rule of sovereignty: Every standpoint has a cost; and 5) The rule of invisibility: Successful claims to invisible phenomena require the assertion of power and the fundamental pluralism of human interaction.

5. An illustration of these relationships may be found in P. Foster (1991), and E.H. Epstein (1991). Foster praises Anderson's contribution to a conservative, "gradualist" strategy for education and modernization, his "Puritan morality" and his ability "to test significant hypotheses in the context of more general theory" (p. 220). Epstein as disciple extols what he sees as the external power of Anderson's ideas and the foundational ethos of the Comparative Education Center at the University of Chicago. As editor of the *Comparative Education Review*, Epstein praises his mentor's "monumental contributions . . . the durability of his wisdom regarding educational policy and planning" (p. 211), and concludes that the *Review* under Epstein's control continues to reflect Anderson's perspective and carry "his imprint" (p. 213).

6. Foucault in Harley (1988), p. 4.

7. Soja advances earlier efforts by Henri Lefebvre to move beyond orthodox Marxian political economy with a new unitary theory of space that ties together the physical, the mental, and the social. Here space is simultaneously a spatial practice (or externalized, material environment), a representation of space (a conceptual model used to direct practice) and a space of representation (the lived social relation of users to the environment). In his classic text, Lefebvre (1991) argues that to change life means to change space and the relations between power and space. This is also Soja's thesis. See also the useful review by Lefebvre's disciple M. Gottdiener (1993).

8. Hampden-Turner sees anti-imagists alive and well today in the Puritan-cum-behaviorist intellectual tradition. "Modern behavioral science is thoroughly infused with Puritan ethics, for example, the idea of a scientist as a predicting and controlling agent for scientific determinism; the dogma of 'immaculate perception;' a preference for visible activity publicly verifiable, and the 'godly discipline' of rigorous ex-

perimental minutiae. There is the same rejection of speculative questions, of the private imaginings of subjective personality and reconciling schema in general" (1981, p. 34).

9. Here our mapping rationale is close to Gadamer's (1975, p. 81) call for a critical hermeneutics able to "raise to a conscious level the prejudices which govern understanding. To realize the possibility that other aims emerge in their own right . . . to realize the possibility that we can understand something in its 'otherness.'"

10. We endorse Kenny's caveat here (1992, pp. 178–179): "Although there may be multiple readings of a text, readings are not in any important sense unique to an individual. Meaning is contained within the limits of language . . . and also has a stability based on the social and historical context of interpretation or discourse."

11. Attempts to create more democratic spaces friendly to elusive insight and multiple perspectives are also central concerns of postmodern architecture. For a wonderful example, see Muschamp (1992):

> The building [like mapping] is generous both in inviting images and refraining from making them explicit . . . acoustics [sounds, like cultures, reverberating in space] and empathy [the understanding of the social cartographer] are the two forces driving the design. It promotes accessibility and erodes the barrier between the inside and outside . . . the building [like the map] achieves unity by encouraging individual viewers to quarry their own figures and abstract images . . . and inviting them to co-exist in one place. The place [like the social-cultural map] will be the sum of the perceptions it involves.

12. Studies for and against postmodern perspectives in educational discourse/practice are burgeoning. We especially like Usher and Edwards (1994) advocacy of a ludic, or playful, notion of postmodernism that resists ideological metanarratives better to mount a critical agenda:

> We argue that resistant and ludic postmodernism are two sides of the same coin that each depends for its effects on the other. We . . . argue for the ludic as a form of resistance needing to always deploy the ludic the better to do its work. Without engaging with the ludic we are left with forms of social analysis which become totalizing despite their intent and remain oppositional but ineffective because as forms they lack the emotional investment of a *desire* for change. (p. 16)

In opposition, Beyer and Liston (1992) voice the puritan's lament—"postmodernism seems to undermine moral responsibility" (p. 371). Closer to our assessment, Lather (1991) is cautious and eclectic. She also asks the central question: How do practices we invent to discover our truth impact our lives?

13. Friedrich Nietzsche makes the existence of the world dependent not on a knowing subject using a single and universal objective standard, but on individual interpretations and perspectives. But non-existence of facts or objective standards does not imply an equality of interpretations or subject positions. Rather interpretations result from a human need to interact with the world. And as our needs vary, we receive conflicting interpretations and must struggle with conflicting perspectives. For Nietzsche, perspectivism, far from describing a notion of liberal tolerance for all viewpoints, is actually more akin to a theory of drives and of the way in which internal struggles are fought out in what we call the process of individual cognition (Holub, 1995, p. 69). Some fascinating recent work explains the cognitive process as global mapping, that is as "a dynamic structure containing multiple reentrant maps (both motor and sensory) that are able to interact with nonmapped parts of the brain. Such a global mapping ensures the creation of a dynamic loop (in the human visual system there are over thirty maps in the visual cortex alone) that continually matches . . . gestures and postures to different kinds of sensory signals" (Edelman

1992, p. 89). From this view categorization does not occur from a computer-like program, but from a dynamic of mapping and re-mapping as first proposed by Nietzsche in 1887 and now advocated here.

REFERENCES

Apter, D. E. (1987). *Rethinking development: Modernization, dependency, and postmodern politics.* Newbury Park, CA: Sage.

Baudrillard, J. (1984). Games with vestiges. *On the Beach, 5,* pp. 19–25.

Baudrillard, J. (1990). *Revenge of the crystal.* London: Pluto.

Beyer, L., and D. Liston. (1992). Discourse or moral action? A critique of postmodernism. *Educational Theory, 42*(4), 371–391.

Burbules, N.C. and S. Rice. (1991). Dialogue across difference: Continuing the conversation. *Harvard Educational Review, 61*(4), 393–416.

Clark, T. (1992). *Derrida, Heidegger, Blanchot: Sources of Derrida's notion and practice of literature.* Oxford: Oxford University Press.

Downs, R.M. and D. Stea. (1973). Cognitive maps and spatial behavior: Processes and products. In R. M. Downs and D. Stea (eds.), *Image and environment: Cognitive mapping and spatial behavior* (8–26). Chicago: Aldine.

Edelman, G.M. (1992). *Bright air, brilliant fire: On the matter of mind.* New York: Basic Books.

Epstein, E.H. (1991). Editorial. *Comparative Education Review, 34*(2), 211–214.

Fenton, J. (1996). "On Statues." *The New York Review of Books, 43* (18), 40.

Foss, P. and J. Perfamis (1990). Translator's introduction to J. Baudrilland, *Revenge of the crystal.* London: Pluto, 4–21.

Foster, P. (1991). C. Arnold Anderson: A personal memoir. *Comparative Education Review, 34*(2), 215–221.

Fritzman, J.M. (1990). Lyotard's paralogy and Rorty's pluralism: Their differences and pedagogical implications. *Educational Theory, 40*(3), 371–380.

Gadamer, H. (1975). *Truth and method.* New York: Seabury.

Geertz, C. (1983). *Local knowledge.* New York: Basic.

Gottdiener, M. (1993). A Marx for our time: Henri Lefebvre and *The Production of Space. Sociological Theory, 11*(1), 129–134.

Hackett, P. (1988). Aesthetics as a dimension for comparative study. *Comparative Education Review, 32*(4), 389–399.

Hampden-Turner, C. (1981). *Maps of the mind.* New York: Collier.

Harley, J.B. (1988). Maps, knowledge, and power. In D. Cosgrove and S. Daniels (eds.), *The iconography of landscape* (277–312). London: Cambridge University Press.

Harley, J.B. (1992). Deconstructing the map. In T. Barnes and J. Duncan (eds.), *Writing worlds: Discourse, text, and metaphor in representation of landscape* (231–247). London: Routledge.

Heidegger, M. (1971). *Poetry, language, thought.* Translated by A. Hofstadter. New York: Harper and Row.

Holmes, B. (1984). Paradigm shifts in comparative education. *Comparative Education Review, 28*(4), 584–604.

Holub, R.C. (1995). *Friedrich Nietzsche.* New York: Twayne.

Kenny, J. (1992). Portland's comprehensive plan as text. In T. Barnes and J. Duncan (eds.), *Writing worlds: Discourse, text, and metaphor in representation of landscape* (176–192). London: Routledge.

Lather, P. (1991). Deconstructing/deconstructive inquiry: The politics of knowing and being known. *Educational Theory, 41*(2), 153–173.

Lefebvre, H. (1991). *The production of space.* Oxford: Blackwell.

Marcus, G.E. (1994). What comes (just) after 'Post'? The case of ethnography. In N.K. Denzin and Y.S. Lincoln (eds.), *Handbook of Qualitative Research* (563–574).

Masemann, V. (1990). Ways of knowing: Implications for comparative education.

Comparative Education Review, 34(4), 464–473.

McNee, R.B. (1981). Perspective—use it or lose it. *Professional Geographer, 33(1)*, 12–15.

Merrell, F. (1995). *Semiosis in the postmodern age.* West Lafayette, IN: Purdue University Press.

Mitchell, W.J.T. (1988). *Iconology: Image, text, and ideology.* Chicago: University of Chicago Press.

Muschamp, H. (1992). Gehry's Disney hall: A Matterhorn for music. *New York Times,* 13 December.

Nietzsche, F. (1966). *On the genealogy of morals.* Translated by W. Kaufmann and R.J. Hollingdale. New York: Vintage.

Paulston, R. (1980). Education as anti-structure. *Comparative Education, 16(1)*, 55–66.

Paulston, R. (1990). From paradigm wars to disputatious community. *Comparative Education Review, 34(3)*, 295–400.

Paulston, R. (1994). Comparative and international education: Paradigms and theories. In T. Husén and N. Postlethwaite (eds.), *International Encyclopedia of Education.* Vol. 2 (923–933). Oxford: Pergamon.

Paulston, R., and M. Liebman. (1994). An invitation to postmodern social cartography. *Comparative Education Review, 38(2)*, 215–232.

Raivola, R. (1985). What is comparison?: Methodological and philosophical considerations. *Comparative Education Review, 29(3)*, 362–374.

Ricoeur, P. (1971). Meaningful action considered as a text. *Social Research, 38(3)*, 529–562.

Ringer, F. (1990). The intellectual field: Intellectual history, and the sociology of knowledge. *Theory and Society, 19*, 269–294.

Rust, V. (1991). Postmodernism and its comparative education implications. *Comparative Education Review, 35(4)*, 610–626.

Sack, R.D. (1980). *Conceptions of space in social thought.* Minneapolis: University of Minnesota Press.

Soja, E. (1989). *Postmodern geographies: The reassertion of space in critical social theory.* London: Verso.

Star, S. (1991). The sociology of the invisible: The primacy of work in the writings of Anselm Strauss. In D.R. Maines (ed.), *Social organization and social process: Essays in honor of Anselm Strauss* (265–283). New York: Aldine de Gruyter.

Thrower, N.J.W. (1966). *Maps and civilization: Cartography in culture and society.* Chicago: University of Chicago Press.

Tomlinson, C. (1992). *Written on water.* Oxford: Oxford University Press. Poem reprinted by permission of Oxford University Press.

Usher, R., and R. Edwards. (1994). *Postmodernism and education.* London: Routledge.

From Modern to Postmodern Ways of Seeing Social and Educational Change[1]

Val D. Rust

Rolland Paulston (1992) and his colleagues at the University of Pittsburgh maintain that visual mapping of comparative and international education contributes to a more open social and intellectual dialogue with theoretical orientations that have heretofore been marginalized and excluded from social discourse. Paulston suggests that social cartography provides a mechanism by which the field can move toward a "postmodernist integration." This is a significant promise and a great challenge to contributors of this volume. My task shall be to explain how postmodernism facilitates dialogue among differing theoretical orientations.

Postmodernism has been conceptualized by various theorists in a number of disciplines. Certain theorists have characterized postmodernism as neoconservative, because they feel it fails to support modern liberal ideals of the Enlightenment (Habermas, 1981), and represents "the suppression of reason and the denial of the possibility of truth" (Bloom, 1987, p. 379). Others see postmodernism as a reinterpretation of American political liberalism (Rorty, 1989). Certain scholars of literary criticism have labeled postmodernism as a "new aesthetic sensibility" in the arts and criticism (Huyssen, 1984), while others characterize it as a symbol of world capitalism (Jameson, 1984). It is seen by some as a new style of discourse and philosophical orientation and by others as a new historical stage following the Modern Age (Rust, 1991). The numerous and often conflicting interpretations of postmodernism contribute to confusion as to its meaning, but conceptual diversity also provides an open-endedness that allows the concepts and theories of postmodernism to be continually reworked and adapted. While postmodernism is a concept that casts its net across a number of orientations, a central orientation is that of poststructuralism. The primary frame of reference of this essay will be the ideas of French poststructuralists Jean-Francois Lyotard and Michel Foucault.

We must begin with language, for French poststructuralists, such as Lyotard and Foucault, claim that knowledge exists only through language.[2] This is not a new notion. Semiotics, the study of signs and sign-using behavior, was introduced by John Locke. French poststructuralists rely heavily on their structuralist predecessors, such as Ferdinand de Saussure (1966), who advanced both the notion that knowledge exists through language and the idea that there is no indissoluble link between words and things. Signifiers are not representations of things and events. A word does not represent a thing but gets its meaning in the context of other words. *White* has meaning only in relationship with language signs such as *black* or *red*. Language is but a metaphor of reality, a process of establishing an identity between dissimilar things. What distinguishes humans is their capacity to create signs, to develop symbolic worlds. To quote Nietzsche, language is ultimately "a mobile army of metaphors." One metaphor points to another in an endless chain, and no means of transcending that chain can be found. We live in what Nietzsche has called "a prison-house of language."[3] Sarchett (1995, p. 22) explains that "language is a self-enclosed system which can have only a very murky relationship to a *real* world without words."

From this perspective neither natural language nor Adamic names exist. There is no transcendental signified, no center. There are no metaphysical certainties, no certified, natural truths. Truth is not synonymous with reality. Rather, truth is a property of language propositions. As Rorty (1989) suggests, "truth is made rather than found." Truth is defined and given meaning by language, which is also made rather than found. No language can express the real nature of the world, the self or the community. Our understanding of them is always tentative, ever contingent.

Umberto Eco's novel, *Foucault's Pendulum* (1989), is instructive concerning the images of postmodernism. Early in his novel, the narrator describes the effect of seeing a pendulum hanging on a long wire set into the ceiling of a choir: "The pendulum told me that everything moved—earth, solar system, nebulae and black holes, all the children of the great cosmic expansion—one single point stood still: a pivot, bolt, or hook around which the universe could move" (p. 5).

In Eco's novel, the narrator and two friends, believing there is no ultimate center around which everything revolves, playfully construct a grand theory that they claim acts as a fixed point. They designate the Knights Templar, Rosicrucians and Freemasons as the occult possessors of the absolute, the ultimate truth. At the end of the novel, the three are captured by the philosophical construction of their own fantasies, and the devotees of these "occult groups," who have taken this construction as truth, try to force this truth from Belbo, one of the constructors. When he refuses (is unable) to reveal this truth, they wrap the pendulum wire around his neck and hang him. At first there is a terrible distortion in the swinging, but then, through some "grisly addition and cancellation of vectors," the wire of the pendulum becomes immobile and his body becomes the pendulum concealing "the location of the World's navel" (p. 597).

Eco's novel provides images for postmodernism. First, it stresses the quest on the part of humankind for metaphysical certainty. Second, it suggests that truth is not grounded in some certified, natural form but is contingent and constructed. Finally, the swinging body of Belbo dramatizes the point that even if a metaphysical truth exists "out there" it is always mediated and distorted through the human condition, particularly human language. In biblical terms, the veil between man and God requires that even the spiritually gifted "see through the glass darkly" (I Cor 13:12).

A main feature of postmodernism is the eschewal or avoidance of what Lyotard calls "grand narratives" or "metanarratives," defined by Cherryholmes (1988) as something "similar to paradigms that guide thought and practice in a discipline or profession." Lyotard (1984) claims postmodernists are those who are "incredulous toward metanarratives," because metanarratives lock civilization into totalitarian and logocentric thought systems. They provide a restrictive, totalizing theory of society and history and are based on abstract principles and theoretical constructs rather than direct, subjective human experience.

Most social science communities have lost the ability to impose metanarratives and are largely fragmented and decentered. This does not mean that groups proclaiming metanarratives have disappeared. Of course, they exist and remain as legitimate as other narratives. In fact, those grand narratives that have inspired the dominant themes of comparative education, evolving from Marx, Freud, Hegel, Adam Smith, Comte, Durkheim, Weber and others, are complex and textually dense, whereas many minor theories and local narratives are textually thin, bland and even banal. In spite of this, the danger of grand narrative developers is that they and their devotees continue to overstate the potential of their narrative. They play a power game in that they would exclude other narratives, claiming that only their narrative is legitimate. Habermas, for example, according to sympathetic critics such as David Held (1980), claims that his concepts of communicative competence provide universalistic criteria for communication and the resolution of differences.

However, according to Foucault (1980, pp. 80–82) "totalitarian theories" have proven to be vulnerable to "local criticisms" and "non-centralized theoretical products," whose validity does not depend on the approval of established regimes of thought, and which allow individuals to move beyond "fixed referents" and "terroristic universals." Lyotard and Foucault were combative and antagonistic toward pervasive narratives, but the sharp style of their discourse characterizes the time when these texts were written, in contrast to the more tempered style of recent discourse. It might be helpful to place the rhetoric of scholars such as Lyotard and Foucault in context. To do this I shall make reference to the changing representations of knowledge in comparative education.

Changing Representations of Knowledge

Paulston (1992) claims that since World War II, textual representations of knowledge in comparative and international education have gone through three major periods: orthodoxy, heterodoxy and heterogeneity. The first period, orthodoxy, occurred in the 1950s and 1960s, and was characterized by the hegemonic and totalizing influences of functionalism and positivism. These closely related theoretical orientations dominated comparative education and adherents confidently claimed to be progressing toward the development of law-like statements and generalizations. The second period, heterodoxy, occurred in the 1970s and 1980s, and was characterized by paradigm clashes where critical and interpretive views successfully competed with and challenged the orthodoxy. The prevailing orthodoxy was under fire. The third period, heterogeneity, occurring in the 1990s, is described by Paulston as consisting of disputatious yet complementary knowledge communities that have come to recognize, tolerate and even appreciate the existence of multiple theoretical realities and perspectives. With the loss of legitimacy of metanarratives, what we have left is what Derrida (1978) conceptualized as *difference*.

The three periods of comparative education described by Paulston mirror developments in the social sciences in general, and the first two periods exemplify two primary political problems with metanarratives. In the first period, adherents of the existing orthodoxy assume their metanarrative contains truth and insights about how progress can be achieved, and they expect adherence on the part of members of the field with regard to what might be considered legitimate research problems, suitable funding of projects, publication of appropriate research and so on. This expectation is so strong that it might even be seen as terrorizing because its adherents force consensus and do not tolerate and appreciate other perspectives. In the second period, metanarratives not falling within the existing orthodoxy are able to compete successfully with the orthodoxy, but the outcome is a struggle for power, an attempt to dethrone the pervasive view and replace it. Because metanarratives, by their very nature, are totalizing, the struggle is one of "either/or" competition, closed defense of the favored paradigm and total disdain for opposing paradigms. Even though Lyotard and Foucault hint at ways to transcend agonistic and partisan dramas of orthodoxy and heterodoxy, the tone of their textual discourse reflects the period in which they were writing, the period of combative heterodoxy.

According to Paulston, comparative education and the social sciences have finally transcended both orthodoxy and heterodoxy and have arrived at a period of theoretical pluralism and heterogeneity. This third period is consistent with postmodern sentiments. Elsewhere, I have stated:

> Postmodernists would reject any claim that one way of knowing is the only legitimate way. Rather they would say our task is to determine which approach to knowing is appropriate to specific interests and needs rather than argue some universal application and validity, which ends up totalizing and confining in its ultimate effect. (Rust, 1991, p. 616)

Of course, defenders of metanarratives continue to exist, but those in the field have generally tempered claims of orthodoxy and heterodoxy and have reconciled themselves to a world that is largely eclectic, ambivalent, open-ended and indeterminate. This is what I like to refer to as a postmodern world. It is the world of Lyotard and Foucault without ideological fangs and claws. It is not always a comfortable world. Paulston (1992) points out that it is filled with ambivalence and nostalgia for the old times of greater certainty, but it is also filled with a sense of adventure, as scholars seek new insights and ways of seeing the world and delight in the adventure of diversity and difference.

Difference and Postmodernism

Intellectual heterogeneity is consistent with postmodernism. We noted above that the meaning of a signifier does not exist in relation with actual things and events. Its meaning is derived in its difference from other signifiers, as black to white, hot to cold, good to bad. Every sign has meaning only as a systematic play of differences (Derrida, 1978). In the system of language only differences are found. The inescapability of difference and the need for its recognition are fundamental and have implications for language, self and community, to which we now turn.

Borrowing from Wittgenstein (1953, sec. 23), scholars such as Lyotard (1985) point out that language cannot be seen as a unifying element of discourse but as a plurality of heterogeneous and incommensurable "language games." A language-game move or utterance has meaning only if there are rules and a move satisfies the rules of a given game. However, postmodernists go beyond Wittgenstein in that they understand language to be a differential system. In any community multiple language games exist, each with its own rules and peculiar moves. This results in a plurality of games, each incommensurable with the other games. Lyotard (1985, p. 94) points out that even in the same language game, there are variants, "a patchwork of language pragmatics that vibrate at all times," so that the partners in the same game "occupy positions that are incommensurable to each other."

These language games are not static, but they are continually changing and shifting, both in terms of new rules and different moves. Language games exist both in verbal and written discourse. Consequently, it is impossible to rely on a text alone as the determinant of the text's meaning. Each reader participates in giving the text meaning. There are few essential, changeless aspects to a text. It is folly, for example, to demand that the Constitution of the United States be interpreted today only in the same way it was interpreted at its inception. It may be continuously and contextually reinterpreted. Any text can be understood as an interplay between the text and the one interpreting the text.

This is not to suggest that a text is completely contingent. Kneller (1994, p. 185) is correct in the following claim: "But in reality any text has a content, and many have dates and other period features that decidedly limit the interpretation they can sustain. Kozol's *Savage Inequalities* and Dewey's *Democracy and Education* cannot be interpreted as if they were alike." In other words, neither the text nor the person reading the text can claim complete autonomy in terms of interpretation. Because Paulston's intellectual map of comparative education is based on textual interpretation, it must be understood that both the texts he selects and their interpretation are fluid and can even be self-contradictory. Paulston is surely not interested in providing a canon of comparative education texts in the sense that Harold Bloom (1994) has attempted to do with regard to Western literature. An authoritative collection of texts, a comparative education canon, would be antithetical to postmodern sensibilities. Paulston appears to have selected some texts, not for the place of their authors in comparative education, but because of their impact (e.g., Bowles and Gintis, 1975) or their potential impact on those in the intertextual field (Althusser, 1990). There are texts in comparative education that might have served his purposes. For example, while he has selected Althusser to represent historical materialism, there are texts written by those in the field who also represent that orientation (e.g., see Hofmann and Malkova, 1990). If he were to engage in such an exercise today, I am certain the texts and even the categories he uses would have shifted. These shifting categories would inevitably come, in part, from his discussions with others about his interpretations. I suspect his students have challenged him as, at a distance, my students have challenged him. My students, for example, inevitably question his decision to include the text of Boli, Ramirez and Meyer (1985) to represent modernization, even though the authors claim to deal with "global patterns of education" and the "world system." Student inquiries about his selection always lead to a healthy discussion and consensus is rarely the outcome.

The Decentered Self

Poststructuralists such as Lyotard have repudiated the notion that the self is independent of language. In fact, they claim the self is structured and defined in the context of language. Just as unchanging language games are difficult to imagine, so too are unchanging self-identities. The self is irreducibly mediated by language, and because language is a differential system the self is also essentially fragmented, decentered, incomplete (Haber, 1994, pp. 11–12). Poststructuralists give overriding importance to language and consequently conclude that a true self, a subject, never exists but constitutes "something like an ideological mirage" (Jameson, 1984, p. 63). Other postmodernists would not be as extreme as the poststructuralists, but they would argue that the self is neither fixed nor monolithic but fluid and even contradictory. Echoing Whitman's "I am large, there is a multitude within," Robert Ornstein (1986, p. 9) argues: "We are not a single person. We are many."

Humans are much more complex and intricate than most of us imagine. The self is neither received nor given but is constructed and reconstructed. In many respects the self and self-identity are dynamic, continually defined in the context of changing relationships between individual and environment. As Trinh T. Minh-ha (1989) asserts, the "I" is not a unified subject or fixed identity, but framed from infinite layers. The basic essence of the self is open-ended; it is also biographical in that it is continually defined and redefined by life and experience. In the mid-1980s, for example, some Japanese high school students returning to Japan following extended residence in the United States formed a group named "Hybrid." They, more than most people, recognized that their identities were neither completely Japanese nor American, but a little of each.

What this means for those of us attempting to understand various theorists in comparative education is that we should not expect a person to remain absolutely consistent over time. A final, ultimate, completed self does not exist in any of us. Our essence is adaptable and flexible rather than timeless and unchanging. Scholars writing texts reflect this fluidity. They are not schizophrenic when they change, shift and even contradict themselves. They manifest one of the adventurous traits of being human. Thus, it should not be alarming to find Paulston (1992, Figure 2) writing texts that fit within three of the four different "root paradigms" or world views he has conceptualized.

Those of us advocating social cartography recognize it represents a novel way for comparative educators to see the intellectual landscape of the field. Visual mapping of the diversity in the field allows not only for the beginning of a more open social dialogue, but for representing a different way of seeing. Comparative education texts are almost always formulated and transmitted, at least in American academia, within the context of what Searle (1995) calls the Western rationalist tradition: The appropriate language for that tradition is linear, objective, and scientific. Since the late 1960s and especially after the appearance of Noah and Eckstein's book, *Towards a Science of Comparative Education* (1969), where they announced the passing of "intuitive" approaches in the field and the initiation of projects such as the International Educational Achievement studies, some comparative educators claimed that the field had finally been raised to the standard of a true science. While postmodernists encourage the continuation of science-oriented studies, they question the arrogance of exclusivity that science-oriented authors usually manifest and call not only for variety in terms of orientation but variety in forms of presentation. As David Turnbull argues in this volume, we ought to explore novel means of conceptual representation, including visual representation, that allow for multidimensionality, nonlinearity and turbulence to be more fully expressed.

Some empirical grounding can be found for the above claim that our selves are decentered. Most people are aware of the dramatic research on the brain in the past two decades. Initial brain studies indicated that two upper brains exist rather than one and that they operate, to some degree, in different mental areas. Tony Buzan (1991) explains: "In most people, the left cortex deals with logic, words, reasoning, number, linearity, and analysis, etc., the so-called 'academic' activities. . . . The right cortex deals with rhythm, images and imagination, color, day-dreaming, face recognition, and pattern or map recognition" (p. 17).

As Buzan points out, the right cortex deals mainly with visual maps. If we restrict ourselves to the conventional academic tools of presentation in comparative education, we limit the potential of our minds. Such an assertion is not antiscience. If one picks up a natural science journal, such as *Plasma Physics*, *Nuclear Physics*, or even something more speculative such as *Philosophical Transactions: Physical Sciences and Engineering*, one finds nearly every article filled with graphic illustrations of the issue under discussion. Those in the natural sciences have long recognized the limitations of the written word. They rely more fully on using the complete mind than those who are dedicated to a "scientific" social science.

Even the claim that the human mind has a dual nature is now recognized as being too simplistic. Two of many complicating factors shall be mentioned. First, the brain operates in a nonlinear, chaotic fashion. According to chaos theorist Ilya Prigogine:

> It's well known that the heart has to be largely regular or you die. But the brain has to be largely irregular; if not you have epilepsy. This shows that irregularity, chaos, leads to complex systems. It's not at all disorder. On the contrary, I would say chaos is what makes life and intelligence possible. The brain has been selected to become so unstable that the smallest effect can lead to the formation of order. (Briggs and Peat, 1989, p. 166)

In other words, for the brain, irregularity in neural firing is entirely normal; different cells oscillate at varying frequencies, setting off surrounding cells, so that traveling and rotating waves of cell action fluctuate through the brain. These chaotic processes are entirely normal and it is only when the wave patterns settle into regular and predictable patterns that the brain appears to dysfunction.

Second, the work of Roger Sperry, Eran Zaidel and Robert Ornstein has demonstrated that both brain hemispheres are intimately linked and each hemisphere has the capacity to engage in a much wider range of mental activities than a dualistic model suggests. Thus, Ornstein (1986) can speak with confidence about multiminds rather than dual minds, of multiple selves rather than a dual self.

Scholars of comparative education have developed a sophisticated intellectual and theoretical base, but the very nature of comparative analysis has become highly specialized, though predictable and conventional, with language the only tool. And the language used must be "appropriate." The language that has dominated the field has been that which conforms to logical, sequential, linear analysis. This means that the models, concepts and theories we express must also be logical, sequential and linear. While these tools are valuable with respect to those aspects of social reality that function according to clockwork predictability, they falter when dealing with conditions of instability and turbulence. Social scientists who have attempted to unravel the collapse of the Soviet Union and its satellites in Eastern Europe have found conventional theories not very helpful. They explain neither why the collapse occurred nor how the pieces are being put back together again. One reason for this is that the analytic models have generally been linear and sequential but the phenomena being studied have been terribly messy and complex. It is difficult to make sense of a complex, nonlinear system with a simple, linear model.

A major difference between linear and nonlinear systems is that in linear systems the whole is always equal to the sum of its parts, while in nonlinear systems the whole is greater than the sum of its parts. In linear systems, the parts exercise a certain independence and their behavior can be precisely measured. Light and sound are linear, which means that when I turn the light on while my television is on, the room light does not interfere with the light from the television screen or the sound from the speaker. The light and sound rays pass through each other and don't affect each other. Much of social life is also linear. If we come to an intersection with a traffic signal, we stop on red and go on green regardless of traffic in the street. Much of economics is based on a notion of linearity, in that the independent judgment of all people supposedly defines the economic system. My decision to build a deck is deemed to be independent of my neighbor's decision to build a swimming pool. Of course, we intuitively sense that there may be some interdependence, but the model does not take that into account. Many school notions are also linear. In the minds of instruction experts, how a math teacher instructs is assumed to have nothing to do with how the literature teacher next door instructs. However, the two may be highly interrelated. The students may attend the math class then move directly into the literature class, bringing the emotional and mental dispositions they gained in math class with them. Our linear models, however, do not take such interrelatedness into account. These systems are highly complex and their parts feed on each other with remarkable results. Even though the individual sounds of the musical instruments of a symphony orchestra behave in a linear fashion, in that we can hear each instrument, the sum of those individual sounds is much greater than the individual parts.

One aspect of certain emerging models of complex phenomena is that feedback loops occur in such a way that small inputs result in enormous changes. In chaos theory, for example, the so-called butterfly effect is a label given to small inputs in weather conditions that cause terrible storms on the other side of the world (e.g., Lorenz, 1993). In the past it has been too difficult to determine such measures, so specialists have not even tried. Computers aren't so particular. They will work endlessly on tasks and specialists have been happy to develop computer simulations that allow the computers to attack problems that are nonlinear and complex in nature.

Those developing nonlinear models are not just attempting to forecast future events nor are they necessarily interested in controlling a particular system. Their preoccupation may be to learn about the system, how it functions, changes, resists change and even collapses. They often wish to understand how to live with the system and harmonize with its basic processes. How much richer the field of comparative education would be if these aspects of analysis and research were included. Unfortunately, the field has been somewhat dormant in terms of its potential to use the full capacity of mind and analysis. A major contribution of Paulston and his colleagues is that they have awakened some scholars in the field to broader possibilities.

The Decentered Community

Poststructuralists insist that social differences be protected and even encouraged. They promote a politics of difference and claim a society without group differences is neither possible nor desirable. Such an emphasis appears, on the surface, to fly in the face of the Enlightenment ideal that group differences, including caste and class, have no place in a just society. The aim of the past two hundred years has been to do away with differential rights, privileges and obligations of various groups. Some postmodernists continue to proclaim such ideals. Richard Rorty (1989, p. 197), for example, advocates communities based on "institutions of liberal democracy." He recognizes a certain ethnocentrism in his point of view but has not attempted to accommodate liberal democracy with pluralism and difference in any fundamental way. According to Rorty (1989, p. 192):

> Solidarity . . . is thought of as the ability to see more and more traditional differences (of tribe, religion, race, customs and the like) as unimportant when compared with their dissimilarity in respect to pain and humiliation—the ability to think of people wildly different from oneself as included in the range of us.

Rorty's liberalism tends to discount difference in favor of a consensual community. Indeed, one could sympathize with the progress that has been made with regard to elimination of social injustice and the development of equality. However, in recent years new postmodern heretics have come onto the scene who challenge the systematic elimination of group difference. They dispute the notion that liberation must be equated with the elimination of group difference. Resisting assimilation, claiming that socially privileged groups are actually setting the norms for assimilation, they reject the idea that groups falling outside those that are privileged are on the margins of full humanity and citizenship. This does not pertain only to social class and race. The women's movement is a good case in point. Recent feminist theory has begun to recognize and acknowledge differences between women. Spokespersons for women admit that no universal category of women can be defined. "White women's feminism" is particularly in question (Giroux, 1991, pp. 31–35). In other words, those in women's studies now recognize that their own norms of women have been universalized and preclude recognition of groups that differ from conventional feminist categories and attributes (Young, 1990, Chapter 6).

Haber (1994, p. 114) claims that most postmodernists are so intent on emphasizing difference that they tend to "universalize difference" and in so doing they create a totalizing metanarrative. In other words, their dedication to difference propels them to fall into their own metanarrative trap. Lyotard (1984, 1985), for example, especially formulates difference so extreme that the self is almost locked into a form of solipsism. Baudrillard (1984) proudly proclaims that nihilism ought to characterize society. When Foucault (Rabinow, 1984) was asked whether a consensus model might have some value in politics, he claimed it was usually linked with power relationships in such a way that it implied coercion. "The farthest I would go," he concluded, "is to say that perhaps one must not be for consensuality, but one must be against nonconsensuality" (p. 379).

A major proposition of poststructuralism, however, is the contention that communities tend, just as metanarratives, toward totalism, and are inclined to become totalitarian in nature, in that they suppress individual self-expression, expect conformity to community norms and exclude certain people from participation (Phillips, 1993, p. 163). Poststructuralists, such as Cherryholmes and Foucault, cast their lot with the individual, the marginalized and the subjugated.

In spite of pushes on the part of many postmodernists, including poststructuralists, toward difference, the self remains inevitably linked with community. Part of the nostalgia we express about prior times has to do with images of personal relationships, shared values, social cohesion in our respective communities. In fact, the self is largely constituted in a world of others, and our individual selves are derived, in large part, from the social world. No easy resolution is possible in the tension between self and community. The contemporary discussion between Liberalism, which stresses individualism, and communitarianism is vital, as is reflected in the works of scholars such as Alasdair MacIntyre (1981, 1988), Michael Sandel (1982), and Charles Taylor (1989) and popularized by Robert N. Bellah and his colleagues (1985). We would only caution against taking an extreme view toward either individualism or communitarianism. This suggests that the self be viewed as *subject-in-community* (Haber, 1994, p. 114), and communities should best be viewed as open, temporary and ever-changing.

The contemporary comparative education community is increasingly characterized by heterogeneity, by people professing a vast array of knowledge constructs. This presents a dilemma, because the field is consumed by a multitude of different discourses with little prospect of connecting the various language games. If all we have is heterogeneity and no unity, however, we have no community. In a political sense, some notion of community must exist if comparative education is to exist, and the question we must face is how that community is constituted. From a postmodern perspective, it would be tragic if the field, in deference to some need for community, reverted to orthodoxy and the totalitarian press of certain metanarratives. We must point toward a community structured in such a way that it can accommodate and even appreciate difference. It must be conceptualized in such a way that individuals in comparative education can act as subjects, but they must always be subjects in community. Of course, comparative educators depend only partly on common intellectual paradigms for their sense of community. It is more an affective outcome of shared values and a common history, of participation in common activities such as conferences, readings, and research, of feelings of solidarity through "fraternal sentiments and fellow feeling" (Sandel, 1982, p. 150).

The fact that we often use adjectives such as "totalizing," "coercive" or "terroristic" to describe metanarratives suggests that there might be certain metanarratives that are not totalizing, coercive, or terroristic. I have already taken a stand that comparative educators sympathetic with postmodernism should not reject all metanarratives; otherwise we would be trapped into localized frameworks that have no general validity, and even be denied the opportunity to engage in comparison. Under these circumstances comparative education would be at a dead end. Useful metanarratives "ought to open the world to individuals and societies, providing forms of analysis that express and articulate differences, that encourage critical thinking without closing off thought and avenues for constructive action" (Rust, 1991, p. 618).

What might such metanarratives be? Lyotard (1984, pp. 60–67) champions "open systems" as one possibility. This is an attractive option for those of us in education.

Systems theory has long been a staple of social theory. Unfortunately, in education, systems theory has too often been tied closely to deterministic and rigidly controlled systems. These are the kinds of social systems Lyotard (1984, p. 11) criticizes for their "totalizing" and "terroristic" qualities, and they are the types of systems Paulston apparently had in mind when he characterized systems theory as part of the equilibrium paradigm (1976). To Lyotard (1988), this type of system is technocratic in concept, in that "it programs itself like a computer, is the optimization of the global relationship between input and output—in other words, performativity." In fact, a wide range of different system types exist. Bela Banathy has suggested four descriptive continua that typically characterize different systems types: mechanistic versus systemic, unitary versus pluralist, restricted versus complex, and closed versus open. These characteristics might be aligned in such a way that they allow us to chart a continuum of social systems types. Those systems that are closed, mechanistic, unitary and restricted would be characterized as rigidly controlled, while those that are open, systemic, pluralist and complex would be characterized as purpose-seeking. Between these two extremes we might chart a series of social system types ranging from those that are *rigidly controlled* to those that are *deterministic*, and beyond to *purposive*, *heuristic* and finally *purpose-seeking*. A visual map of such system types is outlined in Figure 1.

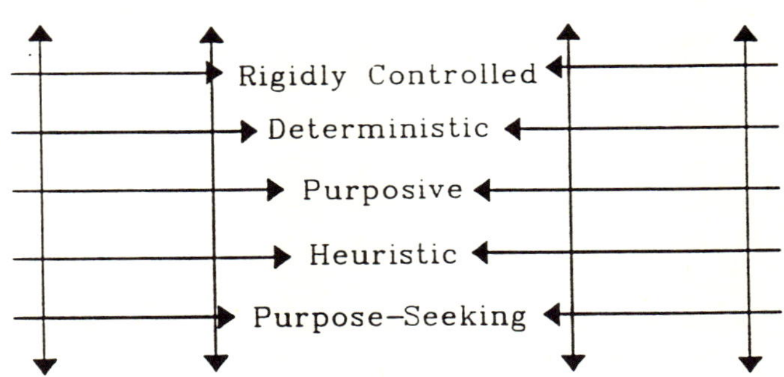

Figure 1. A map of system types

From this scheme, the five system types related to education could be characterized in the following general way, and open educational systems tend toward one end of the system-type continuum.

Rigidly Controlled Systems

These systems are relatively closed in that they have limited interaction with their environments. They operate at a limited number of decision-making levels, have clearly defined goals prescribed to them and have little or no room for self-direction. They behave mechanistically in that their structure is rigid with tightly defined relationships. They exhibit little system dynamics in that they change and innovate with difficulty. Schools in totalitarian states and those sponsored by certain fundamentalist religions typically fit this model.

Deterministic Systems

These systems are a little more open than are rigidly controlled systems, though their boundaries remain closely guarded and controlled. Goals are clearly set for them and operational objectives are externally prescribed. Some minor relational and structural changes might be expected through time; however, the systems remain steady-state in nature. Schools in highly centralized democratic states usually fit this model.

Purposive Systems

These are middle range systems between the two extremes, and are somewhat unitary in terms of goals. Usually strategic goals are externally prescribed while operational objectives and methods are defined by the system itself. Changes do take place but they occur slowly and are coupled with broader environmental evolution. Public schools in decentralized democratic states usually fit this model.

Heuristic Systems

These are relatively pluralistic in that those in the system can formulate their own goals and objectives and they are able to operate within self-selected methods. These systems are characterized by a certain degree of ambiguity. Changes are unpredictable. Nontraditional schools willing to engage in some renewal and reconstruction fit this model.

Purpose-Seeking Systems

These are complex systems guided by images that they shape themselves. They are highly self-directive, shaped by the people in the system itself. Constant change or consideration of change exists in the system. Ambiguity exists, but it is by conscious design to prevent premature closure on decisions. Consciously alternative schools, artistic educational communities and creative schools fit this model.

Lyotard believes open systems, characterized by Banathy as being heuristic and purpose-seeking, are those qualified to serve as appropriate models of science, discourse and knowledge. The contemporary field of comparative education must be characterized as an "open system" and that system has been mapped by Paulston (1992) in such a way as to clarify its multiplicity and complexity.

Comparative Education Cartography

Comparative education cartography is best portrayed by Rolland Paulston's 1992 map of macro intellectual discourse communities (presented as Figure 2 in the preceding chapter). He explains that it portrays his own perception of how the multiple and diverse intellectual communities of the field are situated and interact with each other. Because the map portrays his personal perception, it does not represent "a truth, but a portrait." It represents his interpretation of a "momentary crystallization" of the space claimed by social and ideological ways of seeing (Paulston and Liebman, 1993, p. 15).

Paulston's map of intellectual discourse communities has a metaphorical utility evoking an image of the interconnections between theoretical orientations in comparative education. Such a metaphorical sense has great imaginative utility and represents the way in which social cartography has, to this point, been used. As metaphor, it hints at certain relationships but does not represent other possible connections. I agree with Joseph Seppi, in this volume, that the metaphorical use of cartography is a powerful tool, but it should not prevent us from appreciating the possibility of expanding the metaphor into an analogy. As an analytic tool, a map could take on greater definition and more specific symbols. This would allow those sympathetic to an approach to embellish on the map and to contribute to it and would also allow those critical of an approach to pinpoint their specific criticisms.

I would like to borrow from social network theory in an attempt to suggest a set of possible features that might be helpful in mapping. Maps of social networks typically consist of nodes and lines. In network theory, a node is a point that represents the individuals in a small group, an organization or a community. Nelly P. Stromquist has, in this volume, looked at individuals in various institutional settings as her primary unit of analysis, or node, while Crystal Bartolovich uses transnational corporations as her unit of analysis, or node. A line symbolizes the type of relationships that exist between the individuals. In a Boy Scout troop, for example, each Boy Scout would be seen as a node in the troop network. In mapping the intellectual landscape of comparative education, a node is not necessarily a person, but can be either a text or a particular theoretical orientation. Lines represent the kind of interactions or relationships that exist between different texts or theoretical orientations. I shall draw liberally from J. Clyde Mitchell (1969, Chapter 1), who has provided a number of characteristics and criteria of social networks that are relevant to social cartography. Both morphological characteristics and interactional criteria exist in a network. For illustrative purposes, I shall discuss Paulston's map of intellectual discourse communities (1992) to suggest how the various network features may be defined. Paulston's unit of analysis is a text (e.g., feminist or neo-Marxist) and my unit of analysis will be a theoretical orientation (e.g., feminism or neo-Marxism).

Morphological characteristics of networks are their size, shape, structure and the relationships of their internal parts. I shall outline three possible morphological characteristics of a intellectual map of comparative education: reachability, density, and range.

Reachability

In networks, some nodes are easily accessible to other nodes, while certain nodes are not easily reached. In a map of theoretical orientations, certain of these orientations are easily accessible, while others may be marginalized and not easily accessible. If we were to map Paulston's period of orthodoxy in comparative education, for example, it would be clear that certain orientations, particularly functionalist, modernization, human capital and other evolutionary orientations would be open to each other, while other orientations would not be accessible to the comparative education community. Many orientations would exist, but access to them would not be direct and open.

Density

A community, including the comparative education community, may be close-knit or loose-knit. It may be fragmented or tightly coupled, and a map might provide a mechanism to indicate the degree to which it is either.

Range

In a network, nodes are interlinked with large numbers of other nodes, while other nodes are linked to only a few nodes or may even be outside the range of the community. In the comparative education community, some theoretical orientations are interlinked with large numbers of other orientations while others are linked to only a few other orientations. Certain orientations are also linked via intermediary orientations. For example, in Paulston's map, cultural rationalization appears to have little direct connection with historical materialism, but its linkages, through critical theory and neo-Marxism, are many. Some orientations might be considered isolates in that they don't appear to be well connected with the existing orientations. For example, Paulston gives attention to phenomenography, but it is largely isolated from the main nodes of comparative education. He has not identified two orientations to which I give considerable attention, chaos and self-organization theory, because they currently fall outside the range of the general comparative education community.

Interactional criteria relate to the types of interactions that may occur, and may include content, directedness, intensity and frequency. Paulston's major interests have to do with these criteria.

Content
Probably the most important interactional aspect of networks is the content of the interaction. What exactly is transmitted in an interactive process? It may be feelings, concepts, specific data, techniques, or processes. Paulston has indicated a host of different kinds of content that are being exchanged in comparative education in this period of heterogeneity, and a map of the field could be designed that shows what content is being exchanged by which theoretical orientations.

Directedness
A direction always exists in an exchange relationship. It can be one-way, reciprocal, or multidirectional. For example, according to Paulston, modernization texts have remained largely closed to orientations that do not have very similar perspectives; yet a number of theorists from other orientations have directed their attention, for good or ill, toward modernization. Some orientations, such as feminism and critical theory, modernization and human capital theory, poststructuralism and feminism, enjoy reciprocal interests. It is usually important to clarify on any map the direction of exchanges.

Intensity
An interaction may be of content that suggests a high identity on the part of exchanging parties. On the one hand, great resonance can be found between modernization, neostructuralism and human capital theory, just as resonance exists between historical materialism, neo-Marxism and certain

variations of dependency theory. Of course, critical or negative relationships also exist. For instance, dependency theorists enjoy contrasting themselves with modernization and human capital theorists.

FREQUENCY

A quantifiable characteristic of interaction may be the frequency of contact between two orientations, defined in some reasonable way as frequency of citation-sharing or number of essays and books in which the other orientation is discussed. The frequency of exchange includes both positive or negative content.

It is neither necessary nor productive to include all of the above characteristics and criteria in a single map. The nature of the map might dictate how inclusive it might be. A 3-D map, for example, of the intellectual field of comparative education might indicate concentrations of power, trends, and size of nodes.

A MACRO-MAPPING OF COMPARATIVE EDUCATION AS AN INTELLECTUAL FIELD

In Figure 2, I would like to propose an elaboration and reconstruction of Paulston's macro-map of comparative education, relying in large part on the information he provides in his own 1992 text. In so doing I benefit in that I do not have to engage in an extended review of texts and can simply refer the reader to the sixty texts Paulston interrogated in his mapping enterprise. I also rely on his basic map coordinates, where the vertical dimension relates to textual dispositions toward social and educational change (*transformation* vs. *equilibrium*), and the horizontal dimension relates to textual characterizations of reality (*objective-realist* vs. *idealist-subjectivist*). Finally, in my map I indicate how some of the characteristics and criteria of the mapping scheme, outlined above, might be incorporated.

Paulston located the field of his map in two overlapping circles. I have used the same coordinates as he, but the field of my map consists of a 2 x 2 matrix resulting from two intersecting axes. Borrowing from Paulston, the vertical axis consists of a *transformations-equilibrium* continuum, and the horizontal axis consists of a *personal-scientific* continuum. While the meanings of each of these continua are implicit, it is important to define the term *equilibrium*, because it has at least two meanings. On the one hand, it has the meaning Paulston used in his classic study of social and educational change (Paulston, 1976), which was of evolutionary, incremental and adaptive change. On the other hand, it has the meaning used by students of dissipative systems (e.g., Jantsch, 1980; Eisler and Loye, 1987; Prigogine and Stengers, 1984), where equilibrium is the final period of the dissipative pro-

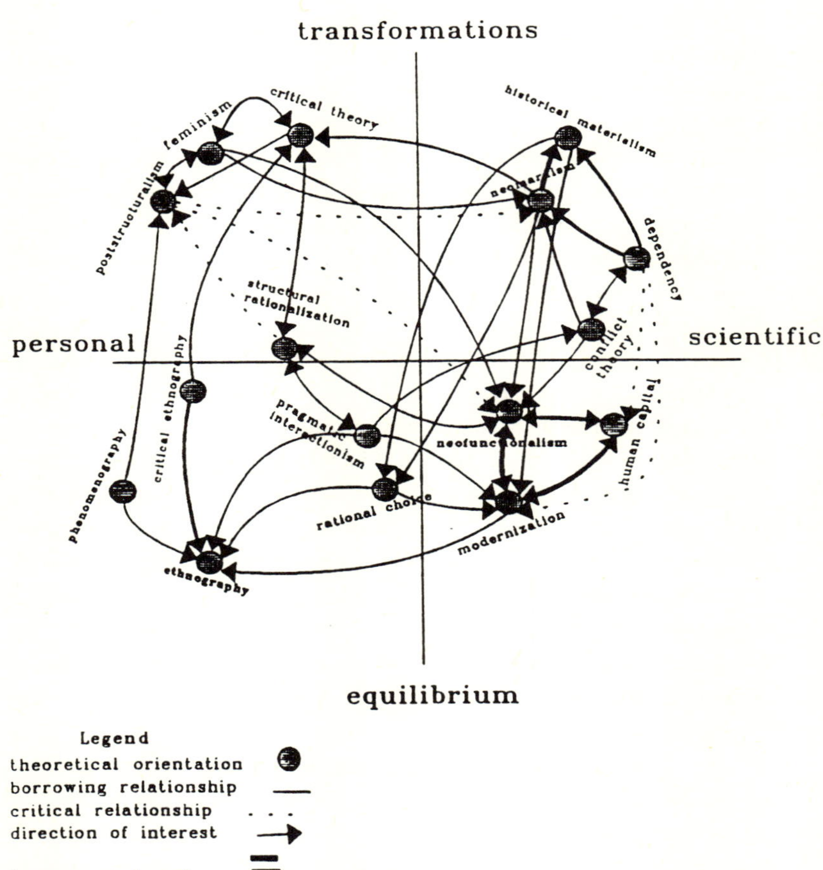

Figure 2. A revised macro-mapping of paradigms and theories in comparative and international education

cess leading to entropy. In my own use of the term, equilibrium is equated with entropy, because it represents one extreme of the continuum.

I have then located sixteen theoretical orientations at an appropriate place in one of the four quadrants of the matrix. They are placed in large part according to the place those representing the orientations locate themselves along the continua, though my own interpretation inevitably plays a significant role in this placement. These orientations form the various nodes of the map. It has not been possible to provide comparative data concerning the various nodes. No indication is given, for example, of the relative size of the research community identifying with each node. Nor is there any indication of the exchange activities at work within each theoretical orientation. A major concern of Paulston has to do with the relationships between theoretical orientations, and I have attempted to indicate in a somewhat more detailed manner the nature of those relationships. A legend has been created to indicate whether the relationship between theoretical orientations, indicated by circles, was of a borrowing or a critical nature. The direction of the relationship, denoted by an arrow, is important. The direction of the arrow indicates that the writer of a text drew from another node or made specific reference to that orientation. The thickness of the line linking two nodes symbolizes the frequency of reference as illustrated in bibliographic citations and number of essays, with a thicker line indicating greater frequency.

Concluding Remarks

In this essay I have provided a postmodern perspective on mapping the field of comparative education, relying primarily on the ideas of Lyotard and Foucault. Comparative education is presently characterized by its heterogeneity rather than orthodoxy or heterodoxy. A postmodern approach to comparative education in this period of heterogeneity is not intended to discredit any orientation, but to see the kinds of relationships that exist. Postmodernists would by and large neither seek to silence or make invisible any of the theoretical orientations in the field, although they would challenge the claims of some that their approach is the only valid perspective. A postmodern approach also encourages a variety of ways of presentation, including social cartography.

Notes

1. Thanks to UCLA students, particularly Mitchell Chang, Megumi Shibuya, and Karen McClafferty, in crafting this chapter.

2. I am not a linguist and I find the allegiance poststructuralists hold to language somewhat confining. My own orientation is more in harmony with that of Ernst

Cassirer, who claims that the physical universe is buffered by a symbolic universe of people's own creation; he maintains that "man lives in a symbolic universe," and not only language but "myth, art, and religion are parts of that universe" (Cassirer, 1944, p. 43).

   3. Fredric Jameson (1972) has written a book using Nietzsche's statement as a title.

REFERENCES

Althusser, L. (1990). *Philosophy and the spontaneous philosophy of the scientists.* London: Verso.

Banathy, B.H. (1988). Matching design methods to system type. *Systems Research, 50*(1), 27–34.

Baudrillard, J. (1984). On nihilism. *On the Beach, 6,* 38–39.

Bellah, R.N., et al. (1985). *Habits of the heart.* New York: Harper and Row.

Bloom, A. (1987). *The closing of the American mind.* New York: Simon and Schuster.

Bloom, H. (1994). *The Western canon: The books and school of the ages.* New York: Harcourt, Brace.

Boli, J., F. Ramirez, and J. Meyer. (1985). Explaining the origins and expansion of mass schooling. *Comparative Education Review, 29,* 145–170.

Bowles, S., and H. Gintis. (1975). *Schooling in capitalist America.* New York: Basic.

Briggs, J., and F.D. Peat. (1989). *Turbulent mirror.* New York: Harper and Row.

Buzan, T. (1991). *Use both sides of your brain.* 3rd ed. New York: Penguin.

Cassirer, E. (1944). *An essay on man.* Garden City, NY: Doubleday.

Cherryholmes, C. H. (1988). *Power and criticism: Poststructuralism investigations in education.* New York: Teachers College, Columbia University.

Derrida, J. (1978). *Writing and difference* (translated by A. Bass). London: Routledge.

de Saussure, F. (1966). *Course in general linguistics.* Translated by W. Baskin. New York: McGraw-Hill.

Eco, U. (1989). *Foucault's pendulum.* San Diego: Harcourt Brace Jovanovich.

Edwards, B. (1979). *Drawing on the right side of the brain.* Los Angeles: J.P. Tarcher.

Eisler, R. and D. Loye. (1987). Chaos and transformation: Implications of nonequilibrium theory for social science and society. *Behavioral Science* 32:53–65.

Foucault, M. (1980). *Power and knowledge: Selected interviews and other writings, 1972–1977.* New York: Pantheon.

Giroux, H.A. (1991). Modernism, postmodernism, and feminism: Rethinking the boundaries of educational discourse. In H.A. Giroux (ed.), *Postmodernism, feminism, and cultural politics* (1–60). Albany: State University of New York Press.

Haber, H.F. (1994). *Beyond postmodern politics: Lyotard, Rorty, Foucault.* New York: Routledge.

Habermas, J. (1981). Modernity versus postmodernity. *New German Critique, 22,* (Winter), 3–14.

Held, D. (1980). *Introduction to critical theory: Horkheimer to Habermas.* Berkeley: University of California Press.

Hofmann, H.-G., and Z. Malkova. (1990). Comparative education around the world: The socialist countries. In W.D. Halls (ed.), *Comparative education: Contemporary issues and trends* (109–144). London: Jessica Kingsley.

Huyssen, A. (1984). Mapping the postmodern. *New German Critique, 33,* 5–52.

Jameson, F. (1972). *The prison-house of language: A critical account of structuralism and Russian formalism.* Princeton, NJ: Princeton University Press.

Jameson, F. (1984). Postmodernism, or the cultural logic of late capitalism. *New Left Review, 146,* 53–92.

Jantsch, E. (1980). *The self-organizing universe.* Oxford: Pergamon.

Jencks, C. (1987). *What is post-modernism?* New York: St. Martin's.

Kneller, G.F. (1994). *Educationists and their vanities.* San Francisco: Caddo Gap.

Lorenz, E.N. (1993). *The essence of chaos*. Seattle: University of Washington Press.

Lyotard, J.-F. (1984). *The postmodern condition*. Translated by G. Macpherson. Oxford University Press.

Lyotard, J.-F. (1985). *Just gaming*. Translated by W. Godzich. Minneapolis: University of Minnesota Press.

Lyotard, J.-F. (1988). *The différend: Phrases in dispute*. Translated by G. van den Abbeele. Minneapolis: University of Minnesota Press.

MacIntyre, A. (1981). *After virtue: A study in moral theory*. London: Duckworth.

MacIntyre, A. (1988). *Whose justice? Which rationality?* London: Duckworth.

Minh-ha, T.T. (1989). *Women, native, other: Writing postcoloniality and feminism*. Bloomington: Indiana University Press.

Mitchell, J.C. (1969). *Social networks in urban situations*. Manchester, England: Manchester University Press.

Noah, H. and M. Eckstein (1969). *Towards a science of comparative education*. New York: Macmillan.

Ornstein, R. (1986). *Multimind: A new way of looking at human behavior*. Boston: Houghton Mifflin.

Parsons, T. (1967). *The social system*. Glencoe, IL: Free Press.

Paulston, R. (1976). *Conflicting theories of social and educational change: A typological review*. Pittsburgh: University Center for International Studies, University of Pittsburgh.

Paulston, R. (1992). Comparative education as an intellectual field: Mapping the theoretical landscape. Paper presented at the 8th World Congress of Comparative Education, July 8–14, Prague, Czechoslovakia. [An abbreviated version was subsequently published in *Compare, 23*(2), 101–114].

Paulston, R., and M. Liebman. (1993). *The promise of a critical postmodern cartography*. Research Report #2 Occasional Paper Series. Pittsburgh: Department of Administrative and Policy Studies, University of Pittsburgh

Phillips, D.L. (1993). *Looking backward: A critical appraisal of communitarian thought*. Princeton, NJ: Princeton University Press.

Prigogine, I. and I. Stengers. (1984). *Order out of chaos: Man's new dialogue with nature*. New York: Bantam.

Rabinow, P., ed. (1984). *Foucault reader*. New York: Pantheon.

Rorty, R. (1989). *Contingency, irony, and solidarity*. Cambridge, England: Cambridge University Press.

Rorty, R. (1990). The dangers of over-philosophization—Reply to Arcilla and Nicholson. *Educational Theory, 40*, 41–45.

Rust, V.D. (1991). Postmodernism and its comparative education implications. *Comparative Education Review, 35*(4), 610–626.

Sandel, M. (1982). *Liberalism and the limits of justice*. Cambridge, England: Cambridge University Press.

Sarchett, B.W. (1995). What's all the fuss about this postmodernist stuff? In J. Arthur and A. Shapiro (eds.), *Campus wars: Multiculturalism and the politics of difference* (19–27). Boulder: Westview.

Searle, J.R. (1995). Postmodernism and the Western rationalist tradition. In J. Arthur and A. Shapiro (eds.), *Campus wars: Multiculturalism and the politics of difference* (28–47). Boulder: Westview.

Taylor, C. (1985). *Philosophy and the human sciences*. Cambridge, MA: Harvard University Press.

Wittgenstein, L. (1953). *Philosophical investigations*. New York: Macmillan.

Young, I.M. (1990). *Justice and the politics of difference*. Princeton, NJ: Princeton University Press.

Constructing Knowledge Spaces and Locating Sites of Resistance in the Modern Cartographic Transformation

David Turnbull

David Turnbull

INTRODUCTION: THE SCIENCE/CARTOGRAPHY RELATIONSHIP

This chapter explores the relationship of the state, science and cartography. It is argued that the state, science and cartography became strongly intermeshed through a complex of sociohistorical processes beginning around 1600. In effect, they served to coproduce one another. The establishment of modern cartography required similar kinds of ordering in both the production of scientific knowledge and in society. The culmination of the first phase of this cartographic transformation was the establishment of the knowledge space within which scientific knowledge is assembled and the state is organized, and which is now taken for granted in Western culture. The continuation of this transformative process in the last decades of the twentieth century has brought about such a major expansion of the scientific knowledge space that it appears from one vantage point to be totally hegemonic. However, the very process of mapping this emergent space reveals the social labor that has gone into its construction and the ways in which local assemblages of knowledge have been suppressed and eliminated in the process. This revelation of the social and the local opens sites for resistance, as may be seen in the other chapters in this volume advocating various approaches to resistance mapping.

We are largely unconscious of the centrality of maps in contemporary Western life, but they are everywhere and profoundly structure our thinking and our culture. We are physically surrounded and bombarded by maps, in our newspapers, on our televisions, in our books and in our getting around in the modern world (Wood, 1992a). The cartographic trope, or figure of speech, is all-pervasive. We talk of cognitive maps, mental maps and genetic maps, or of mapping ways of seeing and mapping the human genome (Downs and Stea, 1973, 1977; Gould and White, 1974, Paulston, 1993; Tolman, 1948). Minds, languages and cultures are described as maps.

The anthropologist Edmund Leach thinks "our whole social environment is map-like" (Leach, 1976, p. 51). The sociologist Pierre Bourdieu notes that "culture is often referred to as a map" (Bourdieu, 1977, p. 2). Postmodernists constantly resort to the cartographic figure in their explorations of the contemporary cultural landscape (Ferrier, 1990; Diprose and Ferrell, 1991). A salient example is a recent newspaper article that portrays the scientific description of the grammatical structure of English as a process of mapping comparable with the mapping of flora and fauna into phyla and species (Sullivan, 1994).

Ludwig Wittgenstein (1958) indirectly invoked the cartographic metaphor and used it to characterize the whole philosophical enterprise when he said, "A philosophical problem has the form 'I don't know my way about'" (p. 49). Stephen Hall has claimed that if there is one thing that best represents the ways in which we as human beings perceive, represent, measure, know and understand reality, it is the map (Hall, 1992, p. 3). The mode of perceiving, representing, measuring, knowing and understanding reality that best represents contemporary Western technoscience is indeed the map. Our intellectual mind-set, our form of life is so intensely cartographic that it is tempting to accept Denis Wood's description of modern Western society as map-immersed (Wood, 1992a, p. 34). However, the scientific map is not yet the universal form of knowledge space, though it is a dominant one. Just as there is more than one way to navigate, there are many ways of mapping. There has, however, been no Kuhnian-style "cartographic revolution." Instead there has been a rather heterogeneous, locally contingent process in which maps have become integrated with science and the state. Different cultures, different periods and different groups in a culture have produced differing knowledge spaces or ways of assembling knowledge that have simultaneously shaped those cultures, eras and people (Belyea, 1992; Edney, 1993; Turnbull, 1991).

As the end of the twentieth century approaches, the cartographic mode of knowledge assembly is undergoing a series of transformations that seem capable of producing total representational hegemony. Technical innovations like the development of an optimal-conformal projection and fractal computer analysis will enable the incorporation of previously unutilized satellite data in topographical maps. But the transformation is not simply a matter of continuous innovation in mapping techniques. There has simultaneously been a radical change in the capacity to assemble and analyze spatially referenced data through the developments in computer processing that permit the visualisation and graphic representation of information (Friedhoff, 1989; Hearnshaw and Unwin, 1994; MacEachren et al., 1992). At the moment, this development in the spatial organization of knowledge goes under the somewhat restrictive and innocuous heading of geographical information systems (GIS), but it is not inherently limited to matters geographic, as it is set to provide a radically new technology for obtaining knowledge generally (Morrison, 1989; Mundell, 1993; Papp-Vary, 1989; Taylor, 1989; Taylor, 1993). Over the past twenty years, there has been a vast increase in the range of spatially portrayed domains (Hall, 1992). Scientists are currently mapping everything from the cosmos to the atom, from the brain to society, from physical objects to the laws of physics (Rubiner, 1995). At the same time, there are moves to expand our linear, individual and causal style explanations to include spatial, relational and systemic explanations. This transition to spatial explanation has occurred not only in the physical and biological sciences, as Hall's *Mapping the Next Millenium* so amply illustrates, but is also well underway in the human sciences:

> Understanding how history is made has been the primary source of emancipatory insight and practical political consciousness, the great variable container for a critical interpretation of social life and practice. Today, however, it may be space more than time that hides consequences from us, the "making of geography" more than the making of history that provides the most revealing tactical and theoretical world. (Soja, 1989, p. 1)

So comprehensive is this spatialization of knowledge that is envisaged that it may become possible to represent any knowledge as a "fuzzy cognitive map" (McNeill and Freiberger, 1993, p. 230).

Just as one would expect in this ever more reflexive world, there is a concomitant transformation underway reshaping both what it is we take maps to be and what we take science to be. One of the most important themes that has emerged in the revised understanding of cartography and science was proposed by Brian Harley, whose extensive but unfortunately curtailed writings have displayed the ways in which maps are texts that can be deconstructed to reveal their concealed power (Harley, 1988a, 1988b, 1989; Harley and Woodward, 1987). Denis Wood, in his iconoclastic and penetrating book *The Power of Maps*, has gone even further to argue that maps are "weapons in the fight for social domination" (Wood, 1992a, p. 66). Similar themes have been taken up in a new understanding of science by such writers as Joseph Rouse, who argues that "the experimental and theoretical practices of science are themselves forms of power" (Rouse, 1987,

p. 248). Likewise, Bruno Latour's examination of the practice of scientists and technologists has shown how facts are constructed and techniques are black-boxed through the elaboration of social networks (Latour, 1987).

A first step in this exploration is to reconsider the role of maps in science. One of the most powerful instances of the use of the map metaphor has been in the description of scientific theories. The map-theory relationship has been commented on by such philosophers as Michael Polanyi, in whose view "all theory may be regarded as a kind of map extended over space and time" (1958, p. 4), and Thomas Kuhn, who extended the point, arguing that:

> as a vehicle for scientific theory, the paradigm functions by telling the scientist about the entities that nature does and does not contain and the ways in which those entities behave. That information provides a map whose details are elucidated by mature scientific research. And since nature is too complex and varied to be explored at random, that map is as essential as observation and experiment to science's continued development. Through the theories they embody, paradigms prove to be constituitive of the research activity. They are also, however, constituitive of science in other respects . . . paradigms provide scientists not only with a map but also with some of the directions essential for map making. In learning a paradigm the scientist acquires theory, methods, and standards together usually in an inextricable mixture. (1970, p. 108)

But it is between cartography and science conceived as total systems of knowledge that there is an especially powerful symbiotic and symbolic synergy. Not only is the most common image of scientific knowledge or theories that of the map, but also the strongest theme running through the history of cartography is that of maps as increasingly scientific and ever more accurate mirrors of nature. In this way, the development of "scientific maps" is taken to be identical with a progressive, cumulative, objective and accurate representation of geographic reality and is also taken to be synonymous with the growth of science itself.[1] An instance of the ways in which the processes of science and mapping are jointly embedded is the concept of "discovery." Territorial discovery and scientific discovery are both conflated with, and mediated by, maps. Both are typically portrayed as natural processes of revelation of the unknown. Both, I would argue—along with Bartolovich in her chapter in this volume—are processes of social ordering that have been coproduced with Western capitalism.

Maps and science are jointly embedded in our forms of life at another level, that of rationality and meaning. Lucy Suchman, in her discussion of situated technology and the possibility of artificial intelligence, suggests that "The view, that purposeful action is determined by plans, is deeply rooted in the Western human sciences as the correct model of the rational actor" (Suchman, 1987, p. ix). Plans, in Suchman's usage, are rules, algorithms or instructions. It is this sense of maps or plans as rules or instructions that finds deep resonances with one strand in Western thought concerning the ideals of rationality and of science itself.

So, the question is, how did all this come about? Why is there such a mutuality among science, cartography and modern society? It is proposed here to initiate the explorations by looking at such questions as: Where and when did this prevalence of maps begin? What is the role of maps in other societies? What do maps and science have in common? What kinds of social processses were involved in their introduction? What sort of function do maps have?

Rethinking the Spatiality of Cartography and Technoscientific Knowledge

The interaction between cartography and science provides an especially opportune site for the application of what Woolgar has called "the lever of irony" in opening up the technoscientific black box (Woolgar, 1983). It is precisely the self-evident nature of the science-cartography relationship that provides the ironic point that makes it possible for technoscience to be conceived as being other than it is. Or, to put it as Marshall McLuhan did over thirty years ago,

> Maps are a prime vehicle for repositioning, reframing, rethinking science because theories are maps, maps are science instantiated, without maps science would not have been possible. The art of making pictorial statements in a precise and repeatable form is one that we have long taken for granted in the West. But it is usually forgotten that without prints and blueprints, without maps and geometry, the world of modern science would hardly exist.[2] (1964, p. 57).

The discursive strategy adopted in this chapter parallels that of Shapin and Schaffer in their seminal work *Leviathan and the Air Pump: Hobbes, Boyle and the Experimental Life*. They argue that we now take it for granted that experiments can produce reliable knowledge, but this was not always so. In order for such artifactual knowledge to be considered reliable and authoritative, a great deal of social labor had to go into the establishment of social and literary technologies that are no longer visible as social constructs. Shapin and Schaffer describe their analysis of the origins of experimental practice in the seventeenth century as showing that:

> the experimental production of matters of fact involved an immense amount of labor, that it rested upon the acceptance of certain social and discursive conventions, and that it depended upon the production and protection of a special form of social organisation. The experimental program was, in Wittgenstein's phrases, a "language game" and a "form of life." (Shapin and Schaffer, 1985, p. 22)

They ground their account by problematizing the self-evidentiality of experiment and in so doing invoke the very cartographic trope, or figure of speech, that I have suggested is so pervasive:

> The success of the experimental program is commonly treated as its own explanation. Even so, the usual way in which the self-evident method presents itself in historical practice is more subtle—not as a set of explicit claims about the rise, acceptance, and institutionalisation of experiment, but as a disposition not to see the point of putting certain questions about the nature of experiment and its status in our overall intellectual map. (Shapin and Schaffer, 1985, pp. 5–6)

Thus, in order to pursue the map-science relationship, we need to ask, "What is it that we are disposed not to question about maps?" This question is especially acute when the doyens of cartographic theory, Robinson and Petchenik, ask what there can be about the map that is so profoundly fundamental? Why should a representational system for space be so basic? Their answer is that the problem in analyzing maps as communication is that the universal metaphor turns out to be the map itself; that maps are surrogates of space:

> As we experience space, and construct representations of it, we know that it will be continuous. Everything is somewhere, and no matter what other characteristics objects do not share, they always share relative location, that is spatiality; hence the desirability of equating knowledge with space, an intellectual space. This assures an organisation and a basis for predictability, which are shared by absolutely everyone. This proposition appears to be so fundamental that apparently it is simply adopted a priori. (Robinson and Petchenik, 1976, p. 4)

The geographers Chorley and Haggett claim a common link in language to argue that "it is characteristic that maps should be likened to language and scientific theories" (Chorley and Haggett, 1967, pp. 48–49). Malcom Lewis, a historical geographer, has similarly argued for an evolutionary relationship between language and spatial consciousness (Lewis, 1987, pp. 51–52). Denis Wood finds a Piagetian development in mapping paralleling the cognitive development of children (Belyea, 1992a, 1992b; Wood, 1992a, 1992b, 1993). Michel de Certeau locates the centrality of space in human consciousness, in the more pedestrian but no less vital role of walking in our everyday lives (de Certeau, 1984, p. 91). Much of the essence of these sorts of claims is summed up in the view of the anthropologist Robert Rundstrom in his work on Inuit maps, that "mapping is fundamental to the process of lending order to the world" (Rundstrom, 1990, p. 155). Wood agrees that mapping, in the sense of deploying mental maps, is a common human trait, but argues that it is not the same as map making, which for him, as we shall see, is largely restricted to societies with a high degree of social complexity (Wood, 1993, p. 2).

A claim for the most fundamental role of maps in our understanding also comes from recent work in neurophysiology, which suggests that the role of the human neocortex, whose extensive development distinguishes homo sapiens, is to create and store memories as maps.[3] The ethologist Talbot Waterman even goes as far as to argue that many animals, birds and insects have a "map sense" (Waterman, 1989, p. 175).

Such claims, while seemingly attractive, are more likely to reflect the pervasiveness of the map metaphor in our scientific culture and questions we are not, therefore, disposed to ask, than they are to reflect the existence of maps in animal or human brains. Whatever the etiology of the spatial in our knowledge, these researchers all assume the metaphysically self-evident spatiality of knowledge. This assumption seems to capture the essence of why it seems so natural to think of knowledge in terms of maps. From Robinson and Petchenik's position, an argument might be sketched out along the following lines: Maps are surrogates of space; knowledge is in some sense spatial in virtue of it being structured; organized knowledge or theories are therefore like maps. Likewise, Lewis's claims for a link between spatiality and linguistic structure are extremely suggestive. However, both such arguments fail to pose this question: How do we come to accept the modes of spatiality we do? They cede too much to another seemingly plausible self-evident intuition: that the underlying reality is the topographical relation of objects, that they "always share relative location." It is this taken-for-granted understanding about objects, their relations and our ability to represent those relations that constitutes part of the forms of life that underlie Western contemporary thought and science. Indeed, our representations of spatial relations are coproduced with our understanding of what spatial relations consist of. Given this dialectic, how do we break into the apparent circle?

Maps as Heterogeneous Assemblages

In order to render visible, vocalizable and questionable that which is otherwise transparent, silent and unquestioned about spatial representation, it is necessary to bring into focus the hidden labor, social organization and discursive formations that constitute the mode of knowledge production and render it map-like. One way to do that is to focus on the contradictions inherent in maps as representations. All representation suffers from such contradictions, so much so that a great deal of the art of the twentieth century has become enmired in a reflexive critique of representation. Maps, however, suffer most acutely insofar as they claim accuracy, scientificity and authority. Of necessity, they cannot avoid the use of some mode of representation that is not itself also a representation; that is to say, they are inherently conventional. Equally, they cannot achieve a full one-to-one correspondence with what they represent; that is to say, they are necessarily selective. Whatever pragmatic resolution of these problems is adopted, it is bound to be incompatible with total truth preservation and consequently, as Monmonier has pointed out, all maps are lies (Monmonier, 1991). Similarly, Nancy Cartwright has argued that the laws of physics are also lies (Cartwright, 1983). The problem is that explanation and truth, the joint aims of scientific inquiry, cannot be simultaneously satisfied (Longino, 1992, p. 284). This trade-off between explanation and truth can be seen as a dynamic tension and the source of change (Kuhn, 1977; Longino, 1992). More typically, the potential for incoherence that it generates is handled by suppression of the conditions under which a given representation, explanation or truth claim is produced.

At one level, the selectivity and conventionality of maps are nonproblematic: We all accept that we cannot represent everything at once and that some mode of representation is necessary. But at another level, we are seldom aware of the ways in which our views of the world are ordered by suppressed social constructs. We are blind to the processes by which the social is naturalized. As Beverley demonstrates in his chapter in this book, maps have boundaries, frames, spaces, centers and silences that structure what it is possible to speak of and also that which it is not possible to speak of. Maps are the product of such transparent processes as "compilation, generalisation, classification, formation into hierarchies, and standardisation of geographical data" (Harley, 1996, p. 5). We are in danger of being prisoners in its social matrix. For cartography as much as for other forms of knowledge, "all social action flows through boundaries determined by classification schemes" (Darnton, 1984, p. 192). Escape from the prison of such seemingly natural boundaries is problematic since they are taken for granted. But as soon as the social construction of such boundaries is made apparent, what were once bars become potential sites of resistance.

Over the past twenty years or so, a great deal of work that falls under the social construction rubric has been done in the sociology of scientific knowledge through examining the actual daily practice of scientists in their laboratories (Woolgar, 1983). The picture of science that has emerged from these empirical investigations of both contemporary and historical scientists is that all knowledge claims are constructed at specific sites through the embodied engagements of particular scientists with particular skills, materials, tools, theories and techniques. Such processes of knowledge production are revealed as thoroughly social and contingent, requiring judgments and negotiations by groups of scientists in specific contexts (Woolgar, 1988). Thus, a fundamental characteristic of scientific knowledge production is its localness (Turnbull, 1993c).

Another motif that links the revised views of both cartography and science is messiness. David Livingstone has recently portrayed the history of the geographical tradition as one of "situated messiness," following Shapin and Schaffer's description of the history of science as a series of messy contingencies (Livingstone, 1992, p. 28). I have elsewhere described the ad hoc nature of the cathedral-building process as analogous to the processes of science and how much of the persuasive power of science lies in the invocation of plans and maps which serve to deny or conceal and hence rationalize the mess (Turnbull, 1993b; Turnbull and Watson-Verran, 1995).

This revelation of the messy, local character of scientific knowledge production then raises this question: How does scientific knowledge become universal? How does it move from the site of production and become accumulated at a center? Following the critiques of Feyerabend and others, it is no longer possible to explain the accumulation of scientific knowledge simply by the application of a specifiable set of rules or methods (Feyerabend, 1978; Schuster and Yeo, 1986). Instead, the movement of scientific knowledge from the local site and moment of its production and application to other places and times is more fruitfully described as requiring the deployment of social strategies and technical devices. Such strategies and devices create the equivalences and connections whereby otherwise heterogeneous and isolated knowledges are enabled to move in space and time (Turnbull, 1993e). All knowledge systems, from no matter what culture or period, have localness in common, and many of the small but significant differences between knowledge systems can be explained in terms of the differing kinds of work involved in creating "assemblages" from the "motley" of practices, instrumentation, theories and people.[4]

Knowledge is not simply local, it is located. It is both situated and situating. It has place and creates a space. An assemblage is made up of linked sites, people and activities; in a very important and profound sense, the creation of an assemblage is the creation of a knowledge space. The motley of scientific practice, its situated messiness, is given a spatial coherence through the social labor of creating equivalences and connections. Such knowledge spaces acquire their taken-for-granted air and seemingly unchallengeable naturalness through the suppression and denial of work involved in their construction. However, since they are essentially motley based in messiness, they are polysemous and provide potential sites of resistance.

The general suppressive effects and silences of cartographic representations have been brought out by Michel de Certeau in his critique of the modern map as a "totalising device."[5] He argues that the application of mathematical principles produces a formal ensemble of abstract places and collates on the same plane heterogeneous places, some received from tradition and others produced by observation. The modern map is, in effect, an homogenization and reification of the rich diversity of spatial itineraries and spatial stories. It eliminates little by little all traces of the practices that produced it.

De Certeau's characterization of the map as a homogenized assemblage that eliminates the local practices that produce it is extremely apposite. But why is it held that a scientific representation of the world needs to be abstract and geometric? According to the French historian of cartography, Francois Jacob,

> Maps allow the assemblage of a multitude of heterogeneous inputs in order to subject them to the same mathematical logic and to erase their differences through the coherence of the visual codes. They delocalize knowledge, and thus render it accessible, in a condensed and synoptic form, to future generations of unknown researchers who can reproduce it according to the same design. (Jacob, 1992, p. 464, my translation)

Hence, it is claimed that in the pursuit of the aims of accuracy and objectivity, all locations need to be rendered equivalent and connectable in a mathematizable framework. In this way, subjectivity and error can be eliminated and all new information about the world can be accumulated systematically. Seppi elaborates a similar argument in his chapter.

Put this way, the self-evident necessity of abstract geometrization seems quite compelling and there seems to be a natural coincidence of internal logics between science and cartography. However, there are three broad kinds of considerations that mitigate against the strength of that compulsion. First, there are alternative ways of creating geographic assemblages. Second, such modes of representation cannot be achieved purely scientifically; large amounts of social work are involved in creating the connections between the heterogeneous elements. Without such work, those elements have no natural relationship, and so self-assembly cannot be achieved by logic or structural necessity. Finally, the connections that maps establish with the social life in which they are embedded are also not natural or self-evident and have power effects that pervade the whole society. Indeed, it is through the social work of creating the assemblages that science and society may be seen to coproduce each other (Mitchell, 1994).

The Sixteenth-Century Cartographic Transformation

These three considerations can be approached by addressing this question: What brought about the cartographic transformation of the sixteenth century? It has been noted by many historians of cartography that there was a marked increase in map production and use in the last half of that century. Buisseret, for example, claims that around 1600, maps suddenly start to be "essential to a wide variety of professions" (1992, p. 1; see also Harvey, 1993, p. 65; Tyacke, 1983, p. 18). It is tempting to dub this transformation a Cartographic Revolution and to see in it the origins of map consciousness. Such a Kuhnian inclination is, I think, to be resisted. Though there was a sudden increase in map usage at this time it was by no means homogenous or ubiquitous. The adoption of mapping as a mode of knowledge assemblage was a local and contingent matter and was not uniform in, for example, estate management, where one might have expected its universal adoption (Fletcher, forthcoming; Bartolovich, 1995, p. 268). By contrast, the use of accurate plats or plans was, for example, rapidly taken up in late sixteenth century naval shipbuilding and in the massive Dover harbor project of the same period. Shipbuilding and fortification had a long history of large-scale, complex projects without the use of detailed drawings, so the question becomes, as Stephen Johnston asks in his illuminating *Making Mathematical Practice* (1994), what was the strategic advantage of their introduction? Johnston persuasively argues that the advantage lay in the facility of maps in bringing together otherwise disparate techniques, interests and needs for control.

The newly expanded history of cartography, which has moved beyond the confines of its earlier Eurocentric focus, shows that from the earliest recorded history to about 1600, maps were produced in some form or another in virtually every society, but only in relatively small amounts (Harley and Woodward, 1987, 1992). After 1600 or so, maps became commonplace in some areas and in some societies. Wood argues that we need to acknowledge that mapping occurs in all societies, but systematic map making is restricted to a small number of complex societies (Wood, 1993, 1995). However, this distinction needs some qualification. Maps are not the only way in which societies assemble knowledge; various societies have assembled large complex bodies of heterogeneous knowledge without making maps. The Inca, for example, sustained an empire without maps or written records using *quipu*—knotted strings—and lines of *huachas*—shrines radiating across the land from the capital Cuzco (Bauer and Dearborn, 1995; Turnbull, 1993c). All the examples of such knowledge systems that I have examined, however, have organized that knowledge spatially. The other qualification is that even in map making societies, maps are not the only way of assembling knowledge. Moreover, there are different map registers (Belyea, 1992b; Edney, 1993; Rundstrom, 1993; Turnbull, 1996).

In one register, information can be assembled spatially without concern for accuracy or quantification. In the other register, represented by such maps as a national survey or a street directory, the concern is fundamentally accurate spatial representation and quantification. The latter corresponds with the universalistic and totalizing style of reasoning that often serves the interests of the state or centralized authority, in that it embodies such possibilities as economic plans and ignores local variation and control. That maps can be drawn in either register is well illustrated by Benjamin Orlove's (1991, 1993) accounts of the Lake Titicaca peasants' struggle with the state over control of the reed beds in the lake. The struggle can be seen in the ways in which the peasant maps and state maps represent, in different terms, the same shallow areas of Lake Titicaca.

> The state maps lay out a narrative of the conception and enactment (prolepsis) of a detailed plan for a specific territory in which state officials control the natural features and the human populations. The maps made by peasants tell quite a different story; that is, of the control of the territory by a set of communities, all equal to each other; of the continuity of this control, and of the ratification of this control by a remote state. (Orlove, 1991, p. 7; 1993)

Maps, then, do not have a simple universal form and, as the Titicaca example shows, they may vary radically in the way they are used. Orlove also points to another salient difference: "Peasant maps do not immediately form pieces of other larger maps." It is this difference in capacity for assemblage that marks the strongest distinction in maps between and within cultures (Lamb, 1993).

Given that map making does not arise as a matter of course as societies reach a certain complexity, nor is a particular register necessarily bound to dominate, let us then explore the sixteenth century cartographic transformation and the eventual emergence of a scientific knowledge space as it took place in Europe. In England, a country that has one of the richest cartographical histories, a classic early example of the state assemblage of knowledge was the Doomsday Book. However, in its compilation (1085–86), William the Conqueror recorded all of England's estates and their resources using a literary, not a cartographic, mode of assemblage.

It was another two hundred years before the first definite signs of an emerging map making culture in England became apparent. In his *Opus Maius* (1266), Roger Bacon proposed an equipollent coordinate system of map projection in which every place is of equal geometric significance: "The frame [of such a map] results from the bounds of a spherical coordinate system with origins based on certain arbitrary lines assumed to be on the earth, it is entirely abstract, geometric and independent of geographic content" (Woodward, 1990, p. 119).

Bacon appears to have been the first European to have advocated a coordinate system before the translation of Ptolemy's *Geographia* became available to European scholars. Woodward points out that there were at the time two other fundamentally different geometries of map construction that existed side by side with Bacon's coordinate system: the *mappaemundi* such as the Ebstorf (1239) and the Hereford (1290) examples, and the portolan charts like the *Carte Pisane* (1275). Each of these map types employs different modes of assembling knowledge and results in markedly different kinds of knowledge spaces. In Woodward's perceptive analysis, the differences are seen to lie in the way the knowledge is centered and framed.

Mappaemundi are "center enhancing." Both the frame and the center are predetermined by the theological/cosmological purposes of the map maker and the scale, rather than being invariant, changes in proportion with the familiarity and centrality of the regions depicted. *Mappaemundi* are not snapshots of the world's geography at a given point in time, but a blending of history and geography, a projection of historical events on a geographical framework (Woodward, 1985, p. 514).

Contrasted with Bacon's proposal, the early portolan charts were classic examples of the assemblage of local knowledge.[6] The frame of the map was usually determined by the size and shape of the medium on which it was drawn and had an arbitrary center varying according to what locations were included. They were compiled from tradition and direct experience by concatenating known routes and itineraries. In Woodward's terms, they constituted a "route enhancing space," precisely the kind that de Certeau claims is displaced by totalizing maps. Interestingly, the portolan chart mode of assemblage of local knowledge, while not having a coordinate system or a fixed system of mensuration, when laid over modern maps, generally has over 90% correspondence. However, Woodward comments that in order for the technique to be rendered as "systematic cartography," it had to be combined with an "astronomically fixed coordinate system in order to curb the instrumental error inherent in the method" (Woodward, 1990, p. 119). It is this concern with instrumental accuracy and the connection with the science of astronomy that provides an essential component in the European cartographic transformation.

Another early mode of knowledege assemblage was the sketch plan presented in a court of law. The historian of cartography Peter Barber dates the beginning of the use of sketch plans in English legal disputes at about 1400, when they were used to "illustrate and when possible to clarify disputes over rights, ownership and jurisdiction. Roughly executed, unsophisticated and undecorative as they usually were, the plans seem to have served their purpose and eventually some were even called for and used as evidence in courts of law" (Barber, 1992, p. 27).

The coordinate system, *mappaemundi*, portolan charts and sketch plans are just four alternative modes of geographic assemblage and do not, of course, exhaust the possibilities. They do, however, serve to show that there coexisted within the Western cartographical tradition profoundly different ways of constructing the spaces within which it was possible to know the world.

The integration of cartography with the affairs of state gained some of its impetus from the political revolutions of the 1530s; "for maps were to become one means by which Cromwell's (and later, Burghley's) objective of enhancing royal authority could be achieved, and it was thus that they made their most profound contribution to Tudor government" (Barber, 1992, p. 33). The military building program that brought about the modernization and expansion of England's defenses under Henry VIII was "once and for all, to establish maps and plans as one of the English government's everyday tools in the formulation of policy and in the process of administration" (Barber, 1992, p. 34). However, this revolution was based, according to Barber, on the "humble and frequently inelegant manuscript plats" (1992, p. 33). Scales began to appear on these plats only after 1547.

"The sort of precision to be found in scale maps was often not required by decision makers, who could make do perfectly adequately on most occasions with rough-and-ready picture or position maps lacking scale or standardized conventional signs. These continued to be produced and used for decades and centuries to come" (Barber, 1992, p. 38).

Hence it may be that the kind of calculative rationality we associate, thanks to Weber, with the modern state is a matter of degree of complexity. Wood, for example, argues that map making

> characterises a degree of organisation in society which is not just a function of size but of differentiation and hierarchical integration . . . the need to keep written records occurs when control of the social process in rapidly expanding groups is at stake. . . . The state in its pre-modern and modern forms evolves together *with* the map as an instrument of polity to assess taxes, wage war, facilitate communications, and exploit strategic resources. (Wood, 1993, p. 5)

He concludes that map use did not come about through the Renaissance discovery of Ptolemy, the scientific revolution, painterly realism, rationalized land management or the emerging nation state. Projections, realism and quantification were secondary. Rather, map making emerges as a rationalizing tool of state control during periods of relative prosperity in capitalist state economies around the world (Wood, 1995).

Wood's arguments, though cogent, are incomplete because they overlook the problem of assemblage. Though maps did start to become important tools in government policy-making during the Tudor period, the modern state and the scientific map did not emerge until appropriate modes of assemblage were created. In Tudor times, maps, surveys, plans and plats were accumulated by elite individuals like Lord Burghley, Queen Elizabeth I's secretary of state from 1550 to 1553 and 1558 to 1570 (Barber, 1992, p. 68). Whenever he wanted to assemble his maps, he simply had them bound together; a classic messy motley. The first gleamings of a need for a more formalized mode of assemblage were recognized with the establishment of the State Paper Office in 1610, but it had a turbulent beginning, with powerful individuals using it as an opportunity to pillage maps for their own private collections (Barber, 1992, p. 83).

The earliest attempt to assemble maps systematically in the interest of the state occurred in Spain and Portugal early in the sixteenth century, when they established the Casa de la Contratación and the Casa de Mina. These Boards of Trade housed the first scientific institutions—hydrographic offices that held the Padron Real, a master template map, on which all the knowledge of the New World was to be assembled. This attempted assemblage was ultimately a failure because the Padron Real and all the maps utilized by ships' captains were portolan charts. Portolan charts lacked a grid, projection or a common measure and were rooted in a maritime tradition. The state, despite assiduous efforts, proved unable to regulate, and control of the flood of new knowledge and power remained with ship captains and navigators (Turnbull, 1996).

To achieve the integrated assemblage of geographical knowledge that we now take for granted, science, cartography and the state had to be aligned—as they were, for the first time, through the process of the triangulated surveys of France and England. By the mid-seventeenth century, the problem of the bureaucratic regulation and taxation of the French state had become recognized as one susceptible to cartographic solution, and the problem of assembling cartographical knowledge was treated as being essentially linked to the astronomical and geophysical sciences. The French Academie Royale was set up in 1666, with the explicit purpose of correcting and improving maps and sailing charts, in light of the recognition that the solution of the major problems of geography, chronology and navigation lay in astronomy. In 1667, work was begun on the Royal Observatory at Fauborg St. Jacques (Brown, 1949, p. 213). While the building of the observatory and the collection of more accurate measurements of latitude and longitude brought with it an enhanced capacity to assemble geographical knowledge, that capacity was coproduced with the emerging demands of a centralized economy, which in turn brought with it a yet greater need for a more fully articulated knowledge space. This space was to be achieved through the expenditure of a great deal of physical labor in building a network that linked all points in the country and, ultimately, the Paris and Greenwich observatories in 1787 (Wood, 1993).

This connection between the two national spaces was established by trigonometric triangulation, using lights and a new theodolite at night to span the Channel. However, the initial problem of creating equivalences was not so easily resolved. It was necessary first of all to convert the French *toise* to the English league. Then, having established the linear distance between the meridians of Paris and Greenwich, there were difficulties converting this value into degrees, since, once again, this depended on agreement about the precise length of a degree and the shape of the earth. In addition, there were problems in establishing the difference in clock time between the meridians. Only when social means could be found to solve these seemingly technical problems could the astronomers measure the differences between their observatories, thereby creating one unified knowledge space and hence a new national and political space. This set in motion the process whereby the whole of the earth's territory could be mapped as one, all sites would be rendered equivalent and all localness would vanish in the homogenization and geometrization of space.

The modern dream of one global, unified, knowledge space within which everything could be located, coordinated and regulated is the ultimate totalizing project that postmodernists fear so much. However, despite all the work and negotiations concerning a universal scale, projection and grid that have gone into the globalization of society, and despite the talk of the global village and communication superhighways, "The International Map," with all countries adopting one uniform standardized system, has yet to be fully realized over fifty years after it was first proposed at the Fifth International Geographic Congress in 1945. Such an international map would, however, only be a relatively minor form of universal representation by comparison with the geographical information systems (GIS) that could provide for the ultimate knowledge space. While GIS seems about to displace all rivals, it is plagued with organizational, political and ethical difficulties (Edney, 1993; Taylor, 1989; Taylor, 1993; Wasowski, 1991).

To produce such an ordering of knowledge on a global transnational scale requires the coproduction of a social ordering that is as yet incomplete, and may never be completed given the scale of social ordering required. It may be that the inherent instability and messiness is not containable in a world that is so economically and politically fragmented. Such messiness offers potential sites of resistance to the establishment of a global universal knowledge space; clearly, there are many sources of instability within the already established knowledge space of Western science.

In order to locate such sites, it is necessary to return to the question of the power of maps. This exploration has, thus far, concentrated on the ways in which maps result from collective social work of assemblage. This emphasis on the work involved in map construction has the effect of making it seem as if there is a natural disjunction between the work of map making and the work that maps do, a polarity paralleled by the distinction that is traditionally drawn between knowledge and power. Power is inherent in the making of all knowledge and all representations. Power, as Foucault has taught us, does not merely reside in the use and articulation of maps; it is inherent in them. However, this inherent power comes in a variety of forms and they cross the boundaries between internal and external power. Not only is power inherent in maps, but maps change the space within which external power can be manifested; external and internal powers are coproduced.

In the orthodox realist view, maps gain their power from their accuracy, their correspondence with external reality, and as their accuracy increases so does their capacity to aid men in their endeavors. There is, of course, a level or scale at which there is some truth in this. However, there have been two major conceptual changes that make it no longer possible to suppose that our representations can have an unequivocal correspondence with reality. The first arrived with Einstein's theory of relativity: "Around 1910 a certain space was shattered. It was the space of common sense, of knowledge, of social practice, of political power, a space hitherto enshrined in everyday discourse, just as in abstract thought" (Lefebvre, 1991, p. 25). This everyday space was that of Euclid. With the discovery of non-Euclidean geometry and the theory of relativity, it was no longer possible to hold that there are frame-independent geometric truths of the kind that still seemed so persuasive to Lloyd Brown when he wrote the first general history of cartography in 1949:

> The mapping of the world, with every coastal outline, every river, lake and mountain, town and city in its proper place with regard to its distance and direction, depends on a simple geometric proposition, namely, that the intersection of two lines is a point. In other words, to locate a spot on the earth, to fix it cartographically, it is only necessary to know two things: its latitude and its longitude, two lines which intersect at the desired spot, and lo and behold, it is done. (Brown, 1949, p. 10)

When it became apparent that there was no natural correspondence between the seemingly self-evident truths of Euclidean geometry and the structure of reality, when the possibility of alternative geometries, and logics for that matter, were combined with the understanding that the absolute space that went with Newtonian physics was no longer an "eternal bucket" in which all events and objects were contained, then space itself became a historical and contingent matter. Not only did it become possible that space had a history of its own, which was one implication of Einstein's conception of the space-time matrix, but also it could no longer provide its own natural reference frame. Any understanding of the structure of space would, if even just the broad implications of relativity theory are accepted, be crucially dependent—as argued by Liebman in his first chapter in this book— on the frame adopted. Hence Brown's remarks would have to be modified to recognize that the location of an object is not absolutely given by the intersection of latitude and longitude, but is only so given in a time-independent frame and at the scale of medium-size, terrestrial, material objects. Within this frame, history and time as dependent processes are excluded and it is taken for granted that the world consists essentially of objects that have a fixed spatial location. Less troublesome, but nonetheless significant, the very small, the very large and the nonterrestrial either have to be excluded or treated as essentially like macro-scale objects on earth.

There has, however, been a second conceptual change that Brown could not be chastised for failing to accommodate; that brought about by the construal of yet another geometry, the fractal geometry of Benoît Mandelbrot. Mandelbrot started by asking himself something that looks superficially like a classic cartographic question: How long is the coastline of Britain? This question came up because of one of the more mundane—but classic—relativistic difficulties that plagues the would-be universal map maker. For example, there is no agreement on the length of the border between Spain and Portugal. It is either 987 or 1214 kilometers depending on who is doing the measuring (Mandelbrot, 1983, p. 33). The classic answer to the question would normally be thought to be a trivial empirical fact that could be ascertained with whatever degree of accuracy was required or permitted by the instrumentation available, provided all the agreements about standardization discussed earlier were in place. In other words, within a Newtonian frame it is assumed that there is a fixed, finite, measurable length of the Spanish/Portuguese border or of the coastline of Britain and that successive generations of measurements will produce increasing degrees of accuracy. But in the case of the real world as opposed to the unreal, formal world of mathematics, according to Mandelbrot's analysis, there is no such thing as the length of the coastline of Britain. Its length is not scale-independent, but a function of the length of the measuring stick; the shorter the stick, the longer the coast. In other words, if you increase accuracy indefinitely, so the length measured tends to infinity.

Much of the analysis that enabled a new understanding of the power of maps has been provided by Brian Harley in applying the insights of Foucault and Derrida, and deconstruction generally, to cartography. While Harley's critical approach to maps is profoundly important, its slightly negative edge needs augmentation with an understanding of the more positive senses in which maps have potentially antihegemonic power (Sibley, 1995). Robert Rundstrom, in writing about the accuracy of Inuit maps, has articulated the sense in which maps have power as devices for attributing order and meaning to the world:

Maps may be considered artifacts composed of signs that materialise a way of experiencing. By transforming a given way of thinking into material reality, maps simultaneously reflect and reinforce the worldview or spatial thought of a culture . . . maps both reflect and reinforce the cultural values and beliefs of the people who make them. (Rundstrom, 1990, pp. 155–56)

This is a view that is compatible with Harley's position that "cartography is seldom what cartographers say it is" (Harley, 1989, p. 1); rather it is "primarily a form of political discourse concerned with the acquisition and maintenance of power" (Harley, 1988a, p. 57). Harley spells out the ways in which he thinks that power is manifested and in so doing makes explicit the link between the ways in which scientific and cartographic knowledge is manufactured in labs and workshops:

> Compilation, generalisation, classification, formation into hierarchies, and standardisation of geographical data, far from being mere "neutral" technical activities, involve power-knowledge relations at work. Just as the disciplinary institutions described by Foucault—prisons, schools, armies, factories—serve to normalise human beings, so too the workshop of the map-maker can be seen as normalizing the phenomena of place and territory in creating a sketch of a made world that society desired. . . . An analogy may be drawn here between the "construction of microworlds" in the scientific laboratory and the construction of geographical knowledge in the cartographer's workshop. In both cases the key is standardisation. (Harley, 1996, pp. 5–6)

The construction of the microworlds of science and cartography are directly analagous; science and cartography are intimately related, the social processes in their construction are the same and together they help to open up the knowledge space within which we produce and represent knowledge. However, they do not provide the only mode of assembling knowledge or the only social order; messiness can always re-emerge. There are—as this book demonstrates—always other ways of mapping.[7]

Sites of Cartographic Resistance

How do we locate sites of resistance? The primary sites are those where local groups have constructed their own form of cartographic representation and used it to bypass or to resist the dominant form. Secondary sites are those provided by the previous analysis, which points to the ways in which all forms of cartographic representation simultaneously require a form of social ordering that can therefore be changed. There have been a number of examples of cartographic resistance. Guaman Poma tried to resist what he claimed was the Spanish misrule of Peru by utilizing a cartographic representation of the Incan empire that blended European and Incan modes of representation (Turnbull, 1993a, pp. 8–10). The Mexican revolutionary leader Emiliano Zapata drew maps of the Tlapa and Anenecuilo areas in the style of preconquest maps to demonstrate how local representations of communal ownership of the land contrasted with modern maps of the sugar plantation owners (Brotherston, 1992, p. 83). In more recent times the Lake Titicaca peasants have resisted by producing their own maps (Orlove, 1991, 1993). Similarly, the Indians of northwestern Canada and Alaska have sought to resist development by drawing maps of their own land usage. These maps and their dream maps have been brilliantly described by Hugh Brody in his book *Maps and Dreams* (1983). The Inuit have resisted by taking government survey maps, removing the grid and scale, naming all significant points with Inuit names and then publishing the revision as the official Inuit map (Rundstrom, 1990, 1991). Examples of maps of resistance can also be found in more urbanized societies, such as the wildly successful Parish maps project begun by the pressure group Common Ground in Britain to encourage villagers to preserve and record their local heritage (Worsley, 1991; King and Clifford, 1988). In a related move, Doug Aberley has proposed local mapping techniques to encourage bioregionalism (Aberley, 1993).

As the West has attempted to expand its knowledge space into the third world, Western modes of knowledge assembly have often proved inadequate and alternate modes are emerging as potential sites of resistance and as more appropriate ways of dealing with, for example, agricultural and ecological problems (Hobart, 1993; Peluso, 1995; Taylor, 1995).

Robert Chambers of the Institute of Development Studies at the University of Sussex has been working on a new approach to development and environmental improvement. His Participatory Rural Appraisal scheme (PRA) finds new ways of assembling local knowledge with cartographic methods that contrast with the usual top-down extension of Western expertise (Chambers, 1994). Local people are encouraged to use their firsthand knowledge of practice and priorities to construct their own maps, charts and matrices. These diagrams drawn in public places, often on the ground in the village square, using whatever materials are available—flowers, sand, seeds, sticks or chalk—often reveal a highly visual pattern language unique to particular communities. Chambers says of such maps:

They are creative, most people take to them readily and find it easy to express what they know. And they are levelling because they can empower the nonliterate and the weak to express what they know and want.

By far the most important attribute of these designs for living, however, is that they work. They provide a framework for people to plan and monitor innovations in farm management, agroforestry, resource distribution, community health care education and so on. By putting local people in the center of the picture—their own picture—PRA helps them set and achieve their own management standards working within local means. (Lamb, 1993, p. 38)

The Green Revolution is a classic instance where the failure of the attempt to expand the knowledge space of the West has created an opportunity for an alternative way of mapping. In Indonesia, for example, local people are mapping their own resources and knowledge. The introduction of high-yield rice turned Indonesia from a net importer of rice incapable of feeding its own population to being one of the biggest rice exporters. This was achieved at the price of using massive amounts of fertilizer and pesticides and the abandonment of indigenous rice strains. That success, however, was short-lived. Insect pests reached plague proportions in the monocrop environment and ever-increased applications of pesticide only made the problem worse. A solution—at least in part—followed from the banning of fertilizer and pesticide imports and the introduction of "integrated pest management," an ecologically based approach to pest control that recognizes there will always be pests and that the best way to manage them is to ensure that the populations of competing insects and wildlife remain in a better balance. In order for this system to work, the expansion of the Western knowledge space had to be reversed and local people had to establish their own space and ways of mapping and assembling knowledge. Using traditional knowledge and techniques, local farmers had to become experts monitoring the insect populations on their farms and reintroducing locally appropriate rice strains. The power and the knowledge were relocalized back into the hands of the farmers.

A similar reversal occurred in Bali, where rice was grown under an irrigation system controlled by water temples. The Indonesian government thought this old-fashioned and superstitious and introduced modern scientific methods of water control and distribution. The result was the same as in Java: Initial success followed by a crash in production. So they brought in more Western experts, but this time they included a rather unusual anthropologist and a computer expert (Lansing, 1991). Between them, they were able to map on the computer screen how the ancient system of temple control and knowledge assembly worked and why it was the most efficient. This is an interesting example of synergy between two different ways of mapping and assembling knowledge, which resulted in the knowledge and power being given back to the local people and traditional institutions.

In Australia, it has been possible for Aborigines to resist colonial domination through their own ways of mapping. They have successfully recovered dispossessed land by producing maps in government courts showing the tracks of the ancestral beings as they created the territory in their journeys (Turnbull and Watson-Verran, 1995; Watson, 1989; Chambers, 1994). But at the same time, the Aborigines most urgently need to find new ways to resist being mapped away. The Aboriginal and Torres Strait Islander Commission Act and the Victorian Heritage Legislation have produced maps creating tribal boundaries; such maps legitimate the territorial claims of some groups but deny or diminish the claims of others. Such a mapping process bears little resemblance to Aboriginal understanding of territory. Inscribing boundaries in this way serves only to incorporate them into the state. One way for them to resist is, I suggest, to renew and re-articulate their own traditional ways of assembling knowledge through myth, ritual, painting and kinship; that is, through their own modes of mapping.[8]

NOTES

1. Harley (1989, p. 4) argues, "The normative history of cartography is a ceaseless massaging of this theme of noble progress. . . . It is the dominant theme." Wood (1992a, pp. 26–35; see also Crone, 1953, p. xi) quotes Crone: 'The history of cartography is largely that of the increase in the accuracy with which . . . elements of distance and direction are determined and the comprehensiveness of the maps' content."

2. This is a point first made by Ivins: "We would not have had modern science and technology without exactly repeatable pictorial statements" (Ivins, 1953, p. 3), and later McLuhan. See also Helen Gardner, cited in Romanyshyn (1989, p. 33). The argument is most persuasively developed by Latour (1986)

3. See, for example, the work of Treisman and Allman cited by Hall (1992, p. 17). For a rather different view and, to my mind, a more plausible one, see the work of Damasio et al. reported by Kinoshita. Damasio proposes a hierarchy of assembly areas or "convergence zones" in which elements of memory can be linked as required: "There is no complete picture in any one part of the brain. You have to reconstruct it each time" (Kinoshita, 1992, p. 48).

4. I have adopted the term "assemblage" from Deleuze and Guattari (1987). The evocative term "motley" is used by Ian Hacking (1992).

5. See de Certeau (1984) and Harley (1989, p. 252). See also Rabasa (1985, p. 2): "The effect of universality or totality is only achieved through blindness to the subjective reconstitution of the fragments. The map is a palimpsest [of multiple inscriptions] subject to ironic reconstitution through bricolage."

6. On portolan charts, see Campbell (1987).

7. See Rundstrom (1993) for a recent effort (albeit "conservative [and] Eurocentric") to open scientific cartography to social and cultural realities.

8. Fischer (1994) provides an apt example here.

REFERENCES

Aberley, D., ed. (1993). *Boundaries of home: Mapping for local empowerment*. Philadelphia: New Society.

Barber, P. (1992). England 1: Pageantry, defense, and government: Maps at court to 1550. In D. Buisseret (ed.), *Monarchs, ministers and maps: The emergence of cartography as a tool of government in early modern Europe* (26–56). Chicago: University of Chicago Press.

Bartolovich, C. (1995). Spatial stories: The surveyor and the politics of transition. In A. Vos, *Place and displacement in the Renaissance* (255–283). Binghamton, NY: SUNY Press.

Bauer, B.S. and D.S. Dearborn. (1995). *Astronomy and empire in the ancient Andes: The cultural origins of Inca sky watching*. Austin: University of Texas Press.

Belyea, B. (1992a). Amerindian maps: The explorer as translator. *Journal of Historical Geography, 18,* 267–277.

Belyea, B. (1992b). Images of power: Derrida/Foucault/Harley. *Cartographica, 29,* (2), 1–9.

Belyea, B. (1992c). Review of Denis Wood's *The power of maps*, and the author's reply. *Cartographica, 29*(3/4), 94–99.

Bourdieu, P. (1977). *Outline of a theory of practice*. Cambridge, MA: Cambridge University Press.

Brody, H. (1983). *Maps and dreams: Indians and the British Columbian frontier*. Harmondsworth, England: Penguin.

Brotherston, G. (1992). *Book of the fourth world: Reading the native Americas through their literature*. Cambridge, MA: Cambridge University Press.

Brown, L.A. (1949). *The story of maps*. Boston: Little, Brown.

Buisseret, D., ed. (1992). *Monarchs, ministers and maps: The emergence of cartography as a tool of government in early modern Europe*. Chicago: University of Chicago Press.

Campbell, T. (1987). Portolan charts from the late thirteenth century to 1500. In J.B. Harley and D. Woodward (eds.), *The history of cartography*. Vol. 1, *Cartography in prehistoric, ancient and medieval Europe and the Mediterranean*. Chicago: University of Chicago Press.

Cartwright, N. (1983). *How the laws of physics lie*. Oxford: Clarendon.

Chambers, R. (1994). Participatory Rural Appraisal (PRA): Analysis of experience. *World Development, 22* (9), 1253–1268.

Chorley, R.J., and P. Haggett. (1967). *Models in geography*. London: Methuen.

Crone, G. (1953). *Maps and their makers*. Folkestone, England: Dawson.

Darnton, R. (1984). *The great cat massacre and other episodes in French cultural history*. New York: Basic Books.

de Certeau, M. (1984). *The practice of everyday life*. Berkeley: University of California Press.

Deleuze, G., and F. Guattari. (1987). *A thousand plateaus: Capitalism and schizophrenia*. Minneapolis: University of Minnesota Press.

Diprose, R., and R. Ferrell, eds. (1991). *Cartographies: Poststructuralism and the mapping of bodies and spaces*. London: Allen and Unwin.

Downs, R.M., and D. Stea. (1973). Cognitive maps and spatial behaviour. In R.M. Downs and D. Stea, (eds.), *Image and environment* (8–26). Chicago: Aldine.

Downs, R.M., and D. Stea, eds. (1977). *Maps in minds: Reflections on cognitive mapping*. New York: Harper and Row.

Edney, M.H. (1993). Cartography without "progress": Reinterpreting the nature and historical development of mapmaking. *Cartographica, 30*(2/3), 54–68.

Ferrier, E. (1990). Mapping power: Cartography and contemporary cultural theory. *Antithesis, 4*(1), 35–52.

Feyerabend, P. (1978). *Against method: Outline of an anarchist theory of knowledge.* London: Verso.

Fischer, A. (1994). Power mapping: New ways of creating maps help people protect their landscapes. *Utne Reader 65,* 32–35.

Fletcher, D. (forthcoming). The development of estate mapping in early modern England. *Transactions of the Institute of British Geographers.*

Friedhoff, R.M., and W. Benson. (1989). *Visualisation: The second computer revolution.* New York: Abrahams.

Gould, P., and R. White. (1974). *Mental maps.* Harmondsworth: Penguin.

Hacking, I. (1992). The self-vindication of the laboratory sciences. In A. Pickering (ed.), *Science as practice and culture* (29–64). Chicago: University of Chicago Press.

Hall, S.S. (1992). *Mapping the next millenium: The discovery of new geographies.* New York: Random House.

Harley, J.B. (1988a). Silences and secrecy: The hidden agenda of cartography in early modern Europe. *Imago Mundi, 40,* 57–76.

Harley, J.B. (1988b). Maps, knowledge and power. In D. Cosgrove and S. Daniels (eds.), *The iconography of landscape* (277–312). Cambridge, England: Cambridge University Press.

Harley, J.B. (1989). Deconstructing the map. *Cartographica, 26,* (2), 1–20.

Harley, J.B. (1990). Cartography, ethics and social theory, *Cartographica, 27,* (2), 1–23.

Harley, J.B. (1996). Power and legitimation in the English geographical atlases of the eighteenth century. In J. Wolter and R.E. Grim (eds.), *Images of the world: Essays on the history of the atlas.* Washington, DC: Library of Congress.

Harley, J.B., and D. Woodward, eds. (1987). *The history of cartography.* Vol. 1, *Cartography in prehistoric, ancient and medieval Europe and the Mediterranean.* Chicago: University of Chicago Press.

Harley, J.B., and D. Woodward, eds. (1992). *The history of cartography,* Vol. 2., *Cartography in the traditional Islamic and South Asian societies.* Chicago: University of Chicago Press.

Harvey, P.D.A. (1993). *Maps in Tudor England.* London: British Library.

Hearnshaw, H.M., and D.J. Unwin, eds. (1994). *Visualisation in geographical information systems.* Chichester, UK: John Wiley and Sons.

Hobart, M., ed. (1993). *An anthropological critique of development.* London: Routledge.

Ivins, W.M. (1953). *Prints and visual communication.* Cambridge, MA: Harvard University Press.

Jacob, C. (1992). *L'Empire des cartes: Approche theorique de la cartographie á travers l'histoire.* Paris: Albin Michel.

Johnston, S.A. (1994). *Making mathematical practice: Gentlemen, practitioners and artisans in Elizabethan England.* Ph.D. dissertation, University of London.

King, A., and S. Clifford. (1988). *Holding your ground: An action guide to local conservation.* Aldershot, England: Wildwood House.

Kinoshita, J. (1992). Mapping the mind. *New York Times Magazine,* 18 October, 44–54.

Kuhn, T. (1970). *The structure of scientific revolutions.* Chicago: University of Chicago Press.

Kuhn, T. (1977). *The essential tension.* Chicago: University of Chicago Press.

Lamb, R. (1993). Designs on life. *New Scientist, 140,* 37–40.

Lansing, J.S. (1991). *Priests and programmers: Technologies of power in the engineered landscapes of Bali.* Princeton, NJ: Princeton University Press.

Latour, B. (1986). Visualisation and cognition: Thinking with eyes and hands. *Knowl-

edge and Society, 6, 1–40.

Latour, B. (1987). *Science in action.* Milton Keynes, England: Open University Press.

Leach, E. (1976). *Culture and communication: The logic by which symbols are connected.* Cambridge, England: Cambridge University Press.

Lefebvre, H. (1991). *The production of space.* Oxford: Blackwell.

Lewis, M. (1987). The origins of cartography. In J. B. Harley and D. Woodward (eds.), *The history of cartography.* Vol. 1, *Cartography in prehistoric, ancient and medieval Europe and the Mediterranean.* Chicago: University of Chicago Press.

Livingstone, D. (1992). *The geographical tradition: Episodes in the history of a contested enterprise.* Oxford: Blackwell.

Longino, H. (1992). Hard, soft, or satisfying. *Social Epistemology, 6,* 281–287.

MacEachren, A.M., et al. (1992). Visualisation. In R.F. Abler, M.G. Marcus, and J.M. Olson, (eds.), *Geography's inner worlds: Themes in contemporary American Geography* (99–137). New Brunswick, NJ: Rutgers University Press.

Mandelbrot, B. (1983). *The fractal geometry of nature.* New York: W.H. Freeman.

McLuhan, M. (1964). *Understanding media: The extensions of man.* New York: McGraw-Hill.

McNeill, D., and P. Freiberger. (1993). *Fuzzy logic.* Melbourne: Bookman.

Mitchell, W.J.T., ed. (1994). *Landscape and power.* Chicago: University of Chicago Press.

Monmonier, M. (1991). *How to lie with maps.* Chicago: University of Chicago Press.

Morrison, J. (1989). The revolution in cartography in the 1980s. In D.W. Rhind and D.R.F. Taylor (eds.), *Cartography, past, present and future* (169–85). London: Elsevier.

Mundell, I. (1993). Maps that shape the world. *New Scientist, 139*(3), 21–23.

Orlove, B. (1991). Mapping reeds and reading maps: The politics of representation in Lake Titicaca. *American Ethnologist, 18,* 3–38.

Orlove, B. (1993). The ethnography of maps. *Cartographica, 30*(1), 29–46.

Papp-Vary, A. (1989). The science of cartography. In D.W. Rhind and D.R.F. Taylor (eds.), *Cartography past, present and future.* London: Elsevier.

Paulston, R. (1993). Mapping ways of seeing in educational studies. *La Educacion: Inter-American Review of Educational Development,* 114(1), 1–18.

Peluso, N.L. (1995). Whose woods are these? Counter-mapping forest territories in Kalimantan, Indonesia. *Antipode* 27(4), 383–406.

Polanyi, M. (1958). *Personal knowledge: Towards a post-critical philosophy.* London: Routledge and Kegan Paul.

Rabasa, J. (1985). Allegories of the ATLAS. In F. Barker (ed.), *Europe and its others* (1–16). Colchester, England: University of Essex.

Robinson, A.H., and B.B. Petchenik. (1976). *The nature of maps: Essays towards understanding maps and mapping.* Chicago: University of Chicago Press.

Romanyshyn, R.A. (1989). *Technology as symptom and dream.* London: Routledge.

Rouse, J. (1987). *Knowledge and power: Towards a political philosophy of science.* Ithaca, NY: Cornell University Press.

Rubiner, M. (1995). A new map of the brain. *New York Times,* 23 April, p. 78.

Rundstrom, R.A. (1990). A cultural interpretation of Inuit map accuracy. *Geographical Review, 80,* 155–168.

Rundstrom, R.A. (1991). Mapping, postmodernism, indigenous people and the changing direction of North American Cartography, *Cartographica, 28,* 1–12.

Rundstrom, R.A., ed. (1993). Introducing cultural and social cartography. *Cartographica* (Special Issue, Monograph 44), 30(1), 1–107.

Schuster, J.A., and R. Yeo, eds. (1986). *The politics and rhetoric of scientific method.* Dordrecht: Reidel.

Shapin, S., and S. Schaffer. (1985). *Leviathan and the air pump: Hobbes, Boyle and the experimental life.* Princeton, NJ: Princeton University Press.

Sibley, D. (1995). *Geographies of exclusion: Society and difference in the West.* New

York: Routledge.

Soja, E.W. (1989). *Postmodern geographies: The reassertion of space in critical social theory*. London: Verso.

Suchman, L. (1987). *Plans and situated actions: The problem of human-machine communication*. Cambridge, MA: Cambridge University Press.

Sullivan, L. (1994). Language map gives few wrong directions. *The Australian* (Melbourne), 18 May, 29.

Taylor, B.R., ed. (1995). *Ecological resistance movements*. Albany: SUNY Press.

Taylor, D.R.F. (1989). New perspectives on cartography. *Cartographic Perspectives*, 4, 24–29.

Taylor, D.R.F. (1993). Geography, GIS and the modern mapping sciences/convergence or divergence. *Cartographica, 30(2/3)*, 47–53.

Tolman, E.C. (1948). Cognitive maps in rats and men. *Psychological Review, 55*, 189–208.

Turnbull, D. (1991). *Mapping the world in the mind: An investigation of the unwritten knowledge of the Micronesian navigators*. Geelong, Australia: Deakin University Press.

Turnbull, D. (1993a). *Maps are territories; science is an atlas*. Chicago: University of Chicago Press.

Turnbull, D. (1993b). The ad hoc collective work of building gothic cathedrals with templates, string, and geometry. *Science Technology and Human Values, 18*, 315–340.

Turnbull, D. (1993c). Local knowledge and comparative scientific traditions. *Knowledge and Policy, 6(3/4)*, 29–54.

Turnbull, D. (1995). Rendering turbulence orderly. *Social Studies of Science, 25*, 9–33.

Turnbull, D. (1996). Cartography and science in early modern Europe: Mapping the construction of knowledge spaces. *Imago Mundi, 47*.

Turnbull, D., and H. Watson-Verran. (1995). Science and other indigenous knowledge systems. In S. Jasanoff, G. Markle, T. Pinch, and J. Petersen (eds.), *Handbook of Science and Technology Studies* (115–139). Thousand Oaks, CA Sage.

Tyacke, S., ed. (1983). *English map-making 1500–1650: Historical essays*. London: British Library.

Wasowski, R.J. (1991). Some ethical aspects of international satellite remote sensing. *Photogrammetric Engineering and Remote Sensing, 57*, 41–48.

Waterman, T.H. (1989). *Animal navigation*. New York: Scientific American Library.

Watson, H. (with the Yolgnu community and D.W. Chambers). (1989). *Singing the land, signing the land*. Geelong, Australia: Deakin University Press.

Wittgenstein, L. (1958). *Philosophical investigations*. Oxford: Blackwell.

Wood, D. (1992a). *The power of maps*. New York: Guilford.

Wood, D. (1992b). How maps work. *Cartographica, 29(3/4)*, 66–74.

Wood, D. (1993). Maps and mapmaking. *Cartographica, 30(1)*, 1–9.

Wood, D. (1995). Mapmaking. In H. Selin (ed.), *Encyclopedia of the history of science, technology and medicine in non-western cultures*. New York: Garland.

Woodward, D. (1985). Reality, symbolism, time and space in medieval world maps. *Annals of the Association of American Geographers, 75*, 510–521.

Woodward, D. (1990). Roger Bacon's terrestrial coordinate system. *Annals of the Association of American Geographers, 80*, 106–122.

Woolgar, S. (1983). Irony in the social study of science. In K.D. Knorr and M. Mulkay (eds.), *Science observed: Perspectives on the social study of science* (235–266). London: Sage.

Woolgar, S., ed. (1988). *Knowledge and reflexivity: New frontiers in the sociology of knowledge*. London: Sage.

Worsley, G. (1991). A sense of place: *Country life*, 29 August, 58–59.

The Timely Emergence of Social Cartography

Thomas W. Mouat IV

Cartography, the making of maps, probably originated when an early homo sapiens scratched the ground with a sharp stick in an effort to communicate a hunting strategy or the location of a needed thing. From the first, cartography used symbolic representation: a squiggly line to represent a water course, perhaps, a short sharp curve for a hill. These first abstract representations of concrete objects laid the discipline's foundations. But cartography has not always and only been used to depict concrete reality.

At least several millennia ago, another inspired human realized that the cartographic technique of mapping symbolic representations could be extended in application from concrete objects to abstractions themselves. Occasionally since then, philosophers and mathematicians have used cartographic techniques to symbolically represent general relationships among abstractions. Aristotle's dialectic cycle of thesis, antithesis and synthesis is a classic example of a generalized abstraction from philosophy that can be communicated at the concrete level through cartography (see Figure 1, below). An early example from mathematics would be the elements of geometry, with trigonometry presenting an important early example of the mapping of one set of generalized abstractions onto another.

These early visual depictions of the general relationship among abstract concepts allowed generalized abstractions to be communicated at the concrete level. This permitted individuals who were utilizing a conceptual framework at the fourth-cycle level of generalizing abstractions (Piaget's formal operations) to communicate elements of fourth cycle concepts to individuals who were operating at the less sophisticated third-cycle level of simple, linear, concrete, reversible logic (Piaget's concrete operations).

Cartography began, then, as the making of visual maps locating concrete objects through symbolic representation. It progressed to the making of visual maps locating abstractions through symbolic representations. Several decades ago cartography was partially reconceptualized once again. Cartography still uses a symbol set to map objects in relation to one another; however, in the same way that the mapped objects need not be concrete, the map itself, now, need not be concrete. Thus, although cartography is descriptive and usually conceptualized as employing a visual symbol set, the symbol set can be words and the mapping can be cognitive. When cartography utilizes language instead of or in addition to purely visual description, then the map is at least partially visualized internally, instead of being transferred there via the optic nerve. Of course, since visual representations, when practical, are more elegant and communicative in their parsimony than is verbal text, mapping abstractions is rarely restricted to text alone.[1] Whether text, text and pictorial representation, or pictorial representation alone, the end result is the same: Cartography is the intentional mapping of the relations among a field of objects, or, one step removed, the mapping of maps onto one another.

Although cartography can be used effectively to map both concrete and abstract relationships, until very recently it has been used almost exclusively for concrete mapping. However, in the last few years, emerging relatively independently across disciplines, these techniques once rarely used for mapping abstractions have found a more general application. Abstract mapping is now being applied to sociology, nonlinear mathematics, political science, literature, educational sociology, psychology, interdisciplinary studies, medicine, cognition research, history and the fine arts, among others. Titles such as *Sociology and Visual Representation* (Chaplin, 1994), *Mapping Ideology* (Zizek, 1995), *Mapping the Nation* (Balskrishnan, 1995), *Mapping the Futures* (1993), *Mapping the Terrain* (Lacy, 1994), "Mapping Changes in Science and Technology" (Tijssen and Van Raan, 1994), "Semantic Cartography in Folklore" (Maranda, 1977) and "Mapping Knowledge Perspectives in Studies of Educational Change" (Paulston, 1995), among others, represent only a small fraction of the publications organized around the concept of mapping abstractions.

It seems from the foregoing that a synchronous development of abstract mapping is underway both across disciplines and as an interdisciplinary activity. Synchronicity, in turn, suggests patterned development, and patterned development arises as an effect in a causal chain. Closer inspection suggests that the interrelated causes of abstract cartography's widespread emergence in the postmodern era are threefold.

The postmodern era began with a dawning awareness that "reality" is composed of disconnected fragments. As early postmoderns sought reconnection, they discovered that the concrete representation of interrelationships between and among fragments often eludes expression. As the struggle to discover and express interrelationships intensified, it became apparent that the *abstract* representation of interrelationships is often possible when their *concrete* representation is not. Therefore, social cartography as mapping abstractions arises initially as a vehicle through which to express, in highly condensed, abstract form, the interrelationships between and among elements of systems which are not amenable to concrete description.

However, once invented, cartography also allows the mapping of abstraction upon abstraction. In this view, social cartography arises to meet the need for reintegration of the fragmented universe bequeathed by modernity. As it happens, full reintegration is impossible: Not only has the universe fragmented, the conceptual framework for knowledge has disintegrated. A new framework for knowledge is required and cartography's techniques for mapping abstractions promise to produce such a neomodern synthesis.

As the following exercise in social cartography demonstrates, abstract mapping is a necessary instrument of the synthesis that, barring dysfunction, should soon occur at the level of mapping generalized abstractions. This demonstration utilizes a neo-Piagetian framework to create abstract maps of the cycle of individual thought and the pattern of cognitive development. These abstractions are overlaid upon a map of Western social history. Through this process the cyclic pattern of individual cognitive development and the cyclic pattern of social development are interrelated, and the necessary purpose of social cartography is explicated.

At this juncture it may be appropriate for me to explain why the examples of sociocognitive development that follow are exclusively drawn from European history. This Eurocentric perspective arises in part, no doubt, from my ignorance of non-Western history. But there is another reason. I am interested in the complex, evolutionary processes that construct the global social web. The cyclic pattern of individual and social cognitive development, while interesting, is but one among several cyclic patterns that play a part in constructing the global web. This psychosocial cyclic pattern is neither the most basic cyclic pattern, nor, perhaps, the most important; it produces but one thread among the many that interweave to construct the global fabric. My particular interest in the European manifestation of the cyclic pattern of psychosocial development arises because of the conjunction between the emergence of a new conceptual framework for knowledge, in Europe, during the Baroque era, and the simultaneous emergence of the capitalist world economy, through which the "Western world" began to dominate the globe. In my view, the emergence in Europe of this conceptual framework legitimated, at least to Europeans, Western hegemony. I offer no further apologies for centering on European history, but beg the reader's indulgence.[2]

Since society comprises individuals, psychology and sociology share all of the linkages between individual and group. Moreover, since the aggregate of individual outcomes defines society, if a pattern exists for individual development, it might be anticipated that the pattern would scale to the social level.

Seen from a neo-Piagetian perspective, the communication of information is the fundamental mechanism linking the individual to society. In other words, within the context of society, norms of thought and other social norms are transmitted to the individual members of society through the process of education. This social transmission of knowledge, whether by mother to child or in any other teacher-learner environment, tends to reproduce extant norms. However, new norms come into being when new ideas produce successful outcomes and as a result are copied by others. According to this perspective, the transmission of norms results in stability producing social-reproduction, while the generation of new ideas leads to changes in individual norms. Over time these changes in individual norms result in the changing of social norms and hence provide the basis for the development of social history. At the same time, since the individual's context is society, it should be equally true that the changing of social norms would result in the changing of individual norms. In other words, the individual and society interact complexly so that each affects the development of the other.

At the individual level, according to Piaget, a cyclical thought process is used to make sense of interactions with the environment. In the process of learning about the environment, the repetitive use of this cycle of thought leads the individual to progress through a series of stages in cognitive development. According to this perspective, all individuals utilize the same cycle of thought and follow the same pattern of development. In addition, as noted, individual and social development are inextricably linked through the social transmission of knowledge. As a result, the pattern of individual development is seen to mandate a pattern of social development. To explore this relationship, it is necessary to begin with a review of the cycle of thought and the pattern of development. Based upon this review, it will be possible to demonstrate that this cyclic pattern has a scalar analogue at the social level.

THE INVARIANT CYCLE OF INDIVIDUAL THOUGHT

As Aristotle, Ramus, Hegel and Marx all noted, human thought progresses through a dialectic cycle. For them, the phases of this cycle are identified with the terms thesis, antithesis and synthesis. Piaget identifies the same cycle,

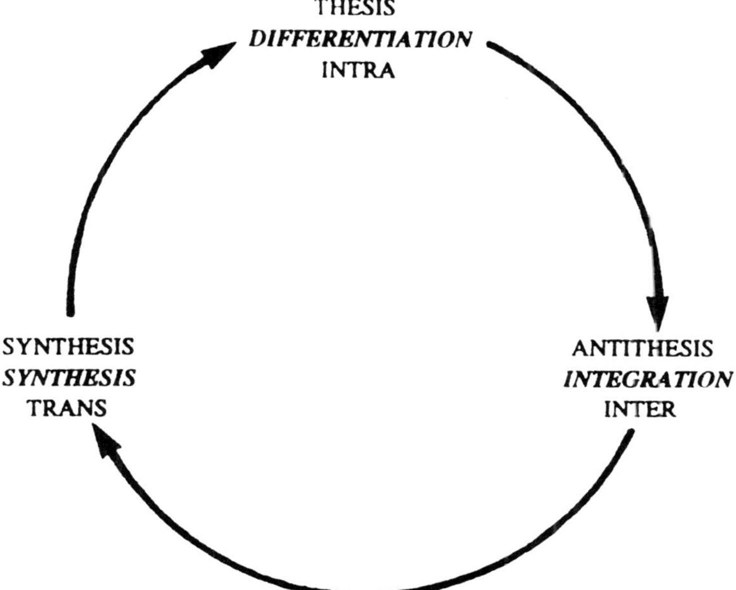

Figure 1. The cycle of thought

but terms the phases intra, inter and trans. However, neither set of terms accurately describes the fundamental operations or processes undertaken during each phase, which are perhaps better expressed by the terms differentiation, integration and synthesis.

Differentiation, the introductory phase of this invariant cycle, occurs when a new thing, idea or event is met. During the process of differentiation, the object of analysis is fragmented or differentiated into its parts in decontextualized isolation for the purpose of discovering defining characteristics. Such local, decontextualized analysis is inward-looking. It results in a high degree of certainty about what has been observed and a high degree of consensus among observers about the defining characteristics of the object under study.

Integration, the second phase of the cycle, follows naturally from the first. Once an object of analysis has been defined through the process of differentiation, the next step is to compare the object with other known objects in an effort to include it in the universe of knowns that the conceptual framework of knowledge comprises. This attempt to assimilate the newly defined object into the conceptual framework is outward-looking and contextualized. The attempt to integrate results in a high degree of uncertainty about how integration should proceed and a high degree of contention between individuals. Uncertainty and contention occur because there are often many possible contexts in which integration could proceed with no context being easily and clearly seen as demonstrably superior. Moreover, there may even be disagreement as to whether the newly met object can be assimilated into the current conceptual framework.

Synthesis, which completes the cycle, occurs when the process of integration discovers that the current conceptual framework for knowledge is incapable of assimilating some newly met objects. In this case the discovered relations between objects either contradict the conceptual framework or no satisfactory relationship can be discovered within the conceptual framework. Regardless, the way in which reality is conceptualized is placed in conflict with reality as it is observed. The result is that either a new conceptual framework must be synthesized that is not in conflict with observed reality or a new framework for observing reality must be conceptualized. Since the production of a new conceptual framework clarifies the relationships among knowns, demonstrates new complexity among relationships and opens new vistas for exploration, synthesis results in a self-confident flowering of knowledge and permits renewed differentiation of objects at the next level of cognitive functioning.

Table 1

Defining Characteristics of the Phases in the Cycle of Thought

Differentiation	*Integration*	*Synthesis*
fragmenting	combining	reconceptualizing
decontextualizing	contextualizing	recontextualizing
isolating	interrelating	restructuring
defining	comparing	reframing
inward-looking	outward-looking	inward and outward looking
consensus	contention	self-confident flowering
certainty	uncertainty	complexity

THE PATTERN OF INDIVIDUAL COGNITIVE DEVELOPMENT

The repetitive application of the cycle of differentiation, integration and synthesis leads over time to a pattern of individual cognitive development. This pattern is characterized by the progressive development of the individual's use of abstraction.

Cycle 1: Birth until the Synthesis of a First Conceptual Framework

The newly born have almost no knowledge about their environments and they have no conceptual frameworks through which to make sense of interactions with their environments. Both of these must be built up by infants through direct interaction with the environment. However, before infants can even begin to build the integrated understandings that permit the eventual synthesis of a conceptual framework, they must first abstract from reality a set of mental symbols with which to represent objects.

This first act of abstraction, which results in a symbolic representation, a permanent mental image of an absent object, is built upon differentiation, as are all subsequent abstractions of this class. The first abstraction may, for instance, involve the care giver, who has a smell, a visual image, a touch, a texture, a voice and many other characteristics. One of these characteristics is differentiated from the others and forms the first abstracted memory, a memory that can be called upon to give the care giver permanence when she or he is absent. Other characteristics are soon integrated, until a complex image or "schema" for representing the absent object can be called forth at will. Once the act of abstraction has been mastered, it is used over and over again, with differentiation capturing the essential characteristics of the person, thing or event, and integration allowing the infant to build up a series of largely disconnected schemata, which represent these objects, as well as later to develop less tangible concepts such as space and time.

Eventually, the infant discovers that she or he can use these symbols not just as mnemonics, but for the purpose of thinking about things, ideas and events that are not immediate. This dawning ability to reflect results both in and from a rapid interconnection of schemata. Thus, as young children implicitly operate on their symbol structures, interconnections between schemata result in the symbols being linked together in the first *synthesis* of a conceptual framework for knowledge.

Cycle 2: The Conceptual Framework of Implicit Irreversible Logic
The conceptual framework for knowledge which the child has synthesized is entirely implicit; that is, the child is not consciously in control or consciously aware of his or her thought processes. As a result, at this stage children's logic is unidirectional: They can reason logically in one direction, but, being unconscious of their own cognition, they are unable to imagine retracing a path. Similarly, since these children are unaware of their thought processes, they can view from only one unconscious and automatic perspective. Moreover, as they are unaware that they are utilizing symbolic representation, children at this stage treat the symbol as if it were the reality, which for them it is. Thus, at this stage, children accept immediate perception as infallibly valid.

Having synthesized a first conceptual framework for knowledge, children continue to seek information about their environments through direct action, but now progressively make sense of that information through reflective symbol manipulation. Through this process and by using successive differentiations and integrations, children fill voids and build interrelationships within their conceptual frameworks for knowledge. As this process continues, children become increasingly aware that immediate perception is, in fact, fallible: Some things are not as they appear to be. They also discover that some schemata are inadequate, while others are linked in ways that lead to paradoxical outcomes, and that as a result their logic is faulty. They begin to test the logic of their thinking in an effort to understand where the discrepancies between cognition and perception arise. This process of testing leads them to look first one step backward in the causal chain of their reasoning, then two, and so on. This testing goes hand-in-hand with a dawning awareness of the process of cognition itself. As children become aware of their own cognitive processes and discover that the testing of logic can be accomplished by reversing the direction of thought, they begin to transfer their trust from the infallibility of perception to the infallibility of reason.

Cycle 3: The Conceptual Framework of Explicit Reversible Logic
As children become explicitly aware of their own cognitive functioning, learn to test their thinking through reversible logic, and transfer trust from perception to cognition, a new framework for knowledge is synthesized and a new flowering occurs. Perhaps the single most important feature of the third cycle in the pattern of cognitive development is that the child who is aware of his or her thought processes, his or her unique perspective on reality, is now for the first time capable of the realization that everyone else has a unique perspective from which he or she sees, too. Thus the child is now potentially able to adopt alternative perspectives or to accept alternative perspectives as possible even when they are too alien for adoption.

An awareness of cognitive processes allows the child to consciously investigate previously implicit methods of mental organization and to explicitly develop and test logical relationships among schemata. As a result, the child rapidly develops the concepts of classification; conservation of serial, ordinal and cardinal; conservation of substance, weight and volume; spatial relations; and relations among distance, time, movement and velocity. Along with reversibility and perspective-taking, these new skills permit a flexibility of thought that is enormously powerful, but still limited.

Although children or adolescents at this stage of development can now consciously manipulate their symbol structure to hypothesize about and mentally test outcomes, their conceptual framework is still limited to considering relationships among objects that are real, observable, and concrete. However, as adolescents continue to explore their increasingly explicit conceptual framework for knowledge, they discover that the symbolic representations used to denote specific instances of relationships between given concrete objects can be generalized to cover similar instances in which the relationship pertains. From this point onward, adolescents begin progressively to operate on and to build their abstractions.

Cycle 4: Generalizing Abstractions and Conceptualizing the Possible

The building of abstractions results in a new (generalizing) way of conceptualizing reality. This synthesis of a new conceptual framework begins when adolescents discover general relationships among specific symbols and schemata which comprise their conceptual framework. At first the individual merely realizes that, having discovered the logical relationship among symbols or schemata, he or she can apply general rules to the logical relationship among the concrete objects which those symbols or schemata represent. However, as the process of generalization continues, the individual discovers that he or she can test certain relationships against the theory embodied in the discovered generalizations without immediate reference to concrete reality; that is, he or she can test mentally against the possible outcomes permitted by theory. Once the realm of hypothesis meets the realm of theoretical possibility at the level of generalized abstraction, individuals can begin to contemplate problems that have no immediate concrete correlates. They now have the conceptual framework necessary for operating at the most advanced level of human cognitive functioning.

Table 2

Defining Characteristics of Stages in Cognitive Development

Cycle 1	*Cycle 2*	*Cycle 3*	*Cycle 4*
first abstraction	*first conceptual*	*awareness of cognition*	*internalizing abstraction*
symbolization	*framework implicit*	*explicit reasoning*	*generalizing relationships*
object permanence	*reasoning*	*reversible logic*	*hypothesizing possibilities*
schema building	*irreversible logic*	*multiperspective*	*testing theoretical*
reflection	*single perspective*	*symbol as abstraction*	*propositions*
	symbol as real	*reason as infallible*	*conceptualizing*
	perception as	*concrete hypotheses*	*complexity*
	infallible	*classification and*	
		conversation	

(The cycle of cognition and the pattern of cognitive development just explicated are common to all of humanity. However, there are exceptional instances in which development does not occur as suggested. Since these instances can affect not only individual development but social development as well, they are discussed in the appendix.)

THE CYCLIC PATTERN OF HISTORICAL DEVELOPMENT

Societies comprise individuals, and many individual instances of differentiation, integration and synthesis are ongoing at all times; therefore, all phases of the cycle of human thought coexist continuously in society. However, since societies operate as an aggregate of individual actions, the three phases composing the cycle of individual cognition are better represented as three continuously flowing streams of process in human social development (see Figure 2). Moreover, although all three streams coexist continuously, in every era of human social development one stream of thought dominates and by dominating provides the central modifier that is society's generally accepted conceptual framework for knowledge generation. Further, the sequence of domination is identical to the sequence established for the cycle of individual cognition: differentiation, integration, synthesis. In addition, because they represent aggregates, these cyclic phases are more distinct for societies than they are for individuals, as is clearly demonstrated by an analysis of Western history.

Although the parallel has not been drawn before, it can be demonstrated that each phase of the cognitive cycle is manifest at the social level with sufficient distinction that historians give names to the phases. Hence, for Western social development the terms Medieval, Renaissance, Baroque, Modern, and Postmodern refer to eras in which a particular phase of the cognitive cycle dominated or dominates the social construction of knowledge. The validity of this observation is demonstrated when the hallmarks of differentiation, integration and synthesis—the phases composing the cycle of human thought— are compared with the hallmarks of the eras in Western development that have been previously identified by historians. When this is done it becomes apparent that the Medieval was primarily a differentiating era, the Renaissance an integrating era, the Baroque was a synthesizing era, the Modern a differentiating era once again, and the Postmodern an integrating era.

Moreover, each complete cycle of synthesis, differentiation and integration forms a stage in social development that finds its direct analogue not only in the cognitive cycle but also in the pattern of individual cognitive development. (Note that I have now placed synthesis at the beginning of the cycle, since it is the synthesis phase that provides the conceptual

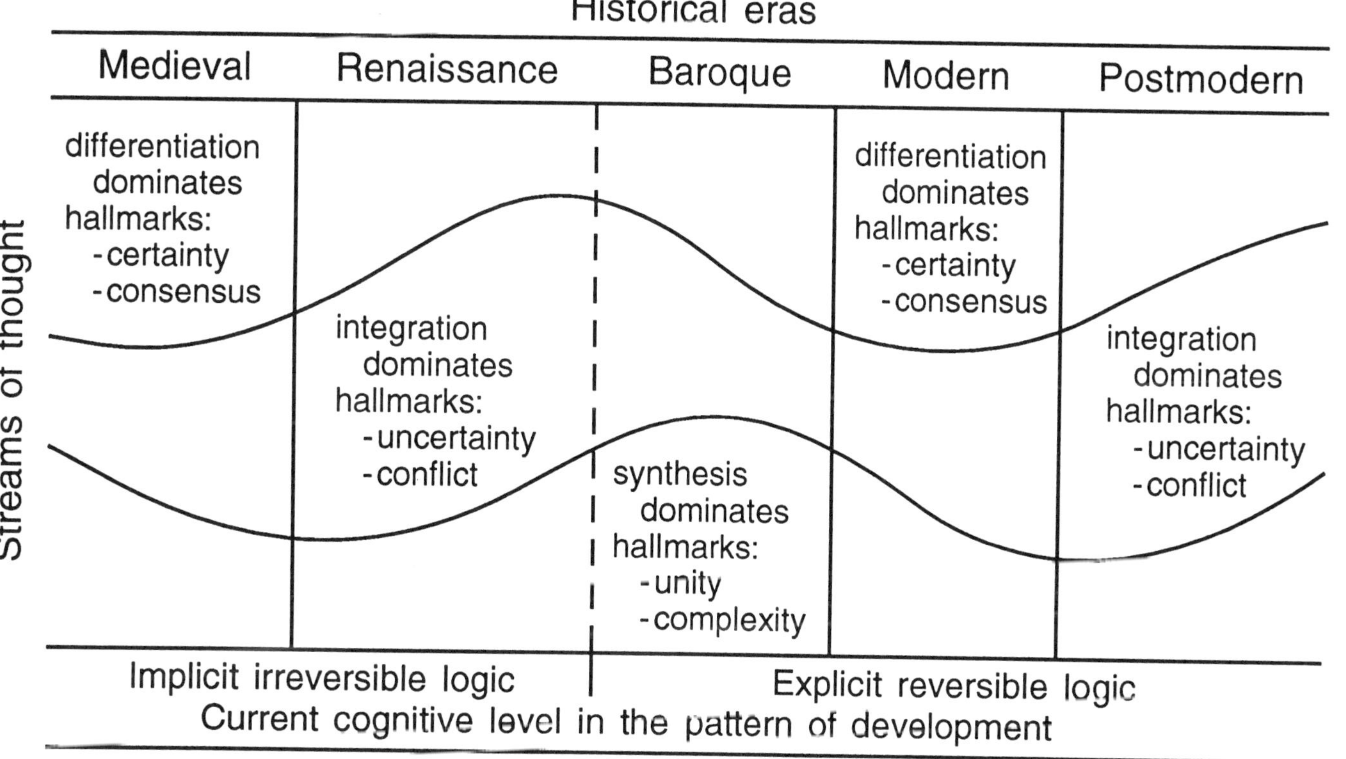

Figure 2. An abstract map of the pattern of cognitive development in Western history

framework that is articulated during the following differentiating and integrating eras.) When Western social history and the pattern of individual cognitive development are compared, the Medieval era and the Renaissance are found to parallel the differentiating and integrating phases of the second cycle in the pattern of cognitive development, while the Baroque synthesis introduces the third cycle which is developed in the Modern age and completed in the Postmodern era.

Although the cyclic pattern of social development can be clearly demonstrated, dating the shift from one phase of a cycle to the next, and from one cycle to another, is more difficult. As will become apparent when attention is focused upon the Postmodern era, during a phase change there is a transitional period during which two modes of knowledge generation compete for dominance. Adding to the problem of dating, as is true for individuals, not all schemata composing the conceptual framework of society make the jump from one cycle to the next at the same time. Therefore, some events appear to occur slightly before the era to which they belong is fully underway, as will be demonstrated for perspective-taking and the timing of the Baroque era.

The Historical Eras and Their Hallmarks

MEDIEVAL EUROPE (TO 1420) — THE DIFFERENTIATING PHASE
According to all accounts, the conceptual framework of medieval Europeans centered on an inward-looking Christianity. Introspective religious certainty and consensus defined the medieval world view. In this highly religious age, the certainty of a Christian God dominated an otherworldly thought process. As a result, self-repression was the norm and the context of thought was local. People conceived the framework of being as a hierarchy with God at the top and the earth at the center, yet the dominant if mainly implicit metaphor was of a unified organism. With God dominant and the church as His shepherd, knowledge generation was religiously directed. Therefore, although universities were founded during the thirteenth century, they taught entirely from a theological basis.

In a differentiating age, certainty means stability, and change is slow to come. However, by the end of the fourteenth century Mediterranean trade had helped the Medici family of Florence to become the bankers of Europe and consequently to make Florence the center of a protocapitalist world economy. As is often the case with the wealthy, the Medici and other wealthy Florentines began to patronize artists and scholars. Along with other goods, they imported a "new" commodity, the knowledge of the ancient Greeks, as well as "modern" ideas from China, India, and the Florentines' Moslem competitors. The result was that, quite suddenly, a new age dawned.

The Renaissance (1420–1610)—The Integrating Phase

Renaissance man had two sets of truths to compare. He had the knowledge of God from Christianity and he had the knowledge of Greek science rediscovered, as well as ideas provided by "modern" Chinese, Indian, and Moslem science. As history records, the age was outward-looking, profane and worldly. In its uncertainty it questioned everything as it turned from self-repression to self-expression—to humanism. For instance, Fra Filippo Lippi (1406–1469) turned from two-dimensional pictures designed to transcendentally focus attention on the act of worship to three-dimensional portrayals, through the natural human form, of God's immanence. Donatello completed David in 1430. Botticelli's Venus was unveiled in 1480.

This was also the era in which Leonardo da Vinci (1452–1519) created his art and conducted his inquiries. Similarly, Copernicus questioned the Aristotelian conceptualization of the solar system and proposed his own version. This was also the era in which printing began and thus the era in which knowledge became more readily available. Mapping flourished. The uncertainty produced by comparing two bodies of knowledge regarding reality led not only to contention regarding the givens of knowledge in science and art but also in religion. As a result of his questioning, Martin Luther began the Protestant Reformation in 1517.

However, as the process of transformation proceeded, the culminating intellectual event of the Renaissance occurred unheralded, circa 1550, when Pietro de la Rammee (Ramus) published his work on the dialectic. Ramus dialectical logic explicitly sought to view thesis and antithesis as complementary parts of a greater whole, which was then sought through the process of synthesis.

The Baroque (1610–1690)—The Synthesis Phase

John Donne, the metaphysical poet and dean of St. Paul's; John Milton, author of *Paradise Lost* and secretary to Oliver Cromwell; Andrew Marvell, metaphysical poet, secretary to Cromwell and parliamentarian extraordinaire; Francis Bacon, the author and scientist who first explicated the scientific method: All studied the dialectic of Ramus, as did many others in the Baroque. Ramus's logic was explicit, and explicitly directed towards synthesis. As such it was a dialectic ready-made for the age.

A synthesizing phase ushers in a new conceptual framework for knowledge. During synthesis, knowledge is reconceptualized, recontextualized, and restructured. Those looking at such an era should find a seeming paradox. The age should at once appear inward- and outward-looking, focusing on the micro and macro, concerned with the global and the local. Also they should find a self-confident flowering of knowledge, and an acceptance of the apparent complexity of interrelationships.

If the synthesis ushering in a third-cycle conceptual framework for society is analogous to the synthesis ushering in a third-cycle conceptual framework for the individual, then social thought should demonstrate a self-awareness of cognitive processes. This awareness, in turn, should lead to a society in which perspective-taking skills flourish. It should also lead to a society that explores nature through empirical science by using those newly acquired skills of the third-cycle conceptual framework: explicit reasoning, reversible logic, trust in reason as infallible, the ability to state concrete hypotheses about the nature of things, an understanding of classification and conservation and the ability to develop concrete relationships between such things as mass, volume and weight, and distance, time, velocity and acceleration. Indeed, these are the hallmarks of the Baroque.

The Baroque era, which is often confused with the Renaissance in histories of science, is clearly identified as a unique era in literature, music, visual art, politics and economics. It is also clearly identified, as will be seen, as marking the beginning of modern experimental science. As Lossky observes, it was a time when new structures were sought for knowledge:

> The seventeenth century in Western Europe was born in chaos and ended in a semblance of order that ushered in the age of Enlightenment. The search for order is the main theme of European history in this period. The chaos at the beginning of the seventeenth century engulfed most aspects of life—religious, intellectual, artistic, economic, social, and political. (1967, p. 3)

So successful was this search for order that Clark states, "Somewhere about the middle of the seventeenth century European life was so completely transformed in many of its aspects that we commonly think of this as one of the great watersheds in modern history" (1947, p. 1).

Perhaps the first clear social signal of the transition to third-cycle thinking was the emergence of complex, multiperspective drama. Throughout the Medieval era and the Renaissance, drama had exerted a powerful influence on society; however, beginning in 1590s London, an entirely new theater experience began to evolve. Over the next twenty years characterization went from one-dimensional to multidimensional; psychological motivation became the reason for action; various characters had their own perspectives; and the interaction of past and present, perspective and motivation among the players created the complex web within which the drama played itself out.

The plays of Shakespeare would be works of genius in any age; however, they could only be written by a being who was cognizant of multiple individual perspectives and of the complex web of interactions that influences outcomes. Perhaps more importantly, the same plays could only be popular with, as a result of being understood by, an audience who was able to appreciate these things; that is, an audience who was able to fully enter into the characters' various perspectives and the plot's complexity. Moreover, a reading of Shakespeare indicates that the author's explicit awareness of both human psychology and the web of complexity increased dramatically over time. It seems that his audience's awareness shared this growth, as his contemporary popularity attests.

It was during this era, also, that metaphysical poetry flourished. This poetry attempts to synthesize a unified understanding from the apparently contradictory elements inherent in things, ideas and events. Most of the metaphysicals formally studied Ramus's dialectical logic during their schooling, and Milton wrote a book on Ramist logic between writing *Paradise Lost* and reorganizing it.

While metaphysical poetry demonstrates synthesis, the birth of the novel during the Baroque, as with Shakespearean theater, demonstrates the emergence of perspective-taking skills. Since the novel relies upon complex plot, theme and characterization, both author and reader must possess the ability to take multiple perspectives before the novel form can flower.

The same rich flowering occurred simultaneously in other European literature, so that, speaking of Baroque French theater, Roaten says:

> The plots fuse and are interwoven; they are not sharply marked off from each other. Plots and subplots are integrated and indivisible. The relations of the plots and subplots are characterized by restless and continuous movement, contrast and a liking for the paradoxical and the bizarre. The several separate plots steadily merge with each other. (1960, p. 28)

Nor was visual art an exception. According to Dupont and Mathey,

There are sound reasons for the great contemporary interest in the seventeenth century; for the artists of that century [who included, among others, Caravaggio, Hals, Hobbema, La Tour, Poussin, Rembrandt, Rubens, Ter Brugghen, Van Dyck, Van Streeck, Velazquez and Vermeer] re-stated and found new solutions to all the problems of creative art. . . . The amazing thing is that all these ends were achieved, not in the course of a slow evolution extending over centuries, but almost simultaneously, within a few years, and in quite different countries. Hence that impression the century gives us of an immense profusion, an abundance of creative genius that seems indeed unique, as compared with earlier or subsequent periods. (1951, p. 7)

Simultaneously, important new structures were appearing in music. As Young states in *A Concise History of Music* (1974):

One generation takes things apart; the next generation puts some of the fragments together again, to form a new pattern. At the end of the sixteenth century men were in a speculative frame of mind. It was the beginning of a new age. Before the Renaissance, ideas had been centered on the principle of God; after the Renaissance, it was the principle of ultimate human responsibility that counted. Music was moved out into the world. (p. 74)

   . . . there needed to be a new relationship between composers and audience, and between them and the solo singers who were to play such a dominant role in the development of music during the *Baroque* era. (p. 76)

In his *History of Music* (1973, pp. 236–69), Worner is more precise when he says,

If one thinks of the time scheme from this point of view [of theory development], one can begin the "modern period in the history of music" around 1600. New developments that occurred around 1600 include:

Opera	Independent orchestral music
Oratorio	Soloist instrumental music

In the three-fold division of the Baroque, individual personalities take their places thus:

Gabrieli	Handel	Bach
Vivaldi	Monteverde	Purcell

The Baroque was as inventive of new structures in the political arena as it was in other areas. As Pole says in *The Seventeenth Century: Sources of Legislative Power*:

> In the early seventeenth century, when the first representative assembly was called to meet on American soil, the English monarchy was believed, by its incumbent, to rule by a divine right that rendered it superior to any popular power. By the end of the century, English kingship had been subordinated to constitutional government. (1969, p. 1)

Although the transition to democracy in England was held in check rather effectively by Cromwell for much of this period, the era did see a monarch who was only able to view from one perspective executed by a populace that was able to view from at least one other perspective. Cromwell, himself, may have died as a result of his adherence to a single perspective. Before the Baroque era ended, governance based on multiple perspectives was a reality.

At the same time that new structures were created in art, music, theater, literature and politics, the seventeenth century saw the consolidation of a new socioeconomic structure: the capitalist world economy (Wallerstein, 1974). Thus, it was at this time that the first global structure solidified. This new structure represents a web of global economic interdependence that ties each to each through a series of commodity chains that cross local, regional and national boundaries.

Clearly, a synthesis of enormous importance occurred during the Baroque. Equally clearly, perspective-taking skills evolved and the classification skills necessary to categorize musical disciplines developed. These are both third-cycle cognitive developments. However, the third cycle in the pattern of cognition involves not only the development of perspective-taking and classification skills, it also involves the development of other skills that permit empirical science to flourish.

The results were astoundingly rich. As Caullery says in his book *French Science*, "It was in the seventeenth century that scientific life was organized and soared away. It was from that time that fundamental discoveries were made in France, and that here, as well as elsewhere, the true foundations of modern science were laid down" (1933, p. 5). Nor does he exaggerate. It was at this time that Kepler (1609), by compiling empirical data about the heavens and comparing various possible theoretical structures with the data, reconceptualized planetary motion and developed the model of the solar system still in use to this day. Simultaneously, Gilbert (1600) investigated the properties of magnetism and developed an understanding of its nature. Soon after, Bacon wrote *Novum Organum* (1620), which made explicit the cognitive process of third-cycle science. He did this in part to develop "the criticisms of Ramus against what he took to be the errors of the Aristotelian schoolmen" (Redwood, 1977, p. 17). In 1616, Harvey completed an understanding of the circulatory system, which he published in 1628. Between the time Harvey realized how blood circulates and published his findings, Galileo Galilei discovered the basis of mechanics, used the telescope to found astronomy, invented the thermometer and discovered the vacuum. Still within the Baroque, Toricelli developed an understanding of pressure and invented the barometer; Huyghens invented the clock; Leeuwenhoek created the microscope; and Boyle founded chemistry.

To reiterate, empirical science requires explicit reasoning, reversible logic, concrete hypotheses and a conceptualization of classification and conservation. These developments permit an understanding of the concrete relationship among things and events. The third cycle of cognitive development provides the conceptual framework that classical sciences requires. Therefore, since empirical science began in the Baroque, the third cognitive cycle in Western European social history commenced with the Baroque.

Thus, the Baroque stands out as the era when the synthesis of an explicit conceptual framework for reversible logic occurred. However, the most important intellectual synthesis of the Baroque era has not yet been touched upon. This synthesis involves the work of Newton and it sets the stage for the modern age.

The Calculus—The Differentiating Process Made Explicit
In Europe, the independent discovery of calculus is claimed for three individuals: Descartes in 1644, Newton in 1665 and Leibniz in 1673. (Note that calculus had also been discovered previously by Chinese, Indian and Arabic mathematicians.) However, Newton gets the primary credit for discovering calculus. His *Philosophae Naturalis Principia Mathematica* (1687) used calculus to demonstrate how the relationships between distance, time, velocity, acceleration and mass can be derived mathematically. In so doing, Newton provided the paradigmatic exemplar (the explicit example) from which arises the Modern era's conceptual framework for third-cycle

knowledge generation. As Guterman and Nitecki state,

> In the late seventeenth century, Isaac Newton in England (1665, 1687) and Gottfried Wilhelm Leibniz in Germany (1673) synthesized several centuries of mathematical thought to create a language and method for describing and predicting the motion of bodies in various physical situations. The invention of calculus was immediately followed by a period of intense mathematical activity, and the effect of these ideas on the development of mathematics, science, and technology makes this event surely one of the most important in the history of Western thought. (1991, p. 1)

In the *Principia Mathematica*, Newton used calculus, in conjunction with his three laws of motion, to derive Kepler's laws of planetary motion and the law of universal gravitation. Kepler had previously discovered his laws by plotting Tycho Brahe's celestial observations against various curves. By doing this Kepler had successfully described planetary motion. Newton, by contrast, demonstrated the principles of physics that determine *why* the motion of bodies in the solar system occurs as it does and derived Kepler's laws from these principles.

While conducting his research, Newton invented calculus, which comprises two complementary branches, differential and integral calculus. Differential calculus, the first branch to be fully developed, permits the rate of change and the limits of change to be calculated. Integral calculus permits the determination of a function from information about its rate of change. Thus, calculus is the mathematics of change and the limits of change. As such it is central to the Newtonian conceptual framework in two distinct ways. First, it supplies the mathematical tools for modeling and thus predicting change in simple, reversible, closed, mechanical systems. Second, the idea upon which differential calculus is based, that a function can be fragmented into the elements that compose it and those elements can be studied in isolation, provides an explicit method for breaking down not only quantifiable mathematical concepts into their component parts, but also qualitative concepts such as ideas and events into their component parts. In addition, the idea upon which integral calculus is based, that the relationship between elements can be built up from their underlying commonalities, provides an explicit methodology for fitting the fragments back into the current conceptual framework for knowledge. Moreover, differentiation led Newton to realize that the relationship between the rates of change of observable quantities is often simpler than the relationship between the quantities themselves. Thus, differentiation provides not only a vehicle for one phase of third-cycle science but also demonstrates a potential method for the simplification of otherwise intractable problems.

So it was that differential calculus supplied the explicit conceptual framework for breaking down not only concepts of physics, but all things and ideas into their component parts. The process of explicit differentiation provides the hallmark of modern empirical science, which has looked more and more closely at things, ideas and events by isolating elements on the basis of their differences, hence differentiating them.

THE MODERN WORLD (1690–1965)—A RETURN TO DIFFERENTIATING

The defining characteristics of the Modern world, the world before 1965, are common currency even to those who grew up as Postmoderns. They include certainty, consensus, fragmentation through differentiation, individuality, materialism, the machine model, empiricism, knowable scientific truth and the belief in foundational knowledge and universal truth.

The important intellectual events are too numerous to mention in any detail, but highlights include Linnaeus's system for classifying plants and animals (1737); Watt's refinement of the steam engine (1776); Jenner's discovery of vaccination (1796); Darwin's *Origin of Species* (1859); Einstein's Special Theory of Relativity (1908); Bohr's extension of Planck's quantum theory (circa 1910); Heisenberg's Uncertainty Principle (1927); Bochner's extension of Radon and Frechet's discovery of integral calculus (circa 1930); and the application of computers (circa 1960).

During the time span in which these discoveries were made, the term "scientist" was invented (1840). University faculties evolved from a core of theology, arts and science, law, and medicine in 1800 to around twenty faculties per university by 1965, with many faculties containing four or more quasi-independent departments where none had existed in 1800. Moreover, research orientations themselves fragmented. This fragmentation results from, in part, the explicit third-cycle ability to classify. Similarly, based upon the explicit classification of fragments, only during Modernity could the sequential divisions of history have been named. Although every age is the modern age for its inhabitants, since history was fragmented into eras during the differentiating phase of the third cycle, it is this era which is termed the "Modern" age.

To continue to develop the equation linking the Modern age with differentiating and the third cycle of cognitive development seems unnecessary. Modern readers are no doubt quite able to further develop the equation on their own. Instead, it may be more valuable to attempt, by looking back from the vantage of Postmodernity, to identify some of the clearest signals indicating that a transition from Modern/Differentiation to Postmodern/Integration was in the offing.

The first clear sign was Einstein's (1908) development of relativity theory, which was soon interpreted to mean that rather than a universal framework for "truth," all spectators at an event see unique events, because their situations in space-time are unique. Thus, while no two perspectives on the same event can be congruent, each perspective has potentially equal validity. It may be worth noting that contemporaneously, the Dadaists were creating art that was intentionally marked by fragmentation, lack of connection and a sense of lost perspective. The second clear sign was Heisenberg's (1927) uncertainty principle, which tells us that we cannot be certain about our objectivity, because the act of measuring, the particular instrumentation and methodology chosen, and therefore our subjective choice, affects the thing being measured and the outcome of measurement. The third clear sign of change occurred when Bochner provided an explicit mathematical tool for recombining or transforming fragments through integration.

A number of mathematicians, including Radon and Frechet, pioneered integral calculus, which became useful as a result of Bochner's (circa 1930) extension of earlier work. Integration takes the differentiated elements and recombines them based upon characteristics which are integral to them all. Thus, integration explicitly provides the hallmark of Postmodern science. Faced with a vast array of apparently dissimilar elements, elements that have been disassociated from each other on the basis of differences, Postmodern science tries to build associations among them. It tries to recognize "transformations" between elements and—as with social cartography—to pattern contested realities.

THE POSTMODERN CONDITION (1965–??)—THE RETURN TO INTEGRATION
From the Baroque era onward, thought has been becoming more and more explicit; therefore, the Postmodern era begins with the dawning, for many people, of a conscious recognition that reality is fragmented, multiperspective, uncertain, decontextualized and differentially valued in a manner that promotes conflict and agonistics. However, the systemic bifurcation (qualitative change of state) that occurred in all realms of thought at about the same time, and ushered in the Postmodern era, began with an instability that gave way to an age of instability and perceptual conflict. As such, many who grew up in the Modern era still cling to the notion

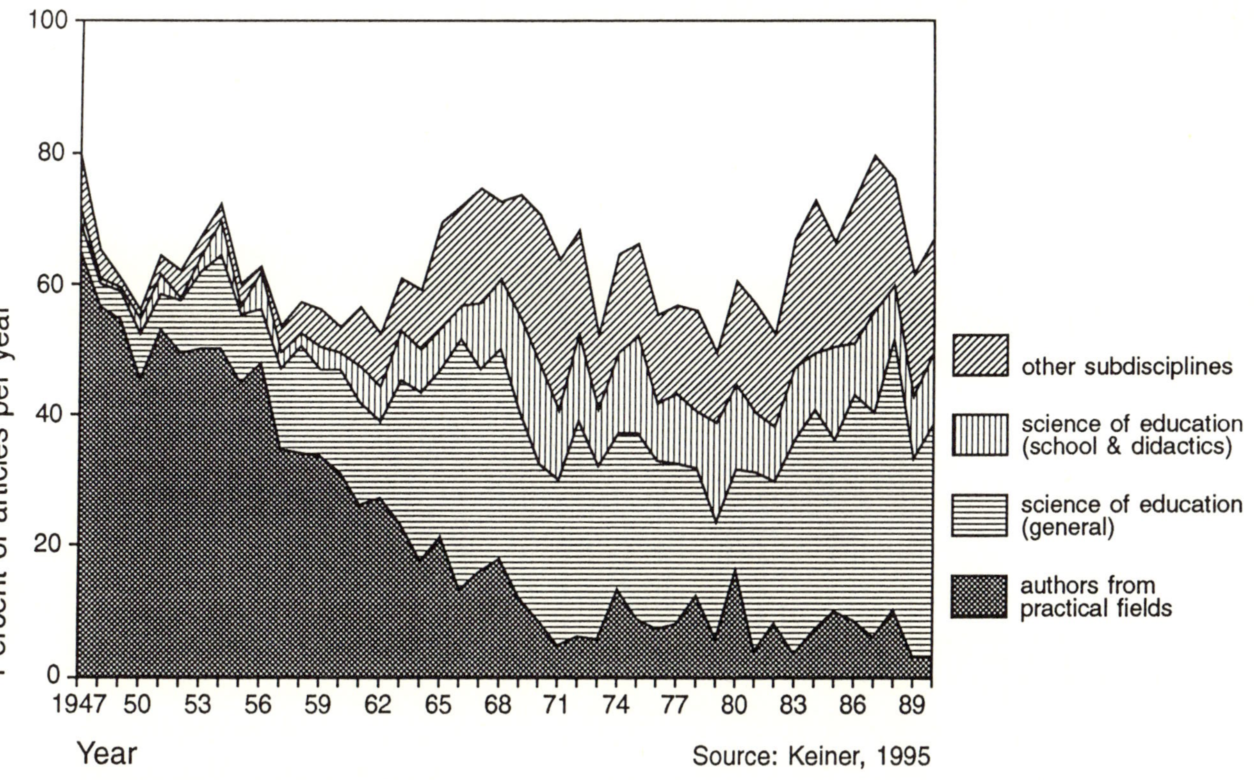

Figure 3. Authors classified according to disciplinary affiliation in three major German educational journals (number of articles = 5,580).

that little has changed. Nonetheless, as the example in Figure 3 shows for educational publications in Germany, recognized or not, a bifurcation did occur. And this bifurcation, this change of state from domination by differentiation to domination by integration, occurred across all disciplines. Since we are immersed in the Postmodern condition, which is apparently multiperspective, comments from and on the critical participants may be most illuminating.

As early as 1955, Merce Cunningham created nonnarrative, emotionless "modern" dance, with abstract or no landscape, no characterization and no situation. He employed differentiated, fragmented elements. Set, costume, music and movement were created in isolation and brought together only at the time of performance, so that no context existed. The action was nonlinear and governed by chance occurrence or randomness. Therefore, it is unsurprising that a Judson Church dancer, one of Cunningham's students, coined the term Postmodern in 1962 to describe her choreography.

As Trachtenberg identifies, early Postmodernism involved an uncertain, almost static questioning of the human condition, though it would be presumptuous to state that this was a search for reconnection:

> [In a review dating from the late 1970s,] the critic Pauline Kael noted that the characters didn't believe in anything at the beginning of the film or at the end either. Critical of existing forms, postmodernism tends to substitute enactment for interpretation—a rhetorical approach for the acknowledgement of history . . .; narrative movement is lateral rather than progressive; figures, even objects, are not depicted. . . . Postmodernism describes a sensibility, a feeling for innovation, for experiment with conventional ways of framing experience so that it is at once removed from recognizable relationships and from the location in which they exist. . . . This movement is variously informed by a skeptical attitude toward illusion, toward a recognizable psychology of human relationships, and towards coherence of any sequence of actions. . . . The self-consciousness of the postmodern novel, as Robert Alder points out, is not pre-occupied with its own artifice but with questioning its premises. (1985, pp. 6–7)

As a conscious recognition grew of the extent to which fragmentation had progressed, direct comparisons between modern and postmodern began, as the following chart indicates.

Table 3

Defining Artistic Symbolizations of Modernism and Postmodernism

Modernism	*Postmodernism*
Romanticism	Paraphysics/Dadism
Form (conjunctive, closed)	Antiform (disjunctive/open)
Design	Chance
Hierarchy	Anarchy
Art object/finished work	Process/performance/happening
Creation/totalization and antithesis	Decreation/deconstruction
Presence	Absence
Genre/boundary	Text/intertext
Semantic	Rhetoric
Paradigm	Syntagm
Metaphor	Metonymy
Selection	Combination
Signified	Signifier
Determinacy	Indeterminacy
Transcendence	Immanence

Source: Hassan, 1985, pp. 91–92.

Once the fragmentary nature of the condition was known to exist, the first attempt was to achieve stability through acceptance. As Collins (1989) contends,

> Within the last decade a wide range of cultural critics representing radically different ideological perspectives have pointed to the crisis within cultural life. Culture is no longer a unitary, fixed category, but a decentred, fragmentary assemblage of conflicting voices and institutions. Whether described as the dissolving of the "mainstream" into a delta by an avant-garde composer like John Cage or the loss of a "common legacy" by the Secretary of Education, a widespread awareness exists that an "official," centralized culture is increasingly difficult to identify in contemporary societies. Although explanations for this development differ quite drastically from theorist to theorist (as do their responses to it), the common denominator remains a recognition that our supposedly "common culture" is continuing to fragment. (pp. 1–2)

The opponents of Post-Modernism have been especially critical of the notion that individual subjects are capable of making selective decisions in the face of this semiotic glut, preferring to dispense with the issue by insisting that this "free to choose" mentality plays directly into the "free market" ideology. But such a move . . . only brackets the central issue—what is the individual to do in such a context? 1) demonstrate that multiple aesthetics are appropriate for different publics within a given culture and time; 2) be heretical in critical practice; 3) strike a balance between involvement and critical distance by recognizing the requirement of choice. (p. 146)

However, as the Postmodern era has progressed, the emphasis has begun to shift from a recognition and initial acceptance of fragmentation to a search for reconnection and revaluation. For instance, Einstein's theory of relativity has been reinterpreted to indicate that not only is every perspective unique, but also partial, so that the understanding of an event requires the negotiation of meaning based upon a multiperspective analysis. As regard, in reference to literature and religion, Collins further notes that:

The postmodern novel, McHale (1987) argues, is characterized by a shift from an "epistemological" to an "ontological" dominant. By this he means a shift from the kind of perspectivism that allowed the modernist to get a better bearing on the meaning of a complex but nevertheless *singular reality*, to the foregrounding of questions as to how radically different realities may coexist, collide, and interpenetrate.

Here, also, no less a person than Pope John Paul II has entered the fray on the side of postmodern. The Pope "does not attack Marxism or liberal secularism because they are the wave of the future" says Rocco Buttiglione, a theologian close to the Pope, but because the "philosophies of the twentieth century have lost their appeal, their time has already passed. The moral crisis of our time is a crisis of Enlightenment thought. For while the latter may indeed have allowed man to emancipate himself "from community and tradition of the Middle Ages in which his individual freedom was *submerged*," the enlightenment affirmation of "self without God" in the end negated itself because reason, a means, was left, in the absence of God's truth, without any spiritual or moral goal (Baltimore *Sun*, 9 September 1987). The *postmodern theological project is to reaffirm God's truth without abandoning the powers of reason.* (1989, p. 41)

It becomes clear that the Postmodern condition should be viewed as comprising two stages. The first stage is transitional, from differentiating to integrating. It involves a reaction against modernity, a dawning awareness that the world comprises dissociated fragments. This realization, in turn, results in an attempt to reassert stability by accepting the fragmented nature of reality. In this stage there is no context; the universe exists as fragments, and the fragments are defined by their unique attributes, their differences from every other fragment. This makes the building of linkages difficult, which results in an initial attempt to accept fragmentation as the natural order. However, people seek stability and a fragmented universe is inherently unstable. In response, the second and definitively integrating stage of the Postmodern era sets in. This stage produces a world reconnecting on the basis of like. The process of integration is embraced and a critical search for reconnection and revaluation is undertaken. The integration of empirical fact and spiritual value, subject and object are pressing concerns. It is at this time and partially to fill this need that abstract cartography emerges across disciplines and as an interdisciplinary connector.

At the same time that transformative reconnections are being sought, Postmodern knowledge cannot be fully integrated within the third-cycle conceptual framework bequeathed by Newton and the Baroque. A new, fourth-cycle conceptual framework for knowledge is required. Already there are signs that indicate that the integrating phase may be short-lived. Significant indications of a shift to synthesis have already appeared. Moreover, the history of the differentiation, integration, and synthesis cycle demonstrates that the time cycle has been collapsing, as the following table demonstrates.

Table 4

The Spiral Pattern of Western Social Development (approximate dates based upon signal historical events)

First Cycle	*Symbolic Representation*	*Interval*
Differentiation	Dawn of humanity until 700 BC	
Integration	700 BC until 460 BC	240 years
Second Cycle	*Implicit Irreversible Logic*	*Interval*
Synthesis	460 BC until 322 BC	140 years
Differentiation	322 BC until 1420 AD	1740 years
Integration	1420 AD until 1610 AD	190 years
Third Cycle	*Explicit Reversible Logic*	*Interval*
Synthesis	1610 AD until 1690 AD	80 years
Differentiation	1610 AD until 1965 AD	355 years
Integration	1965 AD until ?	

The Shift to Complexity—Generalizing Abstractions and the Fourth-Cycle Synthesis

A number of events point to an upcoming or ongoing shift to domination by a synthesizing orientation. The first clear indication of a new web-building or synthesis-oriented framework emerged in the mid- to late 1960s with the birth of environmentalism. Environmentalists tapped into and explicitly recognized the global web of interdependence. Such a recognition is clearly characteristic of the synthesizing and structure-building orientation. Shortly thereafter, in the early 1970s, Wallerstein proposed that there exists a capitalist world economy, global in extent. Wallerstein's argument makes explicit the interconnective social web, and is, thus, a synthesis-oriented endeavor. At much the same time, Prigogine was reinterpreting the "second law of thermodynamics." His reinterpretation prompted the birth of chaos theory, which is the explicit theory of complex systems. This is the clearest indication to date of a shift to the building of structures orientation. Almost simultaneously, global education became a discipline. Global education recognizes the complex web of interrelationships that binds each to each on a global basis. It seeks to reconstruct the web on the basis of collective values. More recently still, the rapid development of global, telecommunications-based computer networks indicates that another structure- or web-building enterprise is underway. Add to this the instrument called abstract mapping, which I believe is the signal instrument of fourth-cycle science, and it might be reasonable to conclude that the shift to synthesis has already begun.

Speculation upon the timing of the next synthesis is also possible on the basis of the time lapse between a precursor and an event. For instance, the dialectic logic of Ramus appeared sixty years before the Baroque began. Similarly, while relativity appeared sixty years before the Postmodern era, the more important precursors—uncertainty and integration—predated the shift by only thirty years. It has now been twenty years since the underpinnings of complexity (chaos) theory won a Nobel Prize for Prigogine.

It is of course impossible to say whether these signs portend an imminent shift from postmodern integration to web-building synthesis. Not only is it impossible for the reasons given above, but because factors germane to other social flows and to individual inputs will affect the timing of this qualitative shift.

Some Further Thoughts

Given an abstract map of this cyclic pattern of psychosocial interaction, other observations are possible. The importance of the social transmission of knowledge through formal education is reinforced. This cyclic pattern also allows a clearer understanding of past, present and future. Taken together, the cyclic patterns of individual and social development demonstrate that human nature and human society are complexly interrelated and that social science cannot be successfully conducted unless the model for analysis recognizes this.

At the individual level, we need to explicitly teach children to use the cycle of thought as a problem-solving mechanism. In order to avoid the development of disconnected schemata and truncation of the pattern of development, we must learn to make interrelationships among schemata explicit for children though, for example, concept mapping. To ensure that coherence and optimization occur, the important schemata and stages of schema development can be identified and children can be encouraged to develop and interrelate these schemata at the appropriate level of development (Bloom, 1995; McIntosh, 1995).

At the same time, the shift from one cycle in the pattern of cognitive development to the next has not been without cost. At each successive level, if the previous level's knowledge is denigrated or abandoned, the tendency occurs for experience to become more cerebral and mediated at the expense of physicality and direct experience. Thus, when the symbolic representation level of immediate physical action is left behind, there is a tendency over time to lose sight of the importance of our physical connection to the external environment and of the physical-mental connection within our own bodies. When the physical connection is lost, we are not whole and we cannot function optimally. Therefore, not only must education seek to maximize everyone's level of cognitive development, but education must also seek to maintain and reinforce the necessary connections across levels of operations so that important learning is not abandoned. This observation holds true for society, the aggregate of individuals, as it does for each individual.

At the theoretical level, the relationship between individual cognition and social development reinforces our developing awareness that the Newtonian conceptual framework for decontextualized knowledge is totally insufficient for modeling the complex interrelationships in society. It becomes apparent that individual cognitive development is complexly interrelated with social development through the social transmission of knowledge. These complex interrelationships cannot be modelled using third-cycle science. Early work in chaos/complexity theory suggests that predictive analysis of social change is, in fact, impossible. Instead, we must reconceptualize so that when social science seeks to analyze large groups it operates much as quantum mechanics does, based upon statistical probabilities that an outcome will occur. However, it is also worth noting that in the same way differential calculus provided a mechanism for moving from complexity to simplicity at the third level of cognitive development, it should be expected that, as society fully enters the level of generalizing abstractions, the mathematics of "complexity" will be developed and that this will permit a new conceptual framework for science. This new conceptual framework will simplify modeling and permit yet another flowering to occur. This simplification may permit the prediction of qualitative outcomes in complex systems.

Using the Baroque example, we should expect that at the beginning of the synthesis phase of a conceptual framework for generalizing abstractions, the hallmarks of synthesis will prevail. In particular, the age will be inward- and outward-looking, seeking to reconceptualize, recontextualize, restructure and reframe knowledge. We should also expect that as development occurs, the Newtonian framework for knowledge will become a special case or concrete example subsumed by the generalizing abstractions' framework. This era should produce innovation at a pace previously unimaginable as a self-confident flowering at the adolescent level of boundless possibilities occurs. At the same time, the beginning of the synthesizing phase brings together two levels of complexity; the complexity inherent in synthesis and the need to conceptualize complexity at the level of generalizing abstractions. As a result, the era will dawn with the world seeming impossibly complex. However, because generalizing abstractions seek the underlying formulas among general relationships, as these formulas are discovered, this feeling of overwhelming complexity will give rise to a sense that complexity is manageable. This era of synthesis will end when a formulaic process for discovering the formulas among general relationships is made explicit, as we might expect from the parallel of Newton's explicit formula for conducting reversible empirical logic as laid out in the *Mathematica Principia*.

During the evolution of a fourth-level conceptual framework, important aspects of the previous conceptual frameworks must be harmonized with generalized abstractions. Thus, for instance, the third-cycle concept of symbolic representation will be reconceptualized at the fourth-cycle level, as will irreversible logic and then reversible logic. I expect that, in its beginning phase, mapping abstractions integrates symbolic representation at the fourth-cycle level, while Prigogine's reinterpretation of the second law of thermodynamics represents the explicit integration of irreversible logic at the fourth-cycle level. I see no signs yet that anyone has begun to theorize regarding the generalized abstraction of reversible logic; but such a theory is required before a formulaic process for generalizing the interrelationships among abstractions can appear. Nonetheless, the movement to fourth-cycle thought is well begun.

However, as discussed in the appendix, it is possible that dysfunction, incoherence, and/or stagnation will prevent society from achieving the level of generalizing abstractions at this time and possibly from ever achieving it. As with the example of ancient Athens, the potential domination of inhibiting factors produces the possibility that the number of humans achieving the level of generalizing abstractions may be too small to lift society to this normative level. Although I do not discount this possibility, I intuit that the world economy functions as a homogenizing mechanism and that as a result of its reliance upon higher formal education as a source of innovation, it uniformly seeks out and raises those who are capable of cognitive advancement to their maximal level of development.

Assuming that human society rises to the level of generalizing abstractions, it remains impossible to predict what the next synthesizing phase after the level of generalizing abstractions will entail. Is there an end? I doubt it. Although it might be expected that, as is apparently the case with some fully mature individuals, all of the phases and stages of development will become operant within a holistic conceptual framework so that each problem is addressed from a fully conscious perspective, even this is speculation. What we can say is that at the next level of cognitive development, the inextricable linkage between individual and social development will produce a system in which the two parts drive each other in such a close relationship that development is relatively concurrent.

Further investigation of this cyclic pattern should probably involve the scalar function touched upon in the appendix. The example of ancient Athens makes it apparent that the interaction is not directly between the individual in particular and society in general, but there are intermediaries or intermediate groups. In the case of second-cycle reasoning, one important intermediate group was probably religious in nature. In the case of third-cycle reasoning, the dominant group is clearly scientists. When these scalar functions are investigated more closely, it should be possible to discover the feedback mechanisms between scales and to understand why and how ideas shift from individual to group and thence achieve social dominance. It seems likely that only people and groups that are operating at an advanced level of cognition are moving society forward. The most probable feedback mechanism would arise when these people achieve success. Their success gives them a competitive advantage and forces others to accept the new knowledge or face relegation.

Another issue that has not been explored but seems central to the development of this understanding involves the mechanisms of communication. It seems reasonable to infer that the development of spoken language coincided with the development of society as we understand it. The development of written language coincided with the first integrative phase in social development and permitted the first synthesis. The development of printing, we know, coincided with the second integrative phase and permitted the dissemination of knowledge, which was necessary before a synthesis to third-cycle reasoning could occur. The development of computers coincides with the third integrative phase. The wealth of knowledge produced by third-cycle differentiation introduces into the action of integration and ultimately into synthesis a complexity that can only be overcome by a mechanism for aiding the process of comparison, transformation and reconceptualization: the computer.

It may be worth noting that both the capitalist world economy and parliamentary democracy arose during the synthesis to third-cycle social functioning. This being the case, it seems not unreasonable to suspect that the synthesis to fourth-cycle social functioning will see a basic reformulation of the world economy, including the demise of the nation-state as an economic unit and in its place the consolidation of the global system under supranational governance. In addition, it also seems likely that parliamentary democracy will be replaced by participatory democracy and central authority by local autonomy.

Finally, it may be worth noting the future is not yet written. It appears that when complex systems become unstable and go through qualitative change, as must occur during a synthesizing phase in the cycle, a small input at the right moment can be so amplified by nonlinear feedback that it redefines the system in a way that is at present wholly unpredictable. In addition, the development of fourth-cycle society that we see occurring is localized in the core regions of the world economy. If this perturbation is sufficiently localized and lacks the "success attributes" that demand that it be copied by the larger global system, then the surrounding system may overwhelm the perturbation entirely, or the perturbation may require several pulses to overwhelm the surrounding system. At the same time, I see early signs that the world economy is decentering. As a result of telecommunication's innovation, high potential "techno-intellectual" human capital is more often being identified, trained, and employed without being relocated to the core of the world economy. This decentering may represent a necessary mechanism for the development of a globally based, fourth-cycle society.

Conclusions

It seems beyond debate, though surely there will be one, that there exists a basic cycle and pattern in human intellectual development and that this ascending spiral has an analogue at the social level. Further, the existence of the cyclic pattern of social development, demonstrated in this examination of the Western experience, necessitates the emergence of social cartography during the fully integrative phase of the Postmodern era. Nor is it likely that the abstract mapping techniques of social cartography, which resulted in the recognition of these mapped and interrelated patterns, could have been identified and utilized, except in such an era. I believe that social cartography arises to fulfill a basic developmental need of society as it approaches fourth-cycle thinking. Abstract cartography is a basic instrument for fourth-cycle thinking and, as such, social cartography's development is central to social development at this time. Hence, the emergence of social cartography during this latter phase of third-cycle integration is indeed timely. It is interesting but unsurprising that an exercise in social cartography should produce an understanding of the purpose of social cartography. It seems inconceivable that in the absence of abstract maps of the interrelationships constructing individual development and social history such an understanding could be achieved. This is reinforced by the need to map one generalized abstraction onto another before the patterns' similarities can be recognized.

Appendix

The cycle of cognition and the pattern of cognitive development just explicated are common to all humanity. However, the cycle can dysfunction and the pattern can fail to develop coherence or it can stagnate. Dysfunction, incoherence and stagnation have the potential to influence both individual and social development; therefore, their influence requires explanation before cognition is related to the construction of history.

The cycle of cognition becomes dysfunctional when individuals choose not to utilize the process of synthesis. This choice is required of individuals who commit to an unchallengeable and unchangeable framework for knowledge. Such commitment can result from the dominance of any inflexible schema that provides a foundation for the individual's construction of reality. Cases of inflexible commitment to a political, scientific or religious framework, among others, run through human history, with perhaps the most insistent inflexibility provided by the framework given "literally" to fundamentalists by God. In all such cases, knowledge must be ignored or misinterpreted if its unaltered inclusion would force reconceptualization.

Lack of a coherent conceptual framework can cause another type of dysfunction. Since much of an individual's cognitive framework is implicit even when he or she is conscious of thought processes, not all schemata are necessarily operating at the same level of cognitive development. When a new conceptual framework for knowledge is synthesized but one or more important schemata for conceptualizing reality are allowed to remain operating at a previous level, the framework fails to achieve coherence. Over time, the continuous articulation of some schemata and the relative stagnation of others makes reconnection progressively more difficult and unlikely. Thus, for instance, in an extreme but not uncommon instance, an individual may be operating at the fourth-cycle level of generalizing abstractions in physics while he or she is still utilizing the implicit irreversible second-cycle logic of a single, unconscious and automatic perspective in his or her social relations, or, equally commonly, an individual may be able to generalize abstracted perspectives in psychology while unable to transform geometric forms to algebraic symbols.

Moreover, even when schema development is coherent, not all individuals are permitted to progress through the entire range of cognitive development. Apparently, some individuals stagnate because they do not possess the mental attributes necessary for further development. Others stagnate because they are not presented with the opportunity to interact with their environment in such a way that development continues to its maximum potential. The opportunity for maximal development may be withheld for a variety of reasons. Since most knowledge is socially transmitted across generations rather than being independently discovered by individuals in each generation, the level of cognitive development attainable by individuals is closely linked to the developmental level current in the society to which they belong. As with individuals, some societies or segments of societies have stagnated through refusal to synthesize a new conceptual framework for religious reasons. Other societies have stagnated because they have struck a harmonious balance with nature and self-structured in such a way that they have not needed nor perhaps considered it intelligent to progress through the entire pattern. They have optimized without maximizing. In such cases it is difficult for even an able individual to achieve full cognitive functioning.

However, although it is difficult for the individual to operate at a cognitive level far in advance of his or her society, some exceptional individuals have done so. When he said, "Do unto others as you would have them do unto you," Christ was operating at the fourth-cycle level of generalizing abstractions. His promulgation was designed to allow individuals operating at the second-cycle level of a single, unconscious and automatic perspective to simulate third-cycle perspective-taking skills.

On the other hand, it is also possible that the opportunity for development may be withheld in even the most advanced society. Since most learning is socially transmitted, the learning environment and teaching strategies employed are often critical to maximizing development, as are other environmental factors. In any case, when an appropriate environment is withheld as a result of social limitations or social deprivation, the result is that the maximization of cognitive development is unlikely.

At the same time, it is possible for an entire society to operate above the cognitive level of surrounding societies. The golden age of Athens presents a well-known anomaly in the pattern of social development that demonstrates this point. Ancient Athens, a city-state of roughly 100,000 inhabitants, achieved synthesis to the second level of cognitive processing in advance of all surrounding groups. The advantage that this gave the Athenians allowed the further flowering of some schemata to the third-cycle level. However, this momentum was insufficient and eventually the anomaly was reabsorbed into the system.

The example of Athens' golden age provides a vehicle for exploring the complexity within which social development takes place. Of Athens' 100,000 residents, only about 5,000 were free citizens. These 5,000 souls set up a social structure that allowed them for a time to produce and maintain an anomaly of great importance. However, the anomaly was too unstable to be self-perpetuating and too small a perturbation relative to the surrounding system to overwhelm the system in a single fluctuation; as a result, the surrounding system overwhelmed the anomaly.

The system was unstable for a variety of reasons, the most important of which is that progress to the third-cycle level of perspective-taking was unacceptable in a society that maintained slaves. Citizens who adopted a multiperspective framework would be unable to support slavery, which was a foundation of the social system. On the other hand, citizens who maintained an unchallengeable schema regarding the right to own slaves would be unable to ascend fully to third-cycle thinking. This paradox alone would be sufficient to doom the Athenian anomaly, since, as we shall see, human society moves first to multiperspective and only then consolidates other schemata at the third level of cognitive functioning.

The foregoing demonstrates, in part, that the interaction between various components of the social matrix produces complex outcomes. It also shows that, rather than there being an invariant relationship between individual cognition and social development, there is a complex, unstable interrelationship, the outcome of which is affected by various individual, natural and cultural inputs. Thus, for instance, the rediscovery of Greek science helped accelerate the development of Western society during the Renaissance, an input wholly outside of that era's social control.

However, overall, if social progress is to occur, it can only be based upon the invariant cycle of cognition and the pattern of individual cognitive development that arises from it.

NOTE

1. The best case for visual representation is made, perhaps, by Tufte (1983 and 1990).

2. By way of justifying this essay's incompleteness, the explanation for the emergence of social cartography from which this chapter grew was not intended for publication in this partial form. The essay originated as a private correspondence written in response to a presentation which Rolland Paulston made in Prague in 1992.

REFERENCES

Balskrishnan, G., ed. (1995). *Mapping the nation.* London: Verso.

Bird, J., B. Curtis, T. Puttnam, and L. Tickner, eds. (1993). *Mapping the futures.* London: Routledge.

Bloom, J.W. (1995). Assessing and extending the scope of children's contexts of meaning: Context maps as a methodological perspective. *International Journal of Science Education* 17(2), 167–187.

Caullery, M. (1933). *French science.* New York: French Institute.

Chaplin, E. (1994). *Sociology and visual representation.* London: Routledge

Clark, G.N. (1947). *The seventeenth century.* 2nd. ed. Oxford: Clarendon.

Collins, J. (1989). *Uncommon cultures.* New York: Routledge.

Dupont, J. and F. Mathey. (1951). *The seventeenth century: Caravaggio to Vermeer.* Translated by S.J. Harrison. New York: Skira.

Gedda, L., P. Parisi, and W.E. Nance, eds. (1981). *Twin research 3.* New York: Liss.

Gleick, J. (1988). *Chaos.* London: Cardinal.

Guterman, M. M., and Z.H. Nitecki. (1991). *Differential equations.* Fort Worth, TX: Saunders.

Harvey, D. (1989). *The condition of post modernity.* Oxford: Blackwell.

Hassan, I. (1982). Postface: Toward a concept of postmodernism. In I. Hassan, (ed.). *The dismemberment of Orpheus: Toward a postmodern literature.* Madison: University of Wisconsin Press.

Hassan, I. (1987). *The postmodern turn.* Columbus: Ohio State University Press.

Keiner, E. (1995). School and schooling in academic discourses. The social contexts of scholarly production: German science of education since the 1950s. In Jürgen Schriewer (chair), *History of educational thought II: The shaping of educational knowledge.* Symposium conducted at the 39th Annual Meeting of the Comparative and International Education Society, Boston, Massachusetts.

Kuhn, T. (1969). *The structure of scientific revolutions.* 2nd ed. Chicago: University of Chicago Press.

Hsu, C.S. (1987). *Cell-to-cell mapping.* New York: Springer.

Lacy, S., ed. (1994). *Mapping the terrain.* Port Townsend, WA: Bay Press.

Lossky, A. (1967). *The seventeenth century.* New York: Free Press.

McIntosh, A. (1995). Conceptual teaching and semantic mapping equal discovering connections. *Perspectives, 14*(1), 11–16.

Maranda, P. (1977). Semantic cartography in folklore. *Recherches-Sociographiques, L8*(2), 247–270.

Paulston, R. (1995). Mapping knowledge perspectives in studies of educational change. In P. Cookson and B. Schneider (eds.), *Transforming schools: Rhetoric and reality* (137–180). New York: Garland.

Piaget, J. (1954). *The construction of reality in the child.* Translated by Margaret Cook.

New York: Basic Books.

Piaget, J. (1970). *The principles of genetic epistemology*. Translated by Wolfe Mays. London: Routledge and Kegan Paul.

Piaget, J. (1976). *The grasp of consciousness*. Translated by Susan Wedgwood. Cambridge, MA: Harvard University Press.

Piaget, J. (1977). *The development of thought*. Translated by Arnold Rosin. New York: Viking.

Piaget, J., and R. Garcia. (1988). *Psychogenesis and the history of science*. Translated by Helga Feider. New York: Columbia University Press.

Pole, J.R. (1969). *The seventeenth century*. Charlottesville: University Press of Virginia.

Prigogine, I., and I. Stengers. (1984). *Order out of chaos*. Toronto: Bantam.

Redwood, J. (1977). *European science in the seventeenth century*. Vancouver: David and Charles.

Roaten, D. (1960). *Structural forms in the French theater*. Philadelphia: University of Pennsylvania.

Stewart, I. (1990). *Does God play dice?* London: Penguin.

Tijssen, R.J., and F.W. Van Raan. (1994). Mapping changes in science and technology. *Evaluation Review, 18(1)*, 98–115.

Trachtenberg, S. (1985). *The postmodern moment*. Westport, CT: Greenwood.

Tufte, E.R. (1983). *The visual display of quantitative information*. Cheshire, CT: Graphics Press.

Tufte, E.R. (1990). *Envisioning information*. Cheshire, CT: Graphics Press.

Wallerstein, I. (1974). *The modern world-system*. London: Academic.

Wallerstein, I. (1984). *The politics of the world-economy*. London: Cambridge University Press.

Worner, K.H. (1973). *History of music*. Translated and supplemented by W. Wager. 5th ed. New York: Free Press.

Young, P.M. (1974). *A concise history of music*. New York: White.

Zizek, S., ed. (1995). *Mapping ideology*. London: Verso.

II Mapping Perspectives

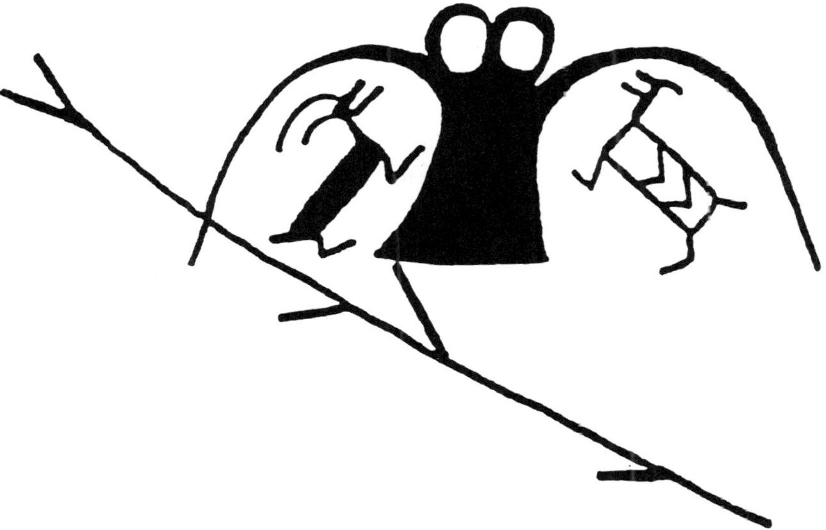

Klu'biíst, the dream teacher. Salish rock painting. From Annie York, Richard Daly, and Chris Arnett (1993), *They write their dreams on the rock forever: Rock painting of the Stein River Valley of British Columbia* (p. 61). Vancouver: Talon. © Talon Books Ltd.

Introduction

Joseph R. Seppi, in his chapter "Spatial Analysis in Social Cartography: Metaphors for Process and Form in Comparative Educational Studies," argues that with the inception of social cartography as an expressionistic medium, an attempt at formalizing the technique must follow. Researchers and even some of the pioneers of social cartography have suggested that the venue must not be constrained by rules or imposed positivist parameters. Social cartography, that is, cognography or free-form mapping, and other narratives used in cultural and comparative studies seem to draw heavily upon cartographic language, as though striving for grounding in some well-founded medium. Extending mapping metaphors used by social cartography into the area of computational spatial analysis may offer a tool for revealing veiled social form and processes. The metaphor begins with a simple and ubiquitous spatial analysis technique that directly maps human thoughts, values and perceptions of their landscape. McHargian suitability analysis is a planner's tool for mapping optimum land uses within a community based on filtering mapped phenomena through the values of various interest groups. The spatial analysis metaphor is expanded upon by showing other techniques for fitting social phenomena into a not-so-rigid binary medium. These new cartographic metaphors may be useful to educators as planning tools to visualize and model social process and form using exploratory techniques.

In her chapter, "Mythopoeic Images of Western Humanism," Anne Buttimer presents a cyclical view of Western humanism and its role in shaping inquiry through the centuries. Humanism is defined as the liberation cry of humanity, voiced at times and places where the integrity of life or thought is threatened or compromised, or when new horizons beckon. The modes whereby the humanist spirit has been negotiated within the changing contexts of Western history reveal a cyclically recurring drama that Buttimer maps using the mythopoeic characters of Phoenix, Faust and Narcissus. It is, she argues, for its potentially emancipatory role that humanism merits attention today as Western scholars seek better communication with colleagues from other cultures about global environmental problems.

This chapter introduces a trilogy of themes and sketches some broad contours of the recent debate on humanism in the arts and sciences. It also sketches some general challenges facing humanism today, and suggests how related fields—such as comparative education—might learn from this intellectual ferment and its mythopoeic representation.

Anne Sigismund Huff's chapter, "Ways of Mapping Strategic Thought," examines from a pragmatic perspective the utility of cognitive mapping for those who wish to understand organizations as interpretive environments where individuals have their own status as "actors." Moving beyond the debate among those who see cognitive mapping as computational models of mental activity, those who see maps as pictorial interpretations of cognition and those who reject cognitive maps, she first examines a continuum of five mapping types: 1) maps that assess the association and importance of concepts; 2) maps that show dimensions of categories and cognitive taxonomies chosen by actors in given settings; 3) maps that search for causal relationships among cognitive elements and system dynamics; 4) maps that show the structure of argument and conclusion—and draw upon philosophy, rhetoric and speech communication; and 5) maps that specify schemas, frames and perceptual codes in efforts to visualize links between thought and actions in structural contexts.

Then, following a discussion of ongoing research issues in the construction of cognitive maps, Huff concludes that mapping is a useful methodological tool for organizational research because it offers a way to distill the many elusive clues of cognition in some reasonably rigorous, reliable and replicable way.

Martin Liebman's chapter, "Envisioning Spatial Metaphors from Wherever We Stand," uses a literary format to discuss several ideas concerning social mapping. Included are sections on metaphor, social incongruity and centers. He then cites examples of contemporary fiction that further draw together these ideas. His thesis advances the claim that social maps are metaphors. A metaphor results from the combining of two incongruities. That is, metaphor takes two incongruous ideas, relates them, and thus creates a new way of seeing. In explaining why social maps are part of metaphor, he extends the form to a new third level, offering that social maps are the metaphor's "face," its tertiary referent. Liebman further recommends that the work of comparative and international education research itself is best seen as an extended metaphor form.

Each particular that a mapper places on a field represents a social incongruity. The reason researchers create maps is to illustrate what stands in one place that is different from what stands next to it. There would be no reason to map if differences did not exist, if we perceived all terrain as flat. These differences are incongruities, and social maps' importance at this time derives from the increase of social incongruities.

In the academic world these incongruities patterned on social maps are called paradigms, theories or intellectual discourse communities. What the social map illustrates, however, are actually people whose worldviews begin from a place on that map that is their center, the point that is the nucleus of their social horizons. While the postmodern era teems with a criticism and censure of centers, individuals seem to retain the notion that they are central to their environments, that they are at the center of a particular universe. Supporting this perspective, Liebman presents a number of examples taken from contemporary novels whose authors recognize and work with the ideas of social space and the metaphors associated with mapping. In an analysis of several randomly selected contemporary novels, he shows that authors, characters, place and space play on mapped fields of intellect, discourse and self-discovery This examination of metaphor shows how the social map is a metaphor identifying previous incongruities now related through research.

SPATIAL ANALYSIS IN SOCIAL CARTOGRAPHY

METAPHORS FOR PROCESS AND FORM IN COMPARATIVE EDUCATIONAL STUDIES

Joseph R. Seppi

INTRODUCTION

With the inception of social cartography as an expressionistic medium, an attempt at formalizing the technique must follow. Researchers and even some of the pioneers of social cartography have suggested that the venue must not be constrained by rules or imposed positivist parameters. Social cartography, that is, cognography or free-form mapping, and other narratives used in comparative and cultural studies seem to draw heavily upon cartographic jargon, as though striving for grounding in some well-founded medium. This study embraces social cartography's need for "furthering comparative investigations through a hermeneutics concerned with extending understanding" (Paulston, 1992, 1993, 1994; Paulston and Liebman, 1994) while developing some new techniques for such investigations. Extending mapping metaphors used by social cartography into the area of computational spatial analysis may offer a tool for revealing veiled social form and processes. The metaphor begins with a simple and ubiquitous spatial analysis technique that directly maps human thoughts, values and perceptions of their landscape. McHargian suitability analysis is a planner's tool for mapping optimum land uses within a community based on filtering mapped phenomena through the values of various interest groups. The spatial analysis metaphor is expanded upon by showing other techniques for fitting social phenomena into a not-so-rigid binary medium. These new cartographic metaphors may be useful to educators as planning tools to visualize and model social process and form using exploratory techniques.

When I first heard about social cartography in 1992, I was dubious to say the least. Rolland Paulston sent me a draft copy of a paper entitled "Comparative and International Education: Paradigms and Theories" (Paulston, 1994). In it were several illustrations referred to as maps (of social phenomena). I was being asked to review the paper; I became interested in the appearance of the "maps" and curious as to how these sketches could be referred to as such. Paulston's map of "knowledge perspectives" (see p. 15), initially seemed nothing more than a schematic diagram that abstracts philosophical benchmarks as shapes in two-dimensional space.

My initial reaction was that Paulston had used the term "map" loosely. As one who is intimate with the mapping sciences of photogrammetry, digital cartography, remote sensing and geographical information systems (GIS), my initial reaction was expectedly positivistic. I decided to put on my relativist hat, however, and take a second look at Paulston's map. It may here be useful to draw the distinction between the term "map" as I would use it and as Paulston has used it. First, I believe that within the context of phenomenography, Paulston has used the term correctly to refer to his drawing. As the cartographer would have it, and in the word's more conventional meaning, a map abstracts real physical features in space as forms or shapes onto a two- or three-dimensional plane and in a Cartesian coordinate system. The cartographer usually maps any number of variables within a given constant frame that is representative of an actual physical space. For example, if a cartographer wanted to map knowledge, he might visualize the human brain as the frame and then classify areas of the brain. On the other hand, he might want to map ways of knowing within a given geographical area, like the United States.

I was faced with a philosophical dilemma. The words "map" or "mapping" do not require a Cartesian approach (Turnbull, 1993), but the word "cartography" does. When I found that there was a movement known as social cartography, I became seriously bothered. Many authors associated with this movement have refrained from using the word "cartography" and instead have opted to use "mapping" (Jameson, 1988; and chapters by Buttimer, Liebman and Mausolff in this volume). Mapping is a process that transcends the Western concept and methodology of scientific cartography (Turnbull, 1993). Many civilizations predating Descartes engaged in mapping their universe or their thoughts (Andrews, 1990) through language writing, song, poem or picture. Today non-Cartesian forms of mapping are found in areas such as data-modeling and CASE (computer-aided software engineering).

If we accept mapping as such a broad method of communication that includes metaphors from just about every other expressionistic medium, we have a real problem on our hands. Furthermore, I am in danger of entering a philosophical hall of mirrors, because I am contributing to a book entitled *Social Cartography*. In 1992, shortly after Rolland Paulston provided me with a bibliography of relevant writings, and still feeling uncomfortable with this use of cartography, I sought further perspective on the matter and found the work of Bertrand Russell (1937) quite helpful. Russell's axiom states (generally) that all logic can be expressed mathematically. If we hold this to be true, then all rational human expression can be decomposed into arithmetic and geometry (Euclidean and non-Euclidean). This changed everything for me. I began to look at free-form mapping metaphors such as Paulston's map of "knowledge perspectives" (see p. 15) in a new and much different way. I saw possibilities that I had not seen before in these schematics.

The concept of Bertrand Russell, expressed in *The Principles of Mathematics*, that all logical premises can be expressed as well as explained with mathematics, is perhaps very useful to justify the need to abstract cognitive/ideological situations. Russell contends that if an entity can be imagined, as variables often are, the influence of the nuances of human thought will not affect logical axioms but can, however, be deduced from those axioms and expressed relative to other synthetic entities as mathematical expressions. If one accepts that knowledge can be expressed by and is in many ways equal to mathematics, then we know that knowledge may be transformed into geometric shapes. A map usually contains geometrical shapes; therefore maps of knowledge are certainly possible. The arrangement of features on a map are thus a direct representation of real phenomena, whether abstracted from real physical subaerial phenomena or subjective cognitive phenomena, or even conjectural metaphysical phenomena, for that matter.

Because real subaerial phenomena are there for the naked eye to see, and because that eye is trained to associate those objects with archetypical (textual) definitions, maps of such objects are more likely to be made to iconify the objects. The outline of the United States is a visual icon synonymous with the words "United States," but it is also a true cartographic map. The US flag is also a visual icon synonymous with the words "United States"; however, it is not a map. It is a diagram with a high degree of internal symbolism.

The type of cognitive phenomena shown in Paulston's map (see p. 15) is not of either of the above types. The cognographer Paulston is transferring or communicating a visual image from his mind's eye that expresses, with a greater amount of freedom, an atextual definition of complex relationships that are conventionally communicated as text or spoken words, thus demonstrating Russell's axiom. The geometrical patterns are real, and once generated, are subject to posteriority and even iconification.

The free-form cognitive map stands as a metaphor for social form and the arrows indicate process. If we normalize this map onto the Cartesian plane and study the geometrical relationships implied, we open the possibility for spatial analysis. We may then have the ability to ask questions about social processes that begin with "what if. . . ." We may also choose to visualize forms that result from modeled processes drawing on the use of computer-aided data visualization. Either way, this can be done while retaining the original spirit of the map or even gaining insight into the author's original intent.

In this chapter, I will build on the cartographic metaphor by discussing some existing methods for mapping human values and perceptions using thematic maps. This will show how planners and landscape architects have superimposed maps of human values and perceptions onto land-based maps. I will also explore the use of computational spatial analysis, or cartographic modeling, to better understand social processes by visualization. While I am not intimately familiar with the field of comparative educational studies, it is my sincere hope that some of what I have to say in this chapter will be found useful by comparative educators, educational planners, and others.

MAPPING HUMAN VALUES AND PERCEPTIONS

Thematic maps are a type of cartographic product that represent classes of a common theme such as vegetation classes, soil types, land use, lithology, demographics, zoning or political boundaries. The concept of the theme has been useful in cartographic modeling because it is analogous to the variable in a mathematical expression. The classes within a theme are usually expressed in nominal terms (e.g., theme = vegetation; class = bottomland hardwood, upland coniferous, scrub/shrub, mixed deciduous, and so on). If classes within a theme are converted to ratio or ordinal terms, we can create even more variables for use in mathematical expressions. Suitability analysis is one form of spatial analysis that tags mapped features with a numerical value that refers to that feature's suitability for a given land use. For example, if a wildlife preserve were proposed for County Cartographica, we would want to create a thematic map of vegetation (Figure 1) and then rank its classes for the number of suitable habitats each could sustain. A habitat suitability index map could be created by reclassifying this map (using those values) into three classes: high, medium and low (Figure 2). Figure 1 represents a *descriptive* map because it attempts to describe, as accurately as possible, the existing conditions of a theme within real geographic framework. Figure 2, on the other hand, represents a *prescriptive* map because it attempts to suggest or model some future scenario based on existing conditions and a set of imposed rules.

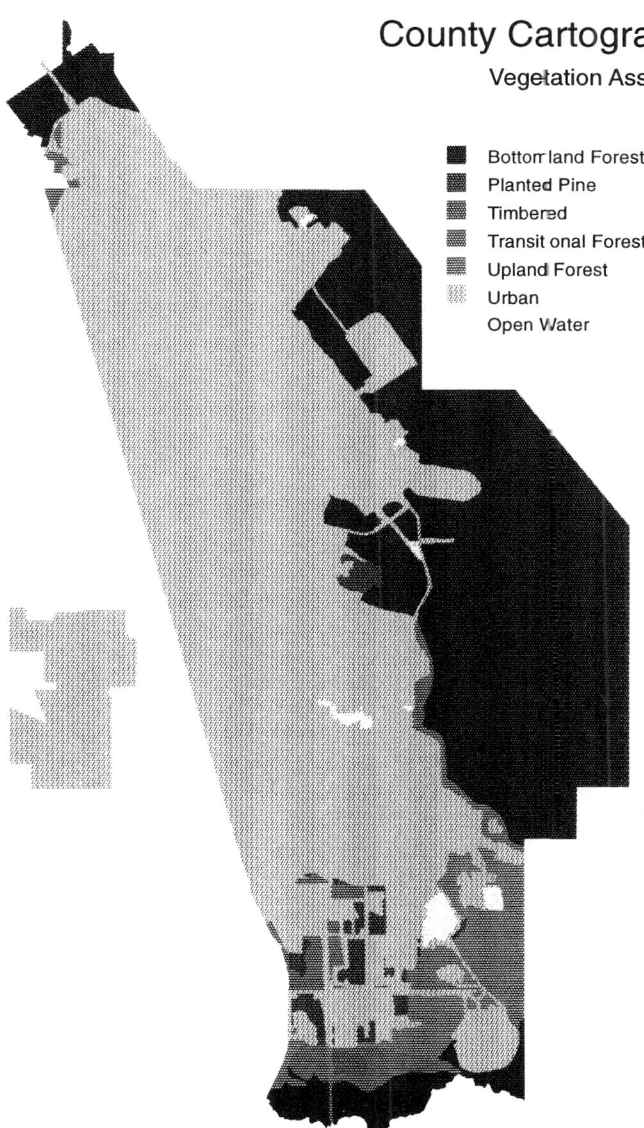

Figure 1. A hypothetical thematic vegetation map

Figure 2. A hypothetical habitat suitability map, showing suitability for a wildlife reserve

If the range of number of sustainable habitats by vegetation class were 1 through 12, the classes would be rated high (9–12), medium (5–8), and low (1–4). Other themes can also be introduced. Suppose we had a road map and knew that no primary roads should be within 1,000 feet of the proposed wildlife preserve. The primary road class from the road map would be "buffered" to a distance of 1,000 feet on either side and given a numerical suitability value of < –12. This new road buffer map would be overlaid with the previous map and would cancel out some of the suitable habitats identified in the previous map.

The technique of applying values to classes within themes allows for the expression of human values and perceptions about those phenomena as well. Ian McHarg, in his book *Design With Nature* (1969) has applied this technique in land-use planning. His approach is both descriptive and prescriptive. McHarg shows how the planner or designer must inventory natural and cultural resources as multiple cartographic themes (Figure 3). The thematic map inventory has often been referred to as the layer-cake or map-overlay planning tool.[1] Human values and perspectives about proposed land-use alterations are identified through interviews with interested parties. A matrix is used to plot interest groups against classes within each mapped theme. Members of each interest group may come together over a table full of maps and matrices and begin to take part in the decision-making process. An interested party places a numerical value in the column of the matrix that pertains to a given phenomenon from a thematic map. In the end, thematic maps are essentially filtered through human values and perceptions.

The result is that each thematic map can be reclassified to show how different people value the phenomena that go into deciding where to develop land (for example). Each reclassified thematic map can be overlaid to produce a composite suitability map for a proposed use or activity. Figure 4 shows such a map indicating young adults' ratings of suitable locations for intensive development for recreation. If a critical disparity arises between two interest groups over a particular use or tract of land, those groups' maps can be isolated in order to find ways of resolving the problem. (It is important to note that often the natural scientist's, planner's, engineer's and politician's vote counts more than the average citizen's vis-a-vis weighting the values.) This exercise results in a *prescriptive* land-use map. *Descriptive* land-use maps can be deconstructed to attempt to determine how certain past human values have prevailed (Harley, 1989 and 1990). Land suitability analysis is a viable metaphor for mapping human values and perceptions toward other social phenomena, such as education. Taking this leap however, requires a higher level of metaphorical substitution, for now the landscape is no longer tangible—part

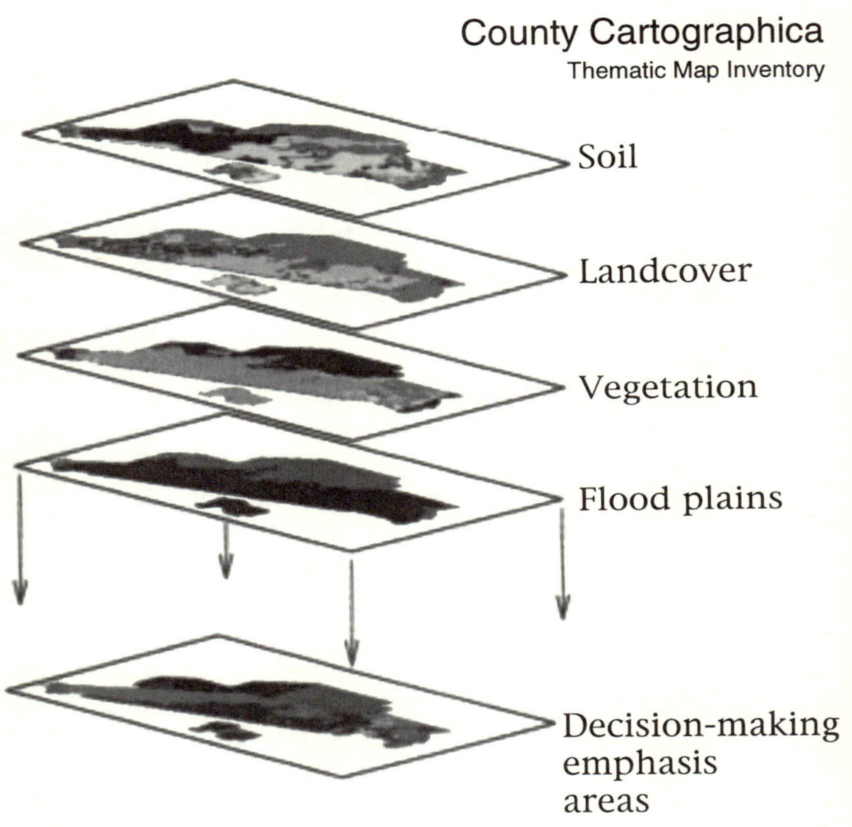

Figure 3. A layer-cake thematic map

of the physical geodetic framework.

The suitability analysis metaphor for social cartography offers a unique way to visualize social form. The metaphor does not lend itself to further modeling. Suitability maps delineate human values and perceptions as they apply to a structured set of real geographic phenomena. The emphasis is on identifying discrete boundaries that everyone can agree on. This type of analysis requires that the extent of landscape and mapping criteria be tangible phenomena from thematic maps. Other mapping metaphors for social cartography may involve a more complete substitution of tangible phenomena with cognitive phenomena. Metaphors for the Cartesian plane itself may yield isopach maps showing the amplitude or volume of a social phenomenon.

COGNOGRAPHY, CARTOGRAPHY, AND SPATIAL ANALYSIS

Mapping human values and perceptions can take place on and off the Cartesian plane. Thoughts, values and perceptions are not always mapped using cartography. Diagrams use symbols to map thoughts, values and perceptions without adhering to the rules of cartography. Such products can be known as "cognographs," and the process of designing them as "cognography." Cognographs can be converted to maps if we treat them as the landscape. Cognographs can be deconstructed to reveal the mapped phenomena within; that is, the internal meaning of the various symbols that make up the cognograph. The hybrid can then be used in a modified suitability analysis to show how different people's thoughts, values and perceptions toward the mapped phenomena compare.

Where no existing diagram or map is found, the next level of metaphorical substitution must be used. In traditional mapping, the first step is the survey. Surveying requires that geodetic control be established prior to any delineation or mensuration. A cognitive control system must be contrived in which we substitute x, y and z with human values and perceptions. Instead of eastings, northings and elevations, we must describe location using a new metaphor. This, in its most fundamental form, is a simple Cartesian graph (Figure 5).

This fundamental data-visualization tool offers a descriptive mapping technique that is a truly Cartesian metaphor for seeing social form. The x, y and the x, y, z graph exploit every aspect of Cartesian coordinate geometry, where x = f(y) or z = f(x, y). Each axis can represent ordinal values about any three phenomena. For example, Figure 6 shows the distribution and number of employees with varying combinations of education and experience in a hypothetical business management structure. A linear or pyrami-

Figure 4. A reclassified thematic map showing suitability for active recreation

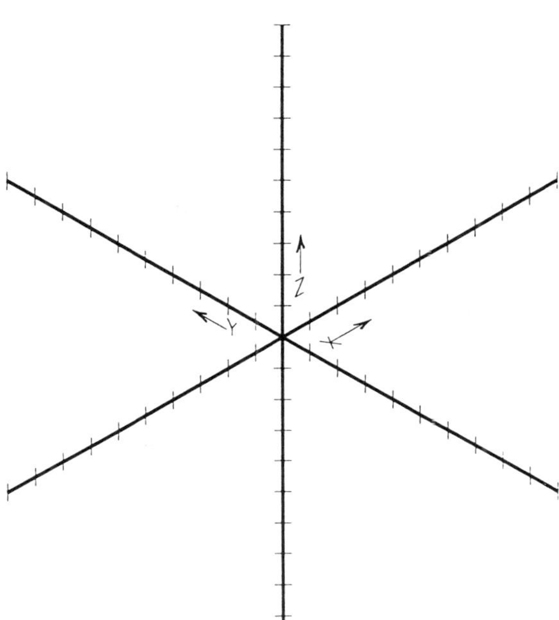

Figure 5. Cartesian planes

dal relationship might be expected; however, we see here that this is not the case. This visualization tool has encapsulated a great deal of data and communicated it with efficiency. In that context it makes a wonderful mapping metaphor.

Three-dimensional graphing software can help to map out a landscape that exists only as a function of social phenomena. If we take the same data set that provided us with Figure 6 and push the cartographic metaphor even further, we can create such a landscape. By applying spatial interpolation methods commonly used in topographical mapping, we can generate a metaphorical terrain model with valleys, hills, streams and a range of other physiographic analogues (Figure 7). Here, we see a true cartographic analogue, where the number of people is represented as elevation (z), education (x) and experience (y) form the planar geographic region metaphor. The individuals that make up the population are analogous to grains of sand blown by the winds of experience and education into dunes.

Figure 7 represents a complete metaphorical landscape and descriptive map generated using only cartographic processes. This map is a digital terrain model to which an incredible number of spatial operators may be applied in order to reveal further metaphorical features. Figure 7 commu-

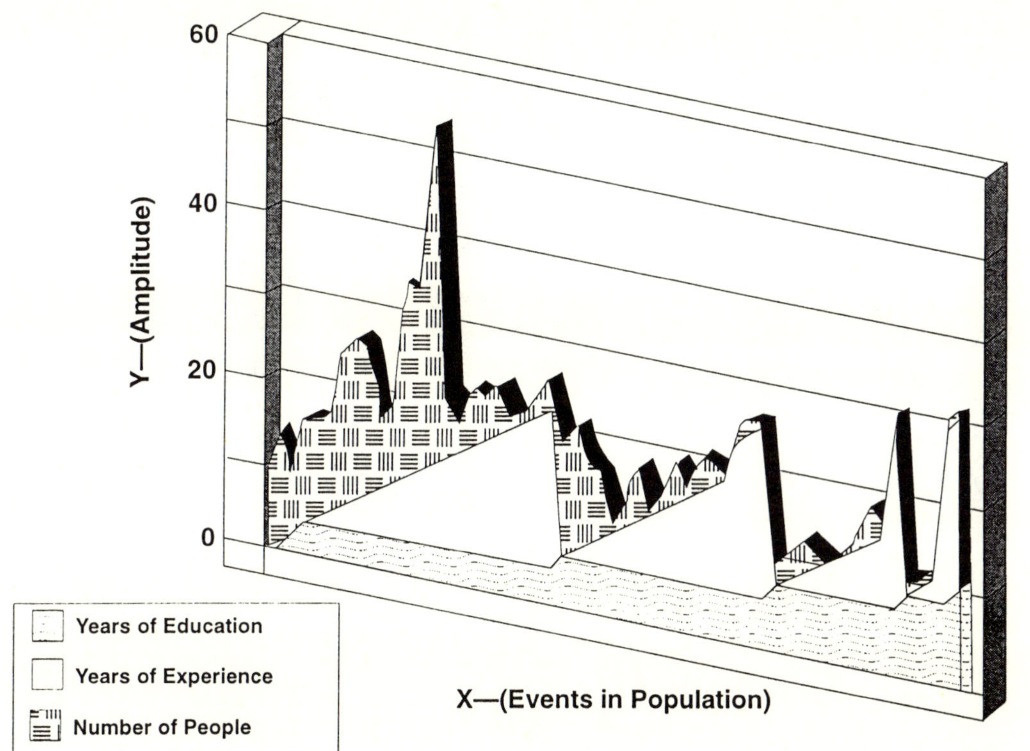

Figure 6. A hypothetical organization showing distributions and number of people by years of experience and education

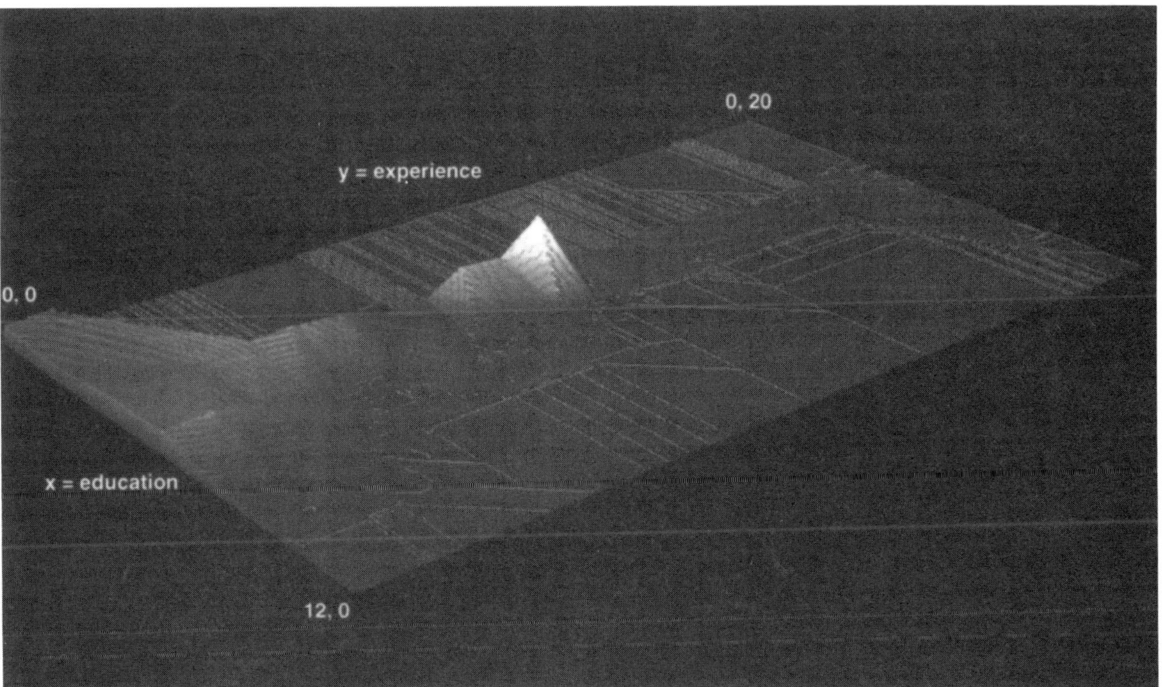

Figure 7. A topographic map of Jobland Corporation showing distribution and number of people by years of experience and education

nicates a fact about an organization—that the majority of the people have around ten years of experience and four years of education. Any other inferences should be left to the map readers. Physiography has much to do with the many complex features of any landscape. To model social process now is a real possibility because the analogues begin to fall into place. We could use algorithms that automatically calculate the expected drainage network that flows across this terrain as rain falls (the drainage network represents paths of least resistance from one place to another). We can drain one drop of water placed anywhere on the model to see where it flows. If we invert the terrain we can run a gravity model to show influence centers. Notice that in Figure 7 the automatically generated drainage network connects clusters by running vectors into the flat plane that represents 0 elevation (or number of people). This could be viewed as a special type of path of least resistance.

Figure 8 represents another terrain or stratum within the Jobland landscape. Here authority within the company (z) defines the elevation changes. The progression from low to high (elevation or authority) shows a relatively linear trend. Notice the small cluster of senior-level staff to the extreme right-middle of the positive map. These staff are Ph.D.s with fewer years at the company but the same authority as some of the longstanding staff.

Other terrains within the same landscape can be generated using the same population. For example, income or time spent in current position as a function of experience and education can be mapped, allowing us to spatially infer two-dimensional themes that can be draped over the authority terrain. For example, we can map gender, race, age and other facts about the population on the experience-education landscape. It is important to hold the position of individuals relative to the original experience-education landscape. We can visualize multivariate equations in a very straightforward way using the terrain mapping metaphor.

Two-dimensional thematic drapings over three-dimensional social terrains can manifest thinly veiled social barriers and provide insight into the processes that have constructed them. These barriers tend to defy expected linear relationships. If rules or laws dictate that linear relationships should strictly define the authority landscape, then running the model for each individual should predict authority. We know that this cannot be true because there are a limited number of positions of authority. This fact in itself becomes a barrier. To map the metaphorical wind gap through which only a select few will be blown, we must explore more variables. Mappable phenomena that apply to individuals of this population might include num-

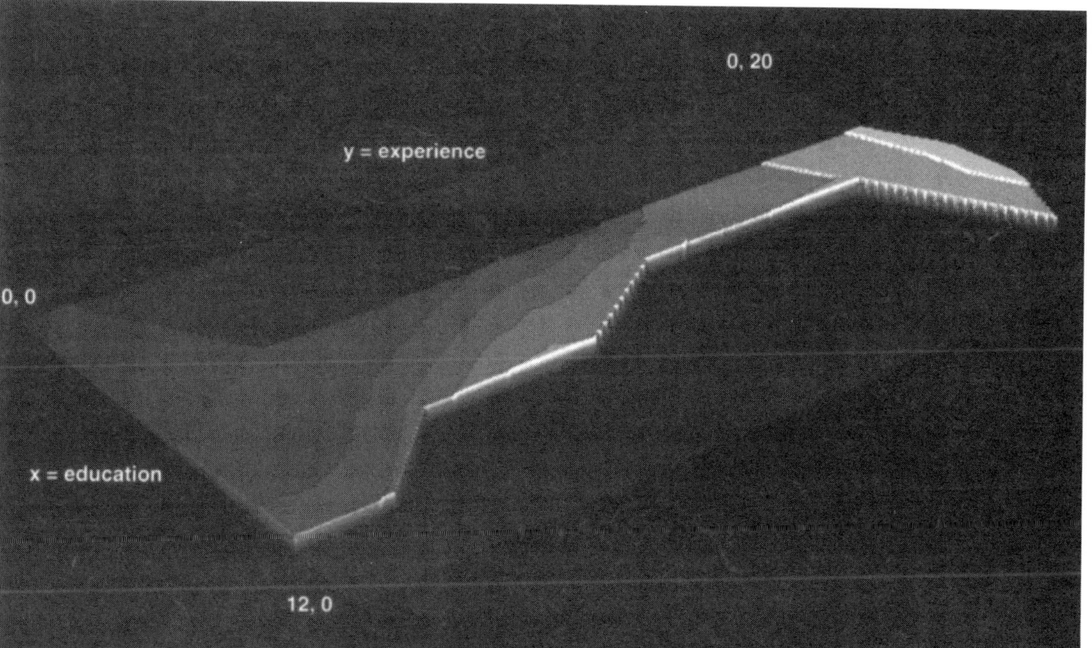

Figure 8. Jobland Corporation: Authority by education and experience, as illustrated by elevation

ber of sick-days taken, number of complaints in one's personnel records, number of commendations, and so on. By cross-plotting the (z) values from each terrain (Figure 9), we can visualize how the two landscapes correlate statistically.

In a sense, the above example could be used as an interactive road map to success for the motivated individual. The model can account for as much or as little reality as is desired. There are certain phenomena, however, that don't readily translate from nominal to ordinal form: Attitude, intuition, and circumstance cannot be easily accounted for. Statistical probability can be used to predict, given an individual's profile, the probability of a given path's being followed. This example, if taken to extremes, can become rather Orwellian. At some point the law of diminishing returns begins to take effect and the mapping process becomes increasingly fruitless.

The Jobland example above uses what is known as a raster modeling approach. Rasters are very similar to computer spreadsheets. A raster is made up of rows and columns; the intersection of row 1, column 2 is defined as cell or pixel 1,2. This simple matrix data structure stores numerical values in each cell. When colors are applied to the values and displayed, we see a map. More sophisticated cartographic modeling software is able to link each cell to a relational data base record capable of storing an infinite number of nominal, ordinal and logical attributes about that cell (Tomlin, 1990). Another way to store additional attributes within a highly structured data model is to use what is known as a voxel. Two-dimensional rasters are made of pixels and three-dimensional rasters are made of voxels. Voxels have the ability to store volume and are commonly used in visualizing geological structure, groundwater bodies or fluid plumes. Using color gradients, concentrations within the volume can be mapped.

There is a wide spectrum of commonly used data models in computational spatial analysis and geographical information systems (GIS) in general. The raster is perhaps the most ubiquitous. The next most widely used data model is known as the vector-based GIS. The vector data model is based on graph theory and is in some ways a more efficient way of storing information digitally. Network modeling is a form of spatial analysis particularly well suited to vector-based maps. If we look at Paulston's map of "knowledge perspectives" (see p. 15), paying particular attention to the arrows he has drawn to connect various orientations, we might have enough information to make a network allocation model. This spatial data model is traditionally used in transportation modeling.

If we think of Paulston's map as a road map in which each arrow is a vector showing directionality, we can expect that navigation within this

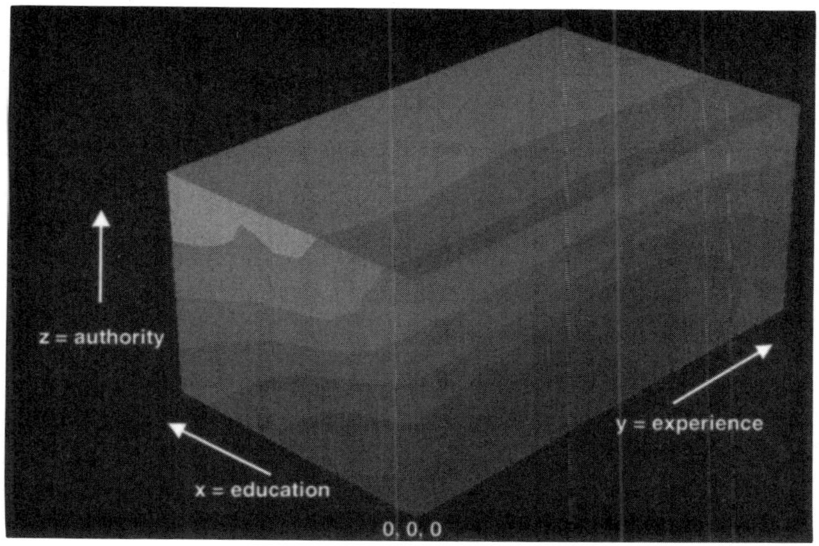

Figure 9. Jobland Corporation:A three-dimensional (voxel) model showing use of multivariate social cartography

network is possible. Under closer examination however, we will find it to be extremely limited. There are, however, some routes that will allow us to traverse multiple orientations, such as from "functionalist" to "pragmatic interactionist" to "cultural rationalization." Using a computer-based network allocation model, Paulston's map could be greatly enhanced to hold much more information. It could become an interactive game that would allow the player to travel through a hypertextual highway. Road signs along the way would affect impedance values such as the speed limit, or caution, one-way or dead-end signs.

A Cognitive Control Framework: Key Concepts

This metaphor for social cartography is extremely challenging. It requires the substitution of physical fact and laws with social facts and laws. The successful identification of three such analogues could yield insight into process and form. Automated three-dimensional visualization could be the next medium for cognitive mapping. Forms that have not been conjured in the mind's eye may take shape using the technology of spatial analysis. Upon seeing such forms, interest groups might experience a sense of deja vu. Such a tool might be useful in developing strategy, anticipating another interest group's viewpoint on as-of-yet nonmanifest social issues. Social modeling, however Orwellian, is perhaps a possibility.

Computational spatial analysis, in order to be accurate, must have an abundance of input data. Input data are usually sampled within a spatial reference system. New surveying metaphors may be required. Instead of linear and angular mensuration, communication and translation may serve as surveying techniques to advance the social cartography project in comparative education studies.

Note

1. Compare the similar imbricated map presented as Figure 3 in the Paulston and Liebman chapter.

References

Andrews, J.H. (1990). Map and language: a metaphor extended. *Cartographica*, 27(1), 1–19.
Harley, J.B. (1989). Deconstructing the map. *Cartographica*, 26(3), 1–20.
Harley, J.B. (1990). Cartography, ethics and social theory. *Cartographica*, 27(2), 1–23.
Jameson, F. (1988). Cognitive mapping. In G. Nelson and L. Grossberg (eds.), *Marxism and the interpretation of culture*. Urbana: University of Illinois Press.
Liebman, M., and R. Paulston. (1994). Social cartography: A new methodology for comparative studies. *Compare*, 24(3), 233–245.
McHarg, I.L. (1969). *Design with nature*. New York: Natural History Press.
Paulston, R. (1992). *Comparative education as an intellectual field: Mapping the theo-*

retical landscape. Paper presented at the 8th World Congress of Comparative Education, Prague, Czechoslovakia.

Paulston, R. (1993). Mapping discourse in comparative education texts. *Compare, 23*(2),101–114.

Paulston, R. (1994). Comparative and international education: Paradigms and theories. In T. Husen and N. Postlethwaite (eds.), *International Encyclopedia of Education*. Vol. 2 (923–933). Oxford: Pergamon.

Paulston, R., and M. Liebman. (1994). An invitation to postmodern social cartography. *Comparative Education Review, 38*(2), 215–232.

Russell, B. (1937). *The principles of mathematics*. London: Allen and Unwin.

Schatzki, T.R. (1991). Spatial ontology and explanation. *Annals of the Association of American Geographers, 81*(4), 650–670.

Taylor, D.R.F. (1991). A conceptual basis for cartography: New directions for the information era. *Cartographica, 28*(4), 1–8.

Tomlin, C.D. (1990). *Geographic information systems and cartographic modelling*. Englewood Cliffs, NJ: Prentice-Hall.

Turnbull, D. (1993). *Maps are territories; science is an atlas*. Chicago: University of Chicago Press.

Mythopoeic Images of Western Humanism[1]

Anne Buttimer

Introduction

This paper frames a perspective on Western humanism and its role in shaping inquiry through the centuries. Humanism is defined as the liberation cry of humanity, voiced at times when the integrity of thought and life is threatened, or when new horizons beckon. The modes whereby the humanist spirit has been negotiated within the changing contexts of Western history reveal a cyclically recurring drama that is here mapped in the mythopoeic characters of Phoenix, Faust and Narcissus. It is for its potentially emancipatory role that humanism merits attention today as Western scholars seek better communication with colleagues from other cultures about global environmental problems.

Waves of concern about global environmental crises and frequently apocalyptic warnings of impending doom sweep through the Western world during these last years of the twentieth century. On the intellectual horizon flash images of Gaia,[2] a living earth, vastly complex and self-regenerating. Voices proclaim the death of the human subject and the apogee of Enlightenment dreams (Derrida, 1972; Ferry, 1995). Postmodernist texts jostle with other texts in a labyrinthine maze of self-reflecting mirrors (Baudrillard, 1990; Anderson, 1990). Thus while "all the king's horses and all the king's men" in Western science and politics press on with Promethean zeal to correct, control and ameliorate humanity's impact on the earth, the human spirit, often with narcissistic longing, sings the song of the absence, seeking to remember forgotten features of human nature and to rediscover wiser ways of dwelling.

Words such as "humanism" have a chameleon ring, revealing as much about their definers and their worlds as they do about any perennial truths. Despite the universalist aspirations of humanist lore, it can be fully appreciated only within specific settings. Proclamations about the essence of humanness, be it *animal rationale, homo sapiens, homo demens, zoon politikon, homo faber* or *homo ludens*, reveal quite as much about the authors of such propositions as they do about human nature.[3] Humanism can scarcely be regarded as an autonomous field of knowledge inquiry. Rather it is a stance on life and place shared by people of diverse walks of life who see in the idea of humanistic wisdom enlarged possibilities for a self-reflective knowledge of culture (Kunze, 1983).

Humanists down the centuries have explored the nature of humanity, its passions and powers, while geographers have studied the earth where humans, among many other life forms, make a terrestrial home. For each facet of humanness—rationality or irrationality, faith, emotion, artistic genius or political prowess—there is a geography. For each spatial interpretation of the earth there are implicit assumptions about the nature of humanness. The common ground is vast indeed. Functional specialization among knowledge fields from the mid-nineteenth century on led to situations where "geography" and "humanities" became distinct and often separate institutions. What stirs excitement today is mutual rediscovery, by individuals from diverse disciplines within the Babel Tower of academia, of common goals and challenges for the future of mankind and earth.

This chapter introduces a trilogy of themes and sketches some broad contours of the recent debate on humanism in the arts and sciences. It also sketches some general challenges facing humanism today, their resonance particularly within the field of human geography, and suggests how related fields—such as comparative education—might learn from this intellectual ferment.

Phoenix, Faust and Narcissus

Phoenix offers a symbol for those emancipatory moments in Western history when new life emerges from the ashes, with prospects for a fresh start (Figure 1). In the careers of individuals, as well as in the course of nations, cultural groups and disciplines, one can identify at least two kinds of emancipatory cry: one seeking freedom from oppression, the other seeking freedom to soar toward new heights of understanding and becoming. Humanism could thus be regarded as the liberation song of humanity, voiced whenever human integrity is threatened or human horizons are dimmed. At times when the academy, church, state, syndicate, or proletariat has tried to exercise monopoly power over thought or life, a humanist protest has appeared. Socrates told parables to baffle the Sophist *virtuosi* and their solipsistic utterances. Pico della Mirandola challenged the ecclesiastical dogmatism of the fifteenth century, paving the way for Giambattista Vico's brilliant interpretations of cultural history two centuries later. Von Herder,

Figure 1. Phoenix, cri-du-coeur of humanity, repeatedly reborn to beckon new horizons for life and thought

Schiller, Blake, Wordsworth and the eighteenth century *literati* inveighed against the rationalistic claims of the Enlightenment, heralding those masterpieces of emancipatory thought associated with Goethe, Kierkegaard, Nietzsche and Dostoyevsky in the nineteenth century. Saint-Exupéry, Camus, Sartre and Havel in the twentieth century, as well, have all pleaded attention to dimensions of humanness, to poetic universals, that were largely forgotten or ignored in Western philosophy and science.

Odes to human freedom have certainly not always been evoked by feelings of protest. The sweetest songs in the humanist's repertoire are those that have often come gratuitously: Novel ideas in art, literature, science, music, spirituality or politics have often been expressions of global concern, generous outpourings of a passion for knowledge and beauty. Bruno, Cervantes, Goethe, Shelley, Teilhard, Neruda and others, for example, have come with messages ahead of their times. Phoenix rises gratuitously, from the ashes of former dreams; if the climate is not right for receiving a new impulse, death through fire may be necessary before a new Phoenix can once again emerge.

The key point with Phoenix, therefore, is its emancipatory message. Its impassioned cry implies far more than pleas for intellectual freedom or social reform. Humanist movements in Western history have sought to reaffirm moral, aesthetic and emotional dimensions of humanness. At times they addressed themselves to those dimensions that were repressed by one tyranny or another. At other times they evoked a nostalgia for those dimensions that were forgotten or silent, or beckoned toward horizons hitherto unexplored. Success in capturing an audience, however, often yielded new orthodoxies and structures, based on the ardent desire to affirm that which had been previously ignored (Buttimer, 1993).

So enters the second symbolic figure: Faust. It is surely characteristic of Western ways that once a fresh idea appears, energies are directed toward the building of supportive structures, institutions, and legal guarantees (Figure 2). Goethe's *Faust* (Part 2 especially) stands as a central symbol for this phase: *eines Menschen Geist in seinem hohen Streben* (a mortal soul in high endeavor). For the pioneering spirit, it is often the idea itself and the new vistas for thought and life that are most precious. Details of how to communicate vision, or render it relevant to ongoing societal interests, may be simply tedious. Phoenix therefore welcomes a helping hand to get "the show on the road," as it were, and it is often in that bond that grows among architects of new movements that the emancipatory *élan* is most tangibly felt (cf. Berman, 1982). And when the idea has "made it," and its legal and institutional bases socially accepted, then some kind of metamorphosis usually occurs. Pioneering spirits recede and a later generation directs its ener-

Figure 2. Faust, a "mortal soul in high endeavour," never ceasing to construct for the progress of humanity

gies quite as much to the maintenance and reproduction of structures as to the sustenance of the initial emancipatory ideal (Turner et al., 1990). But Faust continues to build for humanity's sake—for otherwise, if ever he were to pause and gaze on his achievements and utter *Verweile doch, du bist so schön* (Linger a while, how lovely you are), Mephistopheles would appear ready to claim his soul.[4]

As individuals or groups begin to wonder about these tensions and apparent contradictions between spirit and letter, dream and reality, of their everyday practices, a reflective mood sets in that can be symbolized by Narcissus, pilgrim to the muses of Helicon (Figure 3). One wonders how it came to be, for instance, that enthusiastically laid plans and innovative ideas or movements became unwieldy structures and self-perpetuating bureaucracies. In this mood, several possibilities are conceivable. One points toward the vulgar image of narcissism and the propensity to interpret situations through the lenses of one's own cherished self-image. Another points toward a better understanding of history and an appreciation of the drama of events and their context. Ultimately such understanding could lead to a clearing-ground and the prospect of emancipation from the encrustations of one's Faustian period. Does history not suggest that one may have to pass through a potentially painful death to former certainties before a new Phoenix can take wing?

Such cycles of human experience have undoubtedly occurred throughout Western history, yet there are threads of linear sequence discernible in

Figure 3. Narcissus, bewildered by the conflicts between ideals and reality, reflecting . . .

the story (Figure 4). None of these cycles ever found humanity or earth in the same condition as before. Down the centuries there have been certain persistent melodies, unresolved paradoxes, and typical values that have survived in the twentieth century (Kay, 1989). As Mouat argues in his chapter, both the cyclical and the linear account could help us to place our late twentieth century challenges within a broader historical frame. The risk of attempting such a view, of course, is that of glossing over important nuances and counter movements within and between different periods of Western history: A mythopoeic mapping—as presented here in Figure 4—should not be judged in literal terms!

It is in the context of our twentieth century reflective turn that this account is offered. It presents the narrator's central concern that Western humanity needs to redefine its role and discern more appropriate ways of human cooperation in healing a badly wounded planet. It seeks lessons from history that could elucidate the present challenge facing the scholarly world generally, rather than guidelines for circumscribing the proper domain of scholarly fields.

The Humanistic Turn in Geography: Some Lessons and Caveats

It is only a few decades since the term "humanistic geography" has been widely used in geographic literature, and its connotations are now many. What may be subsumed under this rubric varies from one country or language tradition to another, and in some cases, the terms "social," "cultural" and "humanis-

Figure 4. Phoenix, Faust and Narcissus: Cyclical refrain in the Western story

tic" are used interchangeably (Relph, 1981). For some this turn has been a kind of mission to restore human subjectivity to a field where narrow scientific objectivism is seen as too restrictive (Ley and Samuels, 1978). Some have emphasized human attitudes and values, others cultural patrimony. Some have focused on the aesthetics of landscape (Mitchell, 1994) and architecture, others on the emotional significance of place in human identity (Bowden and Lowenthal, 1975; Meinig, 1976). A substantial number, too, have advocated human compassion and engagement in the resolution of social or environmental problems—reminders about twentieth century pilgrims of peace and early Cassandra voices on environmental destruction (Thomas, 1956). Global horizons beckon in the widespread concern about humanity and earth, the shocking record of environmental destruction and radical transformations in culture and politics (Brunn and Yanarella, 1987). From whatsoever ideological stance it has emerged, the case for humanism has usually been made with the conviction that there must be more to geography than the *danse macabre* of materialistically motivated robots (Folch-Serra, 1989; Jackson, 1994).

Some have looked askance at this humanistic turn as a kind of amnesia, a turning away from problems and retreat into nostalgia and esoterica, or as simply a critique of the status quo. An earlier generation suspected that humanistic concern might eventually undermine the identity of the discipline, taking the "geo" out of geography (Leighly, 1983). Today's misgivings reflect not only the doctrinaire antihumanism of structualism and other philosophical currents, but also the internal fragmentation of geography itself. The institutional separation of its physical and human branches, bemoaned today by advocates of environmental sensitivity, was itself the product of an emancipatory project launched by leaders of the interwar period. One sought freedom from the parental bonds of geology, on the one hand, and history, on the other. Fleeing, too, from ghosts of environmental determinism, one sought a substantive focus on space rather than environment, and livelier interaction with other sciences. Then, after World War II, human geography proclaimed itself a social science.

But by the early 1980s, the zest and optimism that accompanied this humanistic turn had faded. "Realists" found the prose of an earlier generation to be quaintly idealistic. Many, too, disillusioned about all the contradictions and tragedies of the Western legacy, condemned attempts to privilege humanism as wellspring of all hubris and vaunting ambition, as a myth overdue for dismissal (Relph, 1981). One prop for this judgment was the conventional practice of identifying the origin of Western humanism with the early fifteenth century Renaissance in Italy, a movement that also bore the seeds of modernism. Humanism thus earned a "guilt by association" for the Promethean excesses of Western humanity and hence shared its condemnation. Materialist manifestos vented spleen on the perennial plots of capitalism to produce and reproduce those power structures that held sway in the global economy (MacLaughlin, 1986). Postmodernist scenarios of texts reproducing texts in a labyrinth of self-reflecting mirrors precluded any assumptions about human intentionality or meaning in lived reality (Jameson, 1983; Said, 1983; Kearney, 1988). A sense of imprisonment in one's own cultural world, of claustrophobia and nihilism (Baudrillard, 1990), of "hitting one's head against the ceiling of language" (Olsson, 1979) characterized some of the critical thought within geography during the 1980s. One became aware of the many ways in which the entire intellectual heritage of the West had mirrored and been mirrored in the peculiar social history of Western humanity (Mitchell, 1994).

Even over this relatively short stretch of disciplinary history, one could discern some resonance of this cyclical movement in life and thought. The 1960s, in retrospect, seem to have been a Phoenix time for many: new life seemed to be bubbling all around. By decade's end, and throughout the 1970s, one witnessed a Faustian will to create new subdisciplines, societies and specialty groups. The 1980s and 1990s have revealed many contradictions between dream and reality, ethos and structure, and a critically reflective mood has set in. Some have become nostalgic for the past, some have reaffirmed the status quo, and some have even begun to envision prospects for a new dawn (Ley and Samuels, 1989; Hanson, 1992).

At least four distinct strands of inquiry could be discerned in the Western humanist tradition. Since classical times, there have been, for example, proclamations and debates about the nature of humanness (*humanitas*): diverse and recurrent themes about individuality and sociality, freedom and responsibility, rationality and hedonism, conservatism and creativity. There have been claims, too, about humanist modes of knowing, which oppose scientific reductionism, seek to elucidate rather than to explain and emphasize the subjectivity of consciousness and the intersubjective nature of scholarly discourse. There is also the tradition of the humanities—that is, fields of learning deemed appropriate for the cultivation of the arts, classical literature and the nurturing of civic virtue. Finally, there has been an enduring concern about the human condition; for example, in humanitarianism, which has sought to encourage social responsibility and liberal politics. In historical perspective, of course, these different strains intermingle, but each points to a distinct set of scholarly practices whose value and relevance has varied over time. To assess the potential lessons to be gleaned from this survey, it seems feasible to orient inquiry toward these core aca-

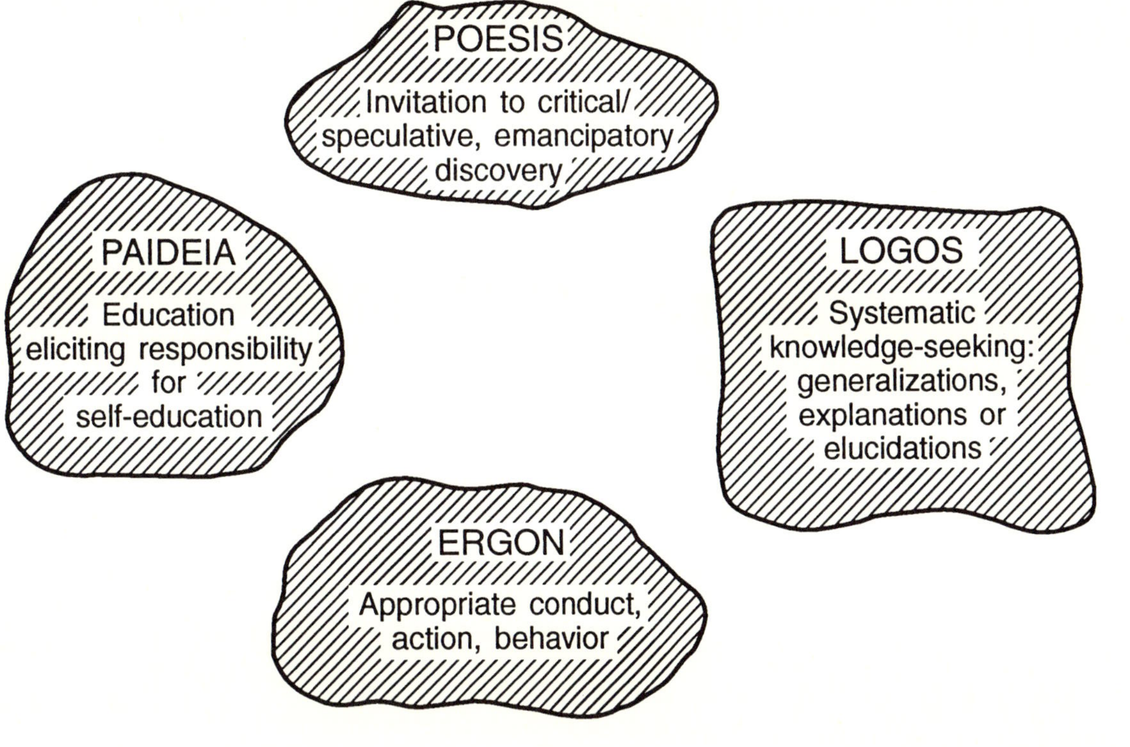

Figure 5. Classical constellations of academic practice. Poesis: *exploration, discovery and critical reflection;* Paideia: *education and learning;* Logos: *systematic analytical inquiry;* Ergon: *practical applications of knowledge to life and conduct.*

demic practices (Figure 5). The ontological questions broached in reflections on the nature of reality, for example, illustrate *poesis*, the art of evoking critical thought and discovery. The epistemological questions entertained in discussions of humanist modes of knowing illustrate *logos*, the human quest for rational knowledge. *Paideia*, or liberal education, has traditionally been the function of the humanities, and *ergon*, appropriate conduct and action, has been a central goal of humanitarianism and concern about the human condition. One of the greatest challenges facing scholars today is to rediscover the bonds among *poesis, paideia, logos,* and *ergon* and to reorchestrate these ideas in practices today.

In all four practices, the cyclical movements of Phoenix, Faust, and Narcissus can surely be discerned, but in none more dramatically than *ergon*, the realm of action and concern about the human condition. Throughout the many and varied movements of humanistic inspiration, it is possible to differentiate an *ethos*, or fundamental spirit, and a *structure* that sought to incarnate that spirit. Socratic seminars became an Academy; catacomb communities became a Vatican; networks of mutual aid among workers and peasants turned into syndicated unions and cooperatives; folk who felt a common sense of ethnic identity became geographically circumscribed nation-states. Malaise over the tensions between spirit and letter, ethos and structure, has generated reflective moments, archival research and the quest for clarifying one's identity. From such reflection, some have emerged with reaffirmations of the status quo, having interpreted history in terms of the preferred aspects of the present. Some have turned parricidal, having found all sorts of grounds on which to condemn their ancestors. Others, unafraid of shedding the harness of routine operational ways, have played key roles in paving the way for some new Phoenix.[5] History affords ample evidence of mankind's incredible resilience and courage to try again.

TWENTIETH-CENTURY CHALLENGES: CONTEXTUAL TURNING

An explicit recognition of context, of environmental and societal circumstances in the shaping of thought, language and action, could perhaps be identified as the major motif of scholarly work during the latter years of the twentieth century. It has confronted each of the "faces" of humanism mapped in this chapter with a major paradox. With respect to *humanitas*, the exaggerated claims of Promethean individualism evoked stormy protest and eventually death warrants on the human subject (Derrida, 1972). To the venerable debates over subjectivity and objectivity, reason and rationality, sociological rather than epistemological modes of discourse were expressed.

Contextual light was eventually cast on that peculiar ambivalence detectable among academic stances on knowledge and life: passionate declarations about the values of intellectual freedom on the one hand, yet an equally passionate commitment to scientific theories designed to prove how determined everything was on the other. Traditional boundaries separating sciences and humanities also eroded and new alliances emerged, reflecting radical reorientations within diverse fields as well as the changing priorities of research funding. Inherited notions about humanitarianism and scientific humanism were also attacked: world wars, the atrocities of so-called civilized peoples, environmental destruction and global terrorism, are surely hard to reconcile with traditional (Western) assumptions about humanitas (Ferry, 1995).

More than forty years ago, Heidegger noted that ever since Roman times the *humanitas* of *homo humanus* has been determined from the view of an already established interpretation of nature, of history, of *Weltgrund*, or being in its totality. Hence humanism has been caught in its own metaphysical stance. "Humanism does not ask . . . for the relation of Being to the essence of man, it (humanism) even impedes this question" (Heidegger, 1947, p. 211). Instead, Heidegger advocated a return to those broader issues of being and becoming that are absent in a knowledge enterprise steered by anthropocentric biases. He called for a sensitive listening to reality—letting reality reveal itself in its own terms—rather than seeking to grasp reality only in the language of preconceived categories and models (Roscoe, 1991). Heidegger's critique of Western humanism roundly satirized the inherited separation of epistemology and ontology (of knowing and being), the intellectual and moral dimensions of thought and action.

The prospectus for the humanities in the late twentieth century re-echoes those classical tensions between Gnostics and Socratics, between images of reality as being in perpetual flux versus images of self-aware human subjects seeking rational understanding of the world. As before, such tensions can scarcely yield a creative outcome until broader horizons for knowledge and life can be perceived by both. The present global crisis, and the daring images projected by Gaia, may well be the catalysts for such an expansion of horizon (Lovelock, 1979). Neither the inherited divisions between intellectual and moral realms of discourse nor those academic fences between "humanities," "divinities" and "natural science" block creative imagination today (Eco, 1986). Education for moral reasoning, Harvey Cox claims, could be a central function for the humanities to "nourish human beings capable of passionate imagination, rigorous reflection, reasoned choice and moral courage" (Cox, 1985).

As the traditional anthropocentrism of Western humanism is attacked from various corners of the academic world and environmental issues hover in public press and debate, the famous Baconian slogan, *natura nisi parendo vincitur* (One only conquers nature by obeying her), had already taken on a new complexion by the 1970s. "Human nature," the "forgotten paradigm," according to Morin, cannot be understood without understanding the complex interplay of culture and nature through history (1973). Exposing the inadequacies of a knowledge enterprise comprising specialized sciences of man and nature—biologism, anthropologism, psychologism—Morin sought to unmask the persistent counterplay of rationality and irrationality (homo *demens* versus *homo sapiens*), order and disorder, trial and error in human history. Pelt, with more explicitly environmental concern, claimed that humans would never discover their own nature until they rediscovered the essential reciprocity of sociality and ecology (1977). And Passet, focusing more directly on the tensions between economics and biology, noted the cardinal differences between mechanist and organicist principles of organization in all elements of the biosphere, including humanity (1979). Common among such authors is the concern to evoke awareness of what is lacking in conventional stances on life and knowledge, on the one hand, and a plea for more holistic and open play of perspectives, on the other (Barnes, 1996; Marchand and Parpart, 1995).

Phoenix, Faust, and Narcissus: Hopes and Hazards of Humanism

Each fresh discovery in humanity's spatial understanding of the world has heralded changes in its self-images, hopes and fears. Phoenix moments in Western intellectual history have been those when new levels of understanding humanity and its terrestrial home have beckoned on the horizon. Classical Greece explored the nature of creation as a whole, observing interconnections between mind, society, nature and the gods, and proclaimed reason as the distinctive quality of humanness (Glacken, 1967). The Renaissance celebrated man as microcosm of the universe, and saw in rational scientific inquiry a way to leash ecclesiastical dogmatism. Romanticism rediscovered nature as a primeval and sacred force, rebelling against the scientific rationality of the Enlightenment. The twentieth century contextual turn echoed many previous turnings to the here-and-now, revealing the intricate interweaving of thought and life and noting particularly the role of society in steering relationships between humanity and world. Each of these periods produced a clearly identifiable emancipatory élan, often seeking a deliverance from confining orthodoxies or structures of thought, politics or material conditions, and at times uttering a prophetic note about future possibilities. Each period also witnessed a will to build structures and institutions, bequeathing its own legacy of Faustian forms and unresolved tensions to its offspring. And in the subsequent attempt to understand and transcend those structures during those dark moments of orthodoxy before the dawn of a new Phoenix, there was the ardent longing of a Narcissus.

Might one not then characterize the challenge of Western humanism in late twentieth century in terms of the "ardent longing of Narcissus"? Frustrations over those Faustian fences that today surround thought and life, disillusionment over the ironic contradictions of word and action in many idealistically motivated social movements of the twentieth century, and maybe a lingering fear about those Spenglerian prophecies about the decline of the West have all been catalysts for critical reflection. Given such an inheritance, one finds expressions of vulgar narcissism, outpourings of satire and brilliantly articulated autopsy on "realized ideals." Yet the experience itself somehow calls for another "face" for Narcissus: that of the pilgrim, already oriented toward horizons that could reveal a new Phoenix. Having passed through a reflective and narcissist phase, the humanist spirit again utters a liberation cry; beckoning, toward a more global comprehension, a mapping of things from the heart of the atom to the mysteries of the universe (Norgaard, 1994).[6]

In retrospect, Phoenix moments for human thought have burst through the Faustian systems within which managerial interests sought to contain them. Fresh understanding of mankind's terrestrial home has led to fresh understanding of mankind itself. In defiance of the traditional Kantian limits to their vision, some scholars have dared to look inward, exploring cultural differences in environmental perceptions and experience,[7] while colleagues in the humanities dare to look outward, freely exploring ground traditionally trod only by natural science and divinities (Cosgrove, 1984; Eco, 1986). The occasion for joint exploration into the mystery of human creativity and wisdom in humanity's modes of dwelling has come. New alliances between physical and biological sciences and between them and the human sciences are taking shape within the academic world (Lovelock, 1979; Prigogine and Stengers, 1984). One can scarcely again consider matter as dead or nature as a complex of blind forces. Rather, one is discovering the dynamic wisdom written into the nature of the universe (Lovelock, 1979) and the basic bonds between humans and fellow living creatures on the earth (Lopez, 1986; Castner, 1990).

What kind of Phoenix then, might one anticipate? In the light of this narrative, the question might be posed metaphorically: What or where could a "Mediterranean" occur in the twenty-first century?

The late twentieth century map of humanity shows that Euro-America and its legacy of Hellenic and Mediterranean models of *humanitas* is but one corner in an evolving noösphere, that is, the sphere of intellect. Yet the analogies with pre-Renaissance Mediterranean worlds is surely striking, as noted so eloquently by Umberto Eco (1986). Today's world map of power and politics has taken on radically new dimensions due to transformations in technology and trade. The explosion of information and transcontinental circulation of people, commodities and ideas raise exciting new possibilities for thought and life, if we dare transcend our inherited orthodoxies. It was just such transformations that inspired Renaissance pioneers, many of whom were jack-of-all-trades poetic types eager to stay abreast of new developments in all realms of human becoming. What the retrospective glance deems as "revolutionary" could scarcely have been possible if these diverse aspects of humanness had not been capable of mutually enriching one another, either in dialogue among human individuals, or within the individual person.

Implications for the human sciences today—and especially for comparative studies—are clear. While each of our practices demands its own level of specialization, so, too, does it need to be orchestrated with the others (Figure 6). Inherited Faustian forms and externally imposed priorities may shed light on how our present fragmentation of effort has come to be, but to suggest that only structural adjustments are needed to clear the way for tomorrow's styles of practice surely evidence a failure of imagination. The humanist's emancipatory hope is that as a scholarly community we begin at home, as it were, to critically evaluate the priorities that have been inherited, and bravely experiment with new ways to see and orchestrate our ideas (Hanson, 1992). Achieving communication and mutual understanding, the very tissue of the noösphere, is one of the greatest challenges facing humans today. And this involves emotion and will, quite as much as cognitive brilliance, technical ingenuity or the design of media. Ultimately, understanding redounds to an expansion of heart and spirit to embrace the twofold challenge facing humanity's dwelling in its terrestrial home in the late twentieth century: sociality and ecology. Sociality, one of the highest art forms of *zoon politikon*, faces the challenge offered today by the spatial juxtaposition of culturally diverse peoples: to understand these often involuntary movements and convergence of humans as springboards for new creativity in politics and social life, rather than problems to be solved by ethnic cleansing or top-down social engineering. Ecology also, in the original meaning

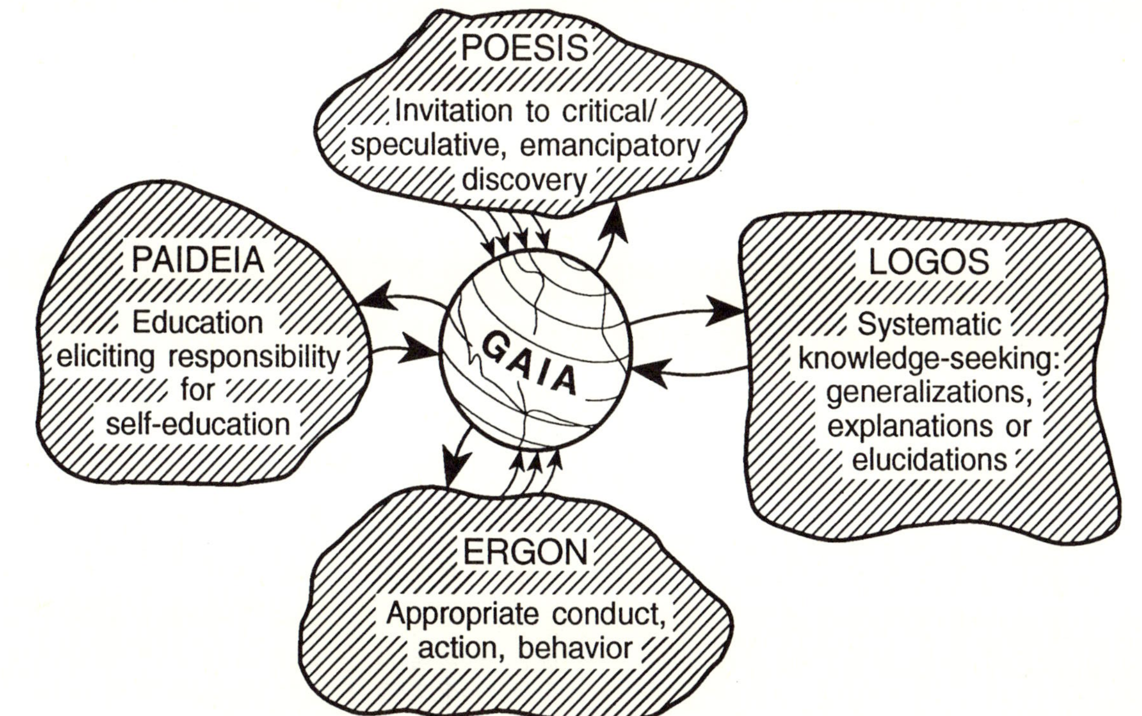

Figure 6. Gaia: *Global challenge for the practice of geography, education and knowledge*

of the term, calls for a sense of how life as a whole functions. Today's chorus of protest against environmental degradation has indeed been heard, but the Faustian fashion in which problems are being addressed often serves more to fragment than to unify efforts toward ameliorative action. Social scientists have much to offer by way of more cross-cultural and comparative evidence from mankind's experiences in sociality and ecology (Wilbanks, 1994).

CONCLUSION

Perhaps the West, far from overestimating *humanitas*, has grossly underestimated it. The recovery of the human subject, the recognition of human agency as an integral part of the lived world and the creativity again apparent on peripheral and previously marginalized regions are all seen as offering fresh potential for human studies. One is challenged not only to regard humanity and earth in global terms, but also to understand the local ecological and social implications of a world humanity now "planetized." The need for cross-cultural and comparative research implies an extension of time horizons on the history of the earth and human occupants to date, as well as reflection on the potential future of Gaia, our living planet. Havel puts it well, saying "The only real hope . . . today is in a renewal of our certainty that we are rooted in the earth and at the same time, the cosmos. This awareness endows us with the capacity for self-transcendence" (1994, p. A15).

As a stance on life, this humanist view is one that welcomes the challenge of discerning the creative potential of individuals and groups to deal with the surface of the earth in responsible and coresponsible ways. Nor is human creativity confined to the intellectual sphere: It involves emotion, aesthetics, memory, faith and will. As Phoenix, then, the humanist turn should refuse to be contained, named and claimed by Faustian structures. It can inspire practitioners across physical, economic, cultural or social studies, and should perhaps not invest too much energy in staking claims for becoming a separate field.

Humanism, this chapter argues, might more appropriately be considered as a leaven in education, as the yeast rather than as a separate loaf. The emancipatory élan recuperable even from our Western traditions could enable humanist perspectives to perform as leaven in the mass of contemporary science and humanities. A Renaissance of humanism calls for an ecumenical rather than a separatist spirit; it calls for excellence in special fields as well as a concern for mapping the whole picture. It beckons sensitivity to what the "barbarism" of our own times might be, and challenges all to responsible action.

1. I thank Bertram Broberg for the excellent illustrations he has prepared for this essay. This chapter was adapted by Rolland G. Paulston from Geography, humanism and global concern, *Annals of the Association of American Geographers* 80:1, 1990 March, 1–33 and is used with permission from the Association of American Geographers and Blackwell Publishers.

2. The Gaia—or living earth—hypothesis, in Lovelock's definition, "postulates that the physical and chemical condition of the surface of the Earth, of the atmosphere, and of the oceans has been and is actively made fit and comfortable by the presence of life itself" (1979, p. 152).

3. Classical proclamations about the nature of humanness (the expression *homo*, "man," intended here, as throughout the paper, in the generic sense) included *animal rationale* = the rational animal, *homo sapiens* = the wise or intelligent one, *homo demens* = the insane or irrational one, *zoon politikon* = the social creature, *homo faber* = the maker, *homo ludens* = the playful or pleasure-seeking one.

4. There are, of course, various versions of this basically Promethean myth. Marlowe's *Dr. Faustus* projected a far more fatalistic outcome than Goethe's *Faust*. Both could find echoes in the history of geographic thought and practice. Elements of the Marlowe version are surely discernible in some contemporary pronouncements about environmental crises. The Goethe version, being somewhat open-ended, seems more inspiring for Western humanity today.

5. In this regard, an early warning that Faustian "environmental science" would result in "global order" and "authoritarian expertocracy" has been sounded by Berry (1972), among others. In an attempt to salvage ecology from this doom, Prakash (1995) advocates support for emerging grassroots ecological perspectives. This view rejects global thinking as an unsituated "impossibility," and invites us instead "to be humble and think little—on a human scale, the scale of our local communities" (pp. 328–29). Prakash offers the example of Gandhi's Tolstoy Farm as an educational means to "conscientize human hands, head and heart" (1993, p. 330). She also contends that such grassroots ecological perspectives

> mark the birth of the first postmodern science: a people's science, replete and rich with concrete knowledge—reflecting experiences drawn from women's lives, from cultures, classes, and persons silenced by the professional norms of modern science. This postmodern science of ecology is strengthening people's resistance to the increasingly oppressive eco-speak, the patriarchal powers and privileges of professional ecologists—now claiming to protect for all of us our planet in peril. (1995, p. 325)

6. David Turnbull's chapter in this volume clearly delineates this mapping transformation in our time.

7. See, for example, Nicholson-Goodman's chapter in this book.

References

Anderson, W.T. (1990). *Reality isn't what it used to be.* San Francisco: Harper Collins.

Barnes, T.J. (1996). *Logics of dislocation: Models, metaphors, and meanings of economic space.* New York: Guilford Press.

Baudrillard, J. (1990). *Revenge of the crystal.* London: Pluto.

Berman, M. (1982). *All that is solid melts into air.* Berkeley: University of California Press.

Berry, W. (1972). *A continuous harmony: Essays, cultural and agricultural.* London: Harvest/HBJ.

Bowden, M., and D. Lowenthal, eds. (1975). *Geographies of the mind.* New York: Oxford University Press.

Brunn, S., and E. Yanarella. (1987). Towards a humanistic political geography. *Studies in Comparative International Development (22)2*, 3–72.

Buttimer, A. (1989). Mirrors, masks, and diverse milieux. In F. Boal and D. Livingston (eds.), *The behavioral environment: Essays in reflection, applications, and re-evaluations* (253–276). London: Routledge.

Buttimer, A. (1993). *Geography and the Human Spirit*. Baltimore: The Johns Hopkins University Press.

Castner, H.W. (1990). *Seeking new horizons: A perceptual approach to geographic education.* Montreal: McGill Queens University Press.

Cosgrove, D. (1984). *Social formation and symbolic landscape.* Totowa, NJ: Barnes and Noble.

Cox, H. (1985). Moral reasoning and the humanities. In T. March et al. (eds.), *Interpreting the humanities* (25–34). Princeton, NJ: Woodrow Wilson National Fellowship Foundation.

Derrida, J. (1972). *Marges de la philosophie.* Paris: Editions du Minuit.

Eco, U. (1986). *Travels in hyper-reality.* New York: Harcourt, Brace.

Ferry, L. (1995). *The new ecological order.* Chicago: University of Chicago Press.

Folch-Serra, M. (1989). Geography and post-modernism: Linking humanism and development studies. *Canadian Geographer 33*, 66–75.

Glacken, C.J. (1967). *Traces on the Rhodean shore.* Berkeley: University of California Press.

Goethe, J.W. von. (1949). *Faust.* Translated by P. Wayne. Harmondsworth, England: Penguin.

Hanson, S. (1992). Geography and feminism: Worlds in collision? *Annals of the Association of American Geographers, 82,* 569–586.

Havel, V. (1994). The new measure of man. *New York Times,* 8 July, p. A15.

Heidegger, M. (1947). *Platons Lehre von der Wahrheit. Mit einen Brief über den "Humanismus."* Bern: A. Francke, 1947.

Jackson, P. (1994). *Maps of meaning.* London: Routledge.

Jameson, F. (1983). Post-modernism and consumer society. In H. Foster (ed.), *The Anti-Aesthetic: Essays on Postmodern Culture* (111–125). Port Townsend, WA: Bay.

Kay, J. (1989). Human dominion over nature in the Hebrew Bible. *Annals of the Association of American Geographers, 79,* 214–32.

Kearney, R. (1988). *The wake of imagination: Ideas of creativity in western culture.* London: Hutchinson.

Kunze, D. (1983). Giambattista Vico as a philosopher of place: *Transactions Institute of British Geographers,* 8(2), 237–248.

Leighly, J. (1983). Memory as mirror. In Anne Buttimer (ed.), *The practice of geography* (80–90). London: Longmans.

Lewis, P. (1985). Beyond description. *Annals of the Association of American Geographers, 75,* 465–77.

Ley, D. and M. Samuels, eds. (1978). *Humanistic geography: Prospects and problems.* Chicago: Maroufa Press.

Ley, D. and M. Samuels. (1989). Fragmentation, coherence, and limits to theory in human geography. In A. Kobayashi and S. MacKenzie (eds.), *Remaking human geography* (223–244). London: Allen and Unwin.

Lopez, B. (1986). *Arctic dreams.* London: Macmillan.

Lovelock, J.E. (1979). *Gaia: A new look at life on earth.* New York: Oxford University Press.

MacLaughlin, J. (1986). State-centered social science and the anarchist critique: Ideology in political geography. *Antipode, 18,* 11–38.

Marchand, M.H. and J.L. Parpart, eds. (1995). *Feminism, postmodernity, development.* London: Routledge.

Meinig, D.W. (1976). The beholding eye: Ten versions of the same scene. *Landscape Architecture, 1,* pp. 47–64.

Mills, W.J. (1982). Metaphorical vision: Changes in Western attitudes toward the environment. *Annals of the Association of American Geographers, 75*, 237–53.

Mitchell, W.J.T., ed. (1994). *Landscapes and power*. Chicago: University of Chicago Press.

Morin, E. (1973). *Le paradigme perdu: La nature humaine*. Paris: Seuil.

Norgaard, R.B. (1994). *Development betrayed: The end of progress and a co-evolutionary revisioning of the future*. New York: Routledge.

Olsson, G. (1979). Social science and human action, or on hitting your head against the ceiling of language. In S. Gale and G. Olsson (eds.), *Philosophy in geography* (287–308). Dordrecht: Reidel.

Passet, R. (1979). *L'économique et le vivant*. Paris: Payot.

Pelt, J.M. (1977). *L'homme re-naturé. Vers la société écologique*. Paris: Seuil.

Prakash, M.S. (1993). Gandhi's postmodern education: Ecology, peace, and multiculturalism relinked. *Holistic Education Review, 6*(3), 178–189.

Prakash, M.S. (1995). Whose ecological perspective? Bringing ecology down to earth. In W. Kohli (ed.), *Critical conversations in philosophy of education* (324–339). London: Routledge.

Prigogine, I., and I. Stengers. (1984). *Order out of chaos: Man's new dialogue with nature*. Boulder, CO: Shambhala New Science Library.

Relph, E. (1981). *Rational landscapes and humanistic geography*. London: Croom Helm.

Roscoe, W. (1991). *The Zuni man-woman*. Albuquerque: University of New Mexico Press.

Said, E.W. (1983). Opponents, audiences, constituencies and community. In H. Foster (ed.), *Anti-aesthetic: Essays on postmodern culture* (135–59). Port Townsend, WA: Bay.

Thomas, R.L., ed. (1956). *Man's role in changing the face of the earth*. Chicago: University of Chicago Press.

Turner, B.L., et al., eds. (1990). *The earth as transformed by human action*. Cambridge, England: Cambridge University Press.

Wilbanks, T.J. (1994). Sustainable development in perspective. *Annals of the Association of American Geographers, 84*, 541–556.

WAYS OF MAPPING STRATEGIC THOUGHT[1]

Anne Sigismund Huff

From a pragmatic perspective, this chapter examines the utility of cognitive mapping for those who wish to understand organizations as interpretive environments where individuals have their own status as "actors." Moving beyond the debate among those who see cognitive mapping as computational models of mental activity, those who see maps as pictorial interpretations of cognition, and those who reject cognitive maps, this chapter first examines a continuum of five mapping types and their preferred methods: 1) maps that assess the association and importance of concepts; 2) maps that show dimensions of categories and cognitive taxonomies chosen by actors in given settings; 3) maps that identify causal relationships among cognitive elements and system dynamics; 4) maps that show the structure of argument and conclusion; and 5) maps that specify schemas, frames and perceptual codes in efforts to visualize links between thought and actions in structural contexts.

Following a discussion of ongoing research issues in the construction of cognitive or mental maps, the chapter concludes that mapping is a useful tool for organizational and comparative research because it offers a provisional way to distill the many but elusive clues of cognition in some reasonably rigorous, reliable and replicable manner.

WHY MENTAL MAPS?

Most academics and practitioners interested in organizations believe that conscious choices can, at least some of the time, affect economic and social outcomes in expected ways. Only recently, however, have we begun to delve into the basic cognitive processes that lie behind deliberate choice. Our interest is part of a growing excitement about cognitive research that is reaching many fields (Barber, 1996). This chapter summarizes some of the leading work in the strategy area that shares this enthusiasm and attempts to make the methods used more accessible to others interested in mapping cognition.

Questions about how people think and how they can understand their world are so basic that they are part of what it means to be human. The literature on the subject, which begins well before Plato, falls under so many perspectives—from poetry to physics—that it is difficult to encapsulate. Relevant research encompasses theory and laboratory experiments on basic neurological and cognitive functions as well as research in fields like anthropology, linguistics and speech communication that study cognition in more "natural" contexts. Clearly this work is of special interest to those who wish to understand and represent organizations as complex interpretive environments.

I believe that it is time for researchers and educators to give more systematic attention to the broad range of subjects that can be approached from a cognitive perspective. Furthermore, our interest in the interacting groups of adults that constitute an organization is consonant with a trend in cognitive psychology, artificial intelligence and other fields to give more attention to real world settings and more challenging tasks. Our interest in the content as well as the processes of strategic thinking puts us in a position to make a real contribution to this basic cognitive science agenda.

One important tool in many cognitive studies is a visual representation, or map, of mental space. On the one hand, the idea of "mental maps" in different forms permeates many discussions of cognition. The status of these images is the subject, however, of considerable debate. One line of thinking closely equates the nature of the world, human representation of the world and the researcher's graphic depiction or model of cognitive activity (Calori, Johnson and Sarnim, 1994). A strong proponent of this view, Johnson-Laird, suggests that:

> At the first level, human beings understand the world by constructing working models of it in their minds. Since these models are incomplete, they are simpler than the entities they represent. In consequence, models contain elements that are merely imitations of reality—there is no working model of how their counterparts in the world operate, but only procedures that mimic their behavior. . . .
>
> At the second level, since cognitive scientists aim to understand the human mind, they, too, must construct a working model. . . . Like other models, however, its utility is not improved by embodying more than a certain amount of knowledge. The crucial aspect of mental processes is their functional organization, and hence a theoretical model of the mind need concern only such matters. (1983, p. 10)

There are cognitive scientists who have a great deal of difficulty with this position. Some observers point out that cognitive scientists like Johnson-Laird are overly enamored with a computational model of mental activity. Other observers feel that all maps invite an inaccurate pictorial interpretation of cognition. Others are disturbed by the inherent simplification of mental models or maps, in whatever form (Martin, 1992).

My own view is that it is not necessary to take sides in this debate (though any given study should be clear about the stance it takes). At the least, pictorial representations are a useful device for summarizing and communicating information (Basu and Blanning, 1994). More expansively, it is possible to point to a good deal of evidence that individuals do encode some information as images rather than words.

The conservative stance treats the map as a tool, subject to the same standards that govern the use of all research tools. My belief is that mental maps can be more than a methodological tool: We can hope to capture something that has the same essential characteristics as thought itself. In this view, the mental map is the knowledge that subjects use themselves (Eden and Ackermann, 1992). Even if current maps fall short of this ideal, we are closer with cognitive mapping to understanding intentional choice than we have been before.

FIVE MAPPING CHOICES

Cognitive maps can be placed on a continuum. At one end are mapping methods that deal with "manifest content" (Berelson, 1952). Here, the underlying, often implicit, model of cognition is relatively simple, and verbal expression is taken as a direct indication of mental activity. Words and related sets of words are counted, and sometimes weighted by placement in the text. Frequent references are taken as indicators of cognitive saliency. The further assumption is that attention is likely to affect action (Cossette and Audet, 1992).

At the other end of the continuum are methods that have been developed in the fields of anthropology, linguistics, literary criticism and artificial intelligence. These methods involve considerable interpretation on the part of the researcher, and they draw on more complicated models of cognition. The assumption, usually made explicit, is that manifest content has to be further analyzed before cognitive structure can be identified. Organizing mental frameworks and the social setting are usually assumed to have strong influences on thought, expression and action.

At a micro level, the diversity of possible mapping methods also comes from the diversity of potential relationships between cognitive elements A and B. Simple connotative association (A reminds me of B); degree of similarity (A and B are different); relative value (A is more important than B); and causal linkage (A causes B) can be identified (Bougon, 1983, p. 177). Or, one might map arguments (since A is true, B is not true); focus on choices (since A, we must do B); and make inferences beyond the relationships present in the text (since A and B are mentioned, the informant must be influenced by C).

The relationships ultimately chosen for mapping depend upon the purpose of the map, and possible subjects of inquiry are as broad as social science itself. It is possible, however, to group the purposes of mapping (and the relationships that help fulfill that purpose) into at least five generic "families" that tend to fall along a continuum demanding increasing interpretive input from the researcher.

1.   *Maps that assess attention, association and importance of concepts*: The emphasis of this kind of map is on inventorying "mental furniture" and its placement. The map maker might search for frequent use of related concepts as indicators of the strategic emphasis of a particular decision maker or group, for example, and then look for the association of these words with other concepts to infer mental connection between important strategic themes. Further judgments might be made about the complexity of these relationships or differences in the use of concepts. Many studies of this type have already been carried out in the social sciences, and these studies provide a source of methodological and theoretic ideas for strategy research. Studies in linguistics help the researcher identify synonyms and compare word use in the data with use of the same concept in more varied sources. Quantitative analysis of the data draws on well-established statistical methods.

2.   *Maps that show dimensions of categories and cognitive taxonomies*: While a great ideal can be done by mapping frequency and simple association of concepts, other map makers wish to investigate more complex relationships among concepts. Empirical research and theory in cognitive psychology supports the map maker in drawing maps that dichotomize concepts and show hierarchical relationships among broad concepts and more specific subcategories. Maps of this type have been used to define the competitive environment and to explore the range and nature of choices perceived by decision makers in a given setting. Since this kind of map usually requires direct inquiry, it offers a means for subjects themselves to collaborate in defining the topics of study.

3.   *Maps that show influence, causality and system dynamics*: Another frequent impetus for map making is the search for causal relationships among

cognitive elements. Causal maps allow the map maker to focus on action—for example, how the respondent explains the current situation in terms of previous events, and what changes he or she expects in the future. This kind of cognitive map is currently the most popular mapping method in organization theory and strategic management. Considerable work along the same lines has been done in political science. As such maps become more complex, systems theory is a natural source of guiding concepts and provides potentially powerful means of analyzing the future growth or decline of nodes on the map.

4.   *Maps that show the structure of argument and conclusion*: More complex yet, in terms of their aims, are the maps that attempt to show the logic behind conclusions and decisions to act. The guiding theory for these maps comes primarily from philosophy, rhetoric and speech communication. The map maker includes causal beliefs in such maps, but looks more broadly at the text as a whole to show the cumulative impact of varied evidence and the links between longer chains of reasoning.

5.   *Maps that specify schemas, frames and perceptual codes*: Work on linguistic structure, cognitive psychology, comparative culture and artificial intelligence provides various supports for the basic notion that cognition is guided by mental frameworks that are not completely accessible to the individual involved. While arguments for the existence of such frameworks are very strong, to infer them (especially in natural settings) requires the greatest leap from text to map of all approaches just described. Those who draw this kind of map, however, claim that an underlying framework affects all of the relationships mapped by other methods. If the map maker wants to understand the link between thought and action, understanding this deeper structure is essential.

There is some range in cartographic techniques available within each of these five generic families, but by and large each basic choice is associated with a circumscribed set of visual representatives. The rest of this chapter assesses these five approaches to mapping in greater detail. Boundaries between each approach are somewhat permeable and in practice map makers often use more than one approach to mapping. By keeping the possibilities initially distinct, however, it will be easier to consider the strengths and limitations of each method. Multiple approaches with compensating features can then be chosen for a specific research project.

Maps That Assess Attention, Association and Importance
Word-based analysis makes one or more of the following assumptions:

- — *frequency* of word use is a reflection of cognitive *centrality*
- — *related* words can be clustered to indicate *themes* of importance
- — *change* in word use can be taken as an indicator of changing *attention*
- — *juxtaposition* of words can be taken as an indicator of mental *connection* between concepts

One of the most influential lines of work on the importance of language was initiated by Sapir (1944) and Whorf (1956), who hypothesized that the way in which people perceive the world is heavily influenced by the categories of their language. Subsequent work showed, for example, that individuals speaking languages with different color categories differed in their ability to recall color chips. Though this work was later qualified by more complicated experiments indicating that speakers from different languages are very consistent in their identification of a few "core" hues (Berlin and Kay, 1969), the basic idea that past experience (and the way in which it is categorized) influences current perception and interpretation remains very influential in cognitive research.

STRENGTHS OF ATTENTION MAPS

Maps focused on word use have several characteristics that recommend them to the map maker. First, the concepts used by organization members form the basic building blocks of cognition. Nelson (1987) reviews basic research in cognitive psychology that supports this view, and quotes Lennenberg to the effect that "concepts . . . are not so much the product of man's cognition, but conceptualization is the cognitive process itself" (1987, p. 223). The nature of these basic cognitive building blocks and changes in their composition over time are too often ignored by management researchers (even those interested in managerial cognition). Work at the concept level would more firmly establish studies of diversification and turnaround, for example, which currently rely on quantitative methods with little sensitivity to connotative subtleties.

The widespread availability of microcomputers and the existence of more and more on-line full-text data bases make mapping methods very accessible to strategic management researchers (Clarke, 1995). Currently available utilities easily identify all uses of key root words and detect changes in the association pattern among key words over time. More sophisticated programs are being developed. Given the quantity of material that can be mapped via the computer, it might be argued that these methods allow the researcher to do the best job of showing "comprehensive intuition" of all the mapping methods described in this chapter.

Another advantage of attention maps is that they require relatively little researcher interpretation and judgment. The data are allowed to speak directly, as evidenced by high intercoder reliability in many studies. Counts of words, and even coding for thematic emphasis, integrative complexity and evaluative assertions, are thus particularly compatible with exploratory studies and grounded theory approaches (Brown, 1992).

POTENTIAL PROBLEMS WITH MAPS OF ATTENTION

Maps that claim to identify attention, association and importance also raise some concerns. A key problem in stimulus-rich organization settings has to do with the limited ability of the human mind to apprehend objects simultaneously. Miller (1956), in an article that has influenced many cognitive scientists, suggests that the "magic number" limiting human attention is seven, plus or minus two. Another line of concern involves the status of the concepts that the researcher ascribes to the subject in this kind of research. The texts most commonly used for analysis (reports, speeches, interviews) are linked to organization purposes and specific audiences. Yet one has to question the underlying assumptions of the method. Does frequency of word use necessarily indicate saliency? Does change in vocabulary indicate change in attention and understanding? Or do shifts in audience and purpose, and the background of the speaker, account for vocabulary change? To what extent does habit and the need for competitive secrecy distort this indicator of cognitive centrality? Can textual proximity of two concepts be taken as an indicator of mental association, or does it primarily reflect rhetorical strategy? Finally, can word use be compared across individuals, much less across different organizational or national cultures?

While all mapping techniques rely on similar kinds of problematic data, direct word counts can do less than other methods to assess the extent to which such things are a problem. Even thematic analysis typically looks only at words in the local context of sentences or paragraphs, or very broadly categorizes large units of text, and is thus less sensitive to nuance than the methods described next. One also might ask if microanalysis of this type is too far from the purpose of strategy research. More complicated mapping techniques, especially maps that allow causal links and argument structure, seem far closer to strategic thought than a thesaurus of the specific words used to make those arguments.

These problems suggest that maps of attention may be of greater value for many purposes when paired with another type of analysis. Word counts can be used to identify basic concepts, review a large set of documents and indicate transition points in the use or association of key ideas; more detailed analysis can then be done with other methods.

Maps That Show Dimensions of Categories and Cognitive Taxonomies
Map makers may attempt to show more specific links among concepts by making the following kinds of assumptions:

— *thinking* involves search and retrieval from *organized memory*
— *learning* involves *categorization*—either the modification of old categories or the formation of new categories
— the *meaning* of any given concept arises primarily from its *contrast* with other concepts

The problem of limited attention can be significantly reduced by categorizing like elements. Nelson notes that recent research basically corroborates much earlier findings:

> In the mid-1800s the question asked was: How many objects can the mind apprehend at one time? . . . "You can easily make the experiment for yourselves," said Sir William Hamilton . . . "but you must beware of grouping the objects into classes." If you throw a handful of marbles on the floor, you will find it difficult to view at once more than six, or seven at most, without confusion; but if you group them into twos, or threes, or fives, you can comprehend as many units; because the mind considers these groups only as units. (1987, p. 6)

Bruner, Goodnow and Austin (1956) were among the first of contemporary scientists to investigate categorization further. Their research, focused on the "strategies" of categorization, suggested among other things that individuals had to successively scan an array of objects to search for hypothesized common attributes.

Map makers in the strategy field who are interested in more closely specifying categories of organizational thought have relied primarily on two approaches. The first of these is personal construct theory, initially proposed by Kelly (1955) as a complete theory of personality but later used primarily as a theory of cognition. Kelly suggests that understanding how individuals organize and map their environments requires that subjects themselves define the relevant dimensions of their environments. A set of techniques known collectively as a "repertory grid" facilitates empirical research guided by the theory.

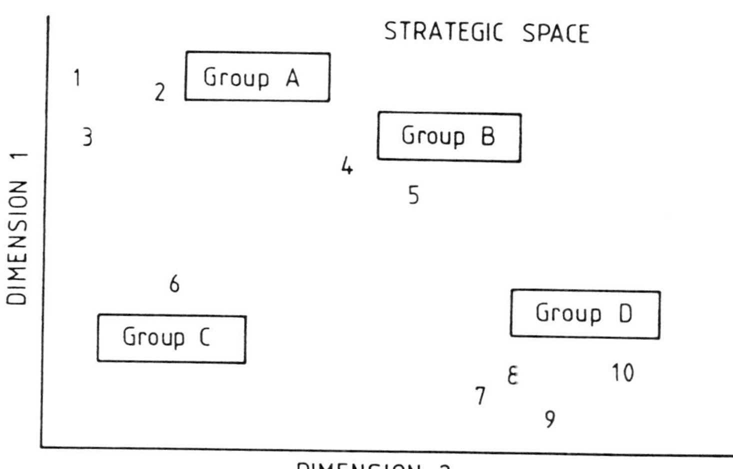

Figure 1. A map of perceived variation in strategic choices

In Figure 1, Kelly's (1972) repertory grid is used to identify the dimensions of competition among regional bank holding companies. Key elements of strategy were grouped by asking twenty-four strategists to identify the two firms most alike in random triads of competitors. Respondents were then asked to place a more extensive list of competitors along the dimensions they had just identified as usefully distinguishing firms in the industry. The results can be mapped along the general lines shown in Figure 1.

Walton (1986), a strategy researcher interested in the ways managers define organizational prototypes, uses multivariate analysis to cluster data drawn from interviews in quite similar ways.

Another approach to defining managerial understanding of the competitive environment involves asking strategists to label the industry they are a part of, and then provide both more specific subdivisions within the industry and a more general label that subsumes the industry. The result is a hierarchical representation, as shown in Figure 2. One interesting finding from research that takes this approach is that managers tend to think of their own business as more prototypic than other similar businesses. Porac and Thomas (1987) suggest that a manager's own company serves as a "cognitive reference point" that guides not only competitive moves but the analysis of others' moves.

Strengths of Maps Defining Concepts
As with maps of attention, one can argue that maps categorizing key concepts provide basic evidence on cognition that is essential to understanding

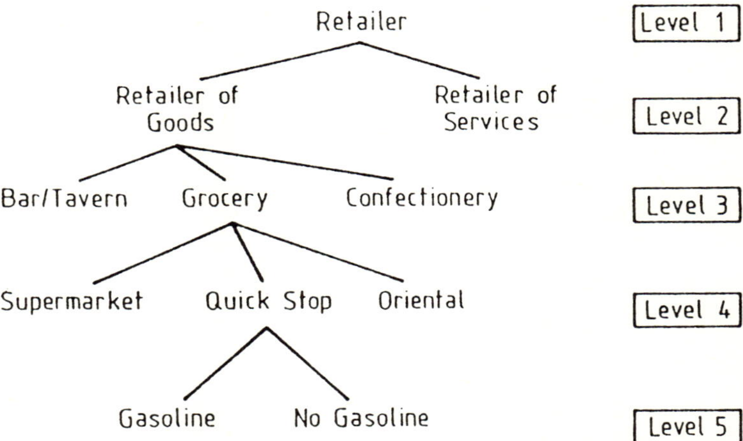

Figure 2. A map of hierarchical relationship

strategic management. The range of subjects that such maps might explore is very large, and studies already completed indicate that these approaches are helpful for understanding perceptions of complex environments. Studies of individual and organizational learning, for example, may also benefit from maps assessing categorization. One might argue that second-order learning (Argyris and Schon, 1978) in particular can be said to have taken place only if concept dimensions and/or relationships can be shown to have changed. More specifically, Hambrick (1981) has suggested that top-level managers in successful companies do not necessarily agree on ends. A map defining concept dimensions and interrelationships might help sort out levels of disagreement among top management or administrative groups.

POTENTIAL PROBLEMS WITH MAPS DEFINING CONCEPTS

While the support for studies of categorization in psychology and related fields is very strong, most empirical work has either involved concept attainment by children or categorization of trivial concepts. The relevance for strategic management of studies that show how people categorize is not yet clear. More complex concepts may not be developed in the same way by knowledgeable adults involved in complex tasks that are of importance to them. Echoing the Sapir-Whorf hypothesis, Murphy and Medlin (1985) suggest that the theories held by the individual give rise to conceptual categories. This approach would seem to be particularly relevant to strategy research, since strategy can itself be seen as analogous to theory.

Hodgkinson and Johnson (1987) offer several criticisms of hierarchical categorization, and suggest that an alternative map of association defines relationships as a network (Anderson, 1983; Sowa, 1984). "Semantic networks" have been proposed as the mental models that allow comprehension of natural language concepts. The network establishes the context-specific way in which a concept with many potential meanings is used, and relates that meaning to other general knowledge. Hodgkinson and Johnson's example mixes different types of associations. This approach to mapping may provide more relevant and complex insights into managerial cognition than maps limited to hierarchical relationships. It also has the advantage of drawing on and potentially contributing to an approach to knowledge representation that is currently central to many studies in artificial intelligence (Sowa, 1984).

One problem with maps focused on categories is that categorization can vary considerably among individuals and may involve entities at different levels of specificity. The basic difficulty is that even a single individual may use overlapping categories that differ in their level of specificity, and different individuals will categorize in different ways. The implications should disturb the potential map maker. It is difficult to think of the decision rules that would allow the map maker to draw a composite map from a matrix of such categorizations.

Mapping mental categories is made more difficult yet by the possibility of multiple nonoverlapping categories. A mug is not only a kind of a cup, it may evoke tea with one's mother and the decor of the local diner. A concept like "capital intensive" may be opposite to both the concept "labor intensive" and the concept "shoestring operation,' even in one person's thoughts. If a given concept does belong to multiple categories and involves multiple dimensions, then one has to ask why this kind of mapping should be carried out at all. Some additional theory is necessary to explain which of many associations will be called upon in specific decision-making situations.

As a final difficulty, one might ask of all maps defining concepts whether the problems of strategy allow the niceties of well-organized categories. Abstract categories (which most analysts assume dominate intended strategy) have shown to have less orderly structure than categories referring to concrete objects. More boldly, members of organizations may only know what they think after they see what they do, extending Karl Weick's well-known aphorism, "How can I know what I think until I see what I say?"

Maps Focused on Causal Reasoning

Causal maps are based on somewhat different assumptions about cognition than the preceding methods, including the ideas that:

— *causal associations* are the major way in which *understanding* about the world is organized
— *causality* is the primary form of *post hoc explanation* of events
— choice among *alternative actions* involves *causal evaluation*

Maps designed to show causal association appear to be the most widely used cognitive maps in the management literature (Roos and Hall, 1980; Ford and Hegarty, 1984; Shrivastava and Lin, 1984; Ramaprasad and Poon, 1985; Fahey and Narayanan, 1986). The nature of causal reasoning has also been the subject of attention in political science (Bryant, 1983) and psychology (Shaklee and Fischhoff, 1982).

Attribution, or post hoc causal explanation for events, is one focus of these studies. The basic idea—as applied by Mausolff in his case study in this volume—is that, in a world of incomplete data, individuals nonetheless make causal inferences that allow interpretation. Fiske and Taylor (1984, pp. 20–99) review some of the extensive work that relates this general notion to emotional states, self-perception, achievement, stereotyping and other behaviors.

An important aspect of this research stream has to do with biases in attribution, and the influence of attribution on the subsequent propensity to act. A good deal of evidence exists to show that individuals tend to claim that their own efforts produced positive results, but that other causes led to poor results (Staw, McKechnie and Puffer, 1983). A variety of explanations have been given for this phenomenon but analysis of the content of causal statements by Huff and Schwerik (1990) suggests a new explanation for the now anticipated pattern of attribution. Their data suggest

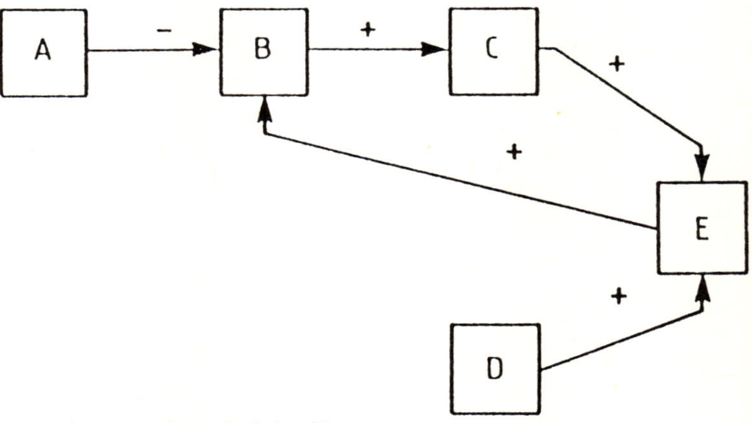

Figure 3. A map of causal relationships

that poor performance challenges the assumptions under which the organization has been in operation. When performance is poor, the search for new understanding of the environment and its impact on the organization leads to more references to external events and actors.

The advantages of causal mapping methods for comparative study include the fact that cause maps provide evidence on the input side of strategy formulation, rather than the outputs normally used to trace strategic changes. Further, mapping holds the promise of capturing strategy as a *coordinated* set of actions. Other methods of data analysis can capture simultaneity, but give less evidence of coordination itself.

Graphic depiction of causal relationships among concepts depends upon identifying variable nodes, or at least nodes that can take on bipolar values. Huff and Schwenk (1990) identified casual statements first, and then combined closely related concepts. The direction of causality between two concepts maybe indicated by an arrow, which can be signed to suggest the direction of causality, as shown in Figure 3.

The information contained in a cause map can also be represented by a matrix in which all elements are listed along the horizontal and vertical axes and the cells show the nature of the causal link (Stubbart and Ramaprasad, 1985). A matrix can also be used to show more subtle casual relationships, as Kelley (1972) suggests. Figure 4 shows a simple example in which factors A and B must jointly reach some threshold level before causing E. Additional examples of this type of map may also be found in the chapters by Paulston and Liebman, and Rust in this volume.

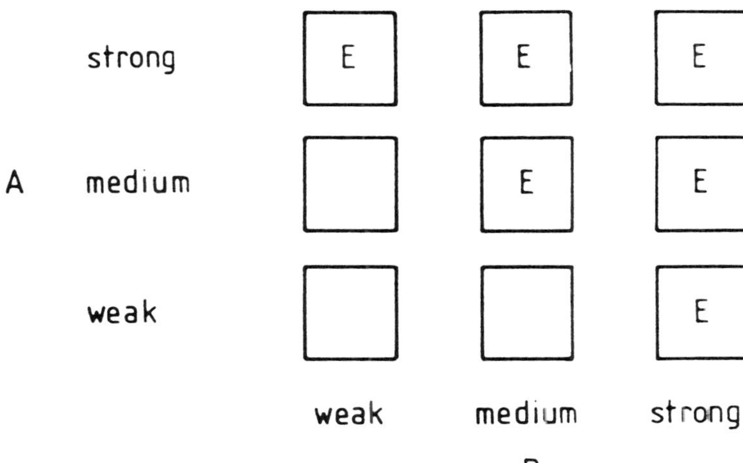

Figure 4. A matrix map suggesting causal relationships

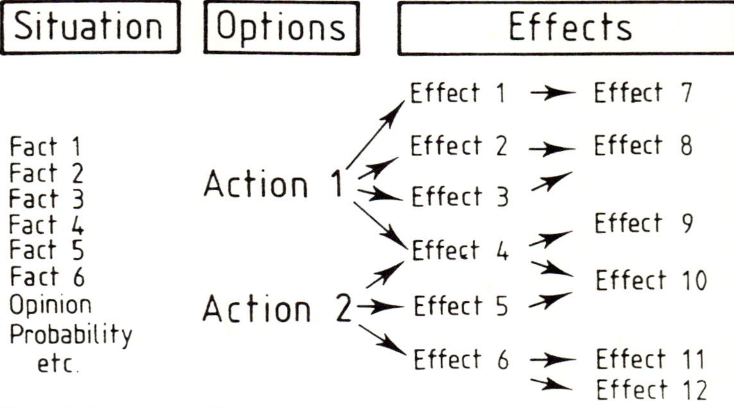

Figure 5. A situation analysis map

Finally, interactively generated maps that focus on causal relationships are also attractive decision aids. Chen (1979) reports on a procedure that begins by mapping individual perceptions of a policy issue. These "situation analysis maps" may take the form shown in Figure 5.

STRENGTHS OF CAUSAL MAPS

The relative simplicity of many causal coding schemes makes it possible to achieve good intercoder reliability (Axelrod, 1976). Despite the fact that decision makers' causal assertions about a given situation can be very complex, the evidence to date also indicates that internal inconsistencies are quite rare, at least in maps drawn from actors whose utterances are subject to close, often adversarial, public scrutiny (Axelrod, 1976). Evidence suggests that the predictive power of the maps is quite good; that is, further decisions follow logically from cause maps. The internal consistency and stability of an actor's cause map over time, and its usefulness in both predicting future actions and explaining those past, are clearly areas of great interest to strategic management, and argue for the continued use of causal mapping in strategic management research.

Furthermore, the range of research questions that can be investigated with causal analysis is demonstrably large. And cause maps invite further analysis. Quantitative manipulation using matrix algebra is one common approach. Control theory forms the basis for another type of analysis that concerns itself with the structure of the cause maps (Axelrod, 1976; Hall, 1984; and Stubbart and Ramaprasad, 1985). Finally, a causal map is a good précis of a long text for some purposes, offering far more information in a relatively compact format than the two types of maps previously described (Paulston, 1994).

Weaknesses of Causal Maps

Axelrod (1976, pp. 251–265) identifies a number of potential limitations of cause maps. Among the problems more uniquely applicable to cause maps is the evidence that maps drawn from documentary evidence tend to show few or no feedback loops. It is not clear whether or not these attributes of the map reflect cognition, but Hall (1984) notes that his subjects had limited ability to perceive feedback loops, even when such loops were deliberately designed into the system used in his experiments.

A second disturbing simplification is that cause maps tend to be "balanced" such that "none of the paths between two given points, A and B, has an indirect effect opposite to any other path" (Axelrod, 1976, p. 264). The result is that the impact of any change within the map can be unambiguously specified. I believe that more realistic maps of strategy-related materials should show points of uncertainty and exhibit contradictory forces. Perhaps communication sets that are less public than the ones studied to date would reveal more complex relationships and less balance. Alternatively, cognitive simplification may be dampening evidence about the very relationships that most interest many strategy researchers.

Another set of problems is more directly related to limitations of the cause map itself. The maps currently being generated show all causal relations at the same level of certainty. In other words, there is no way to distinguish among speculative or "experimentally fitted" relations, relations taken on faith and those that have proven themselves over time.

Influence Diagrams and Systems Maps

Most of the above discussion of cause maps also applies to influence diagrams. Although the two are alike in many ways, there are some essential differences between cause maps and influence maps. Influence diagrams, described as a "qualitative adaptation of systems dynamics modelling" (Diffenbach, 1982, p. 134), have been used primarily as an interactive tool by researchers and consultants to map perceptions of organizational systems (Roos and Hall, 1980; Hall, 1984). More attention is given to defining each variable as a rate in this mapping method, and an attempt is made to define as many feedback loops as possible. The work that attempts to map these complex relationships produces maps of the type shown in Figure 6.

Bougon (1983) uses this kind of map as the basis for conceptualizing and implementing strategic change. He contends that organizations should be defined as "systems of loops" of cause and effect (Weick, 1967). This definition does not require a distinction between participants or orga-

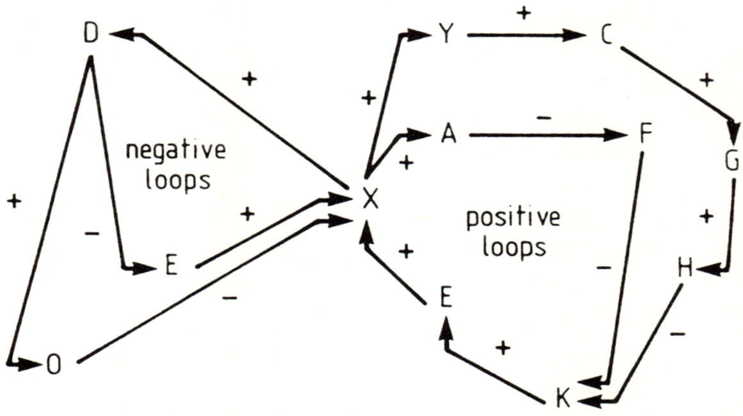

Figure 6. Mapping complex relationships as a system

nizations and their environments, which he argues are artificial distinctions that frustrate change efforts.

Other authors are also in basic accord with other ideas about organizations as systems. Terry Winograd, whose work in artificial intelligence is well known, co-authored a book describing the organization as a "network of commitments" (Winograd and Flores, 1987, p. 150). Finding better ways to deliberately change organizations may rest upon such redefinitions, which mapping can represent and help to clarify.

HYPERMAPS

Another elaboration of basic cause mapping pays more attention to variations in causal perception among individuals. In the context of game theory, Bryant (1983) proposes the use of "hypermaps" as a way of analyzing international conflicts. First-order cause maps show the views of parties involved in conflict (say X and Y); second-order cause maps show what X thinks Y is thinking and vice versa. Third- and higher-order maps are conceivable (how X maps Y's map of X's [or Z's] thinking). But, as with mirrors facing each other across a hall, the details discernible in the reflections of reflections rapidly diminish. (As an interesting sidelight, post hoc explanations of a decision maker's own past actions require second-order maps.)

Eden, Jones and Sims (1981) also use a second-order map to improve decision makers' understanding of a "significant other." The idea behind hypermaps also ties in very closely to stakeholder analysis (Mason and Mitroff, 1981). More complicated maps may, however, involve the problem of incommensurate levels of concept definition.

Researchers interested in reasoning and problem solving sometimes construct cause maps, but they also consider mapping techniques that focus more explicitly on arguments. Four assumptions seem to be particularly salient in these efforts:

— *decisions to act* involve weighing *evidence* for and against an action
— *the evidence* is almost always *inconclusive* and therefore strategists must search for the argument that is strong enough to warrant action
— *disagreements* within a decision-making group are rarely over facts, but are often rooted in the implicit *assumptions* that lead to choosing and interpreting facts
— the power of human *cognition* involves the ability to "nest" arguments so that one chain of logic becomes the basis for additional conclusions

An important basis for argument mapping in strategy research comes from philosophy and rhetoric. Golden, Berquist and Coleman, in their review of "rhetoric as a way of knowing," suggest that "through the critical interaction of arguments where rhetors 'seek' and listeners 'judge' what they hear, knowledge is generated, tested, and acted upon" (1976, p. 285). The philosopher Stephen Toulmin (1958) proposed a scheme for dividing such arguments into five parts, as shown in Figure 7. This approach, along with several others, is an important research tool in the field of speech commu-

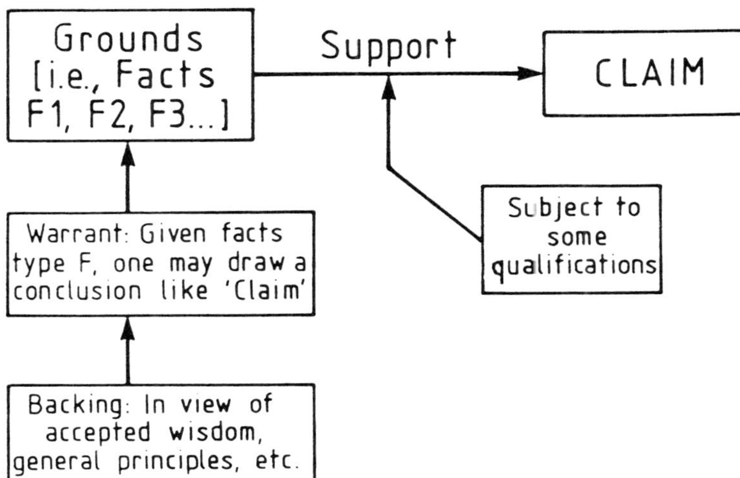

Figure 7. Mapping strategic assumptions in policy analysis

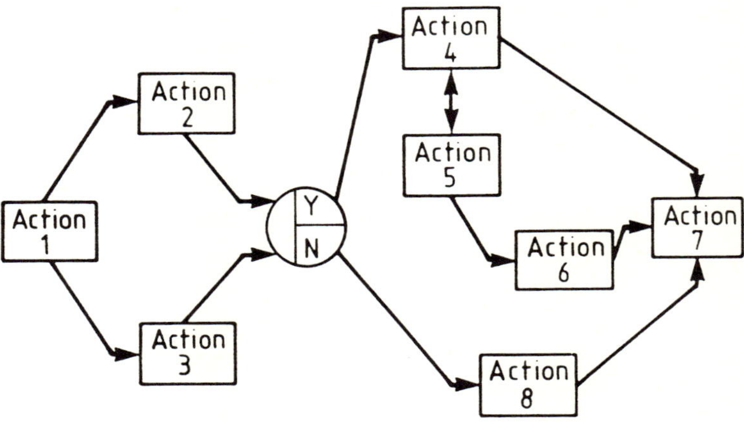

Figure 8. Mapping group decisions in sequencing of actions

nication, and has been frequently applied to organizational settings. Putnam, for example, used argument mapping to study negotiation (Putnam and Geist, 1985). Thompkins and Cheney (1985) looked at organizational identification and unobtrusive control. Within management, Mitroff and Mason (1980) use Toulmin's framework to analyze strategic decision making. Maps of the basic type shown in Figure 7 are used as a tool to examine strategic assumptions and to generate and analyze policy arguments in a dialectical inquiry setting.

A body of literature with somewhat different aims uses argument mapping to focus on the logic used by a speaker. One group of researchers has been interested in the extent to which decision makers are logical in their statements (e.g., Isenberg, 1986). A second group has worked to make decision makers' thoughts more logically complete. Hart et al. (1985), for example, describe a procedure that extends the nominal group technique to promote consensus about the nature and sequencing of actions. The resulting map takes the form shown in Figure 8.

STRENGTHS OF ARGUMENT MAPPING

Having tried almost all of the mapping methods presented here, I feel that one of the greatest strengths of argument maps is their complexity and breadth. This form of data analysis "feels" closest to the mental processes that generate and test strategy, and thus argument mapping methods are a strong candidate for at least part of the research design of many cognitive studies.

The contribution of argument mapping may be clarified by comparing cause mapping and argument mapping. Both methods can serve as an excellent précis of long and involved texts. Cause mapping is less context-dependent in some ways, since a causal assertion is coded regardless of how obvious it is. Argument mapping includes causal assertions, but only those that are related to "potentially controversial" claims. These causal statements are classified precisely because they are put into context. When the conclusion of an argument is a causal assertion, for example, argument mapping provides a far more detailed picture of the mental associations involved than a cause map could. The type of causal relationship embedded in the claim is clear, and not limited to linear-monotonic relationships. The time frame is simple to determine; that is, a speculation about future consequences is clearly distinguishable from post hoc causal reasoning. In addition, argument mapping provides the evidence that leads to causal claims, shows qualifiers and anticipates rebuttals, and to some extent exhibits the force with which the claim is made through the length and detail of the total argument.

Weaknesses of Argument Mapping
While argument mapping provides more context for causal and other types of claims, it is not clear that argument maps are closer to basic cognitive processes. One may argue in different ways, for different purposes, and the additional detail this method picks up may have more to do with communication and persuasion than cognition (Golden, Berquist and Coleman, 1976).

This is also a very time-consuming method of mapping, and the coding techniques required to produce a strategic argument map are not easy to master. The method requires far more judgment than previous mapping methods. A particular problem is to account for nested claims. Longer texts are often designed to flow from topic to topic by providing some linking bridge. It is not always clear whether this is a rhetorical device or intrinsic to the territory to be mapped. Some steps can be taken to improve reliability, such as making multiple passes through the text and reducing the number of coding categories to be coded in each pass. It is also helpful to code all apparent claims, debatable or not. The bottom line, however, is that strategic argument mapping involves a great deal of researcher interpretation, and this may be unacceptable to some researchers.

Maps of Schemas, Frames and Linguistic Codes
The basic assumptions of the last approach to cognitive mapping described in this chapter are that:

—   *expectations,* based on previous experience, *structure* perception

— these expectations provide complex, hierarchical *frameworks* within which *decisions* are made

— *language* can the taken as a *sign* of this underlying structure

Basic research in a variety of fields, from cognitive psychology to artificial intelligence, coincides in the suggestions that cognition is highly conditioned by previous experience. Several different literatures have converged upon this perspective. In artificial intelligence, for example, Winston (1984) suggests that the production and interpretation of news stories draws upon a set of nested expectations.

The notion of an organizing frame or schema—developed in both cognitive psychology (Taylor and Crocker, 1981) and artificial intelligence (Schank and Abelson, 1975)—elaborates upon this basic idea. These researchers suggest that interpreting often fragmentary clues from any given stimulus depends upon more complete details from a schema or frame to "fill in the gaps" and make understanding possible.

Thinking about patterned expectations leads to the literature on semiotics. This perspective, with roots in linguistics and anthropology, emphasizes the principles by which signs (including words) come to have meaning for a community. Barley (1983), for example, asked people working in a funeral home to sort words from their domain into "piles that make sense." The focus of the study is on discovering a core set of "codes" (furnishing the chapel, preparing the body, etc.) that map systems of common meaning for organization participants. Barley's analysis suggests that these codes metaphorically "intimate that perceived opposites are similar" (1983, p. 410): that the chapel is a home, and that the dead body is asleep. They also summarize the "culture" of the organization.

Fiol (1990) also uses semiotics to analyze underlying perceptual codes of CEOs in the chemical industry. She too draws on the basic idea that opposition generates signification. Fiol's study compares annual reports of five medium-sized companies in the chemical industry that formed two or more joint ventures between 1977 and 1979, and five companies that did not form joint ventures during this period or in the preceding ten years. She looks at three levels of structure in the texts: surface structure (the words used as data in the previous four approaches to cognitive mapping); narrative structure (identified by following a set of rules that highlight oppositional dynamics implicit in the text); and deep structure (using the semiotic square to derive fundamental opposition from the narrative structure). Her analysis uncovers significant differences between the two groups of firms, including systematic differences in environmental assessment. The analytic structure is a semiotic

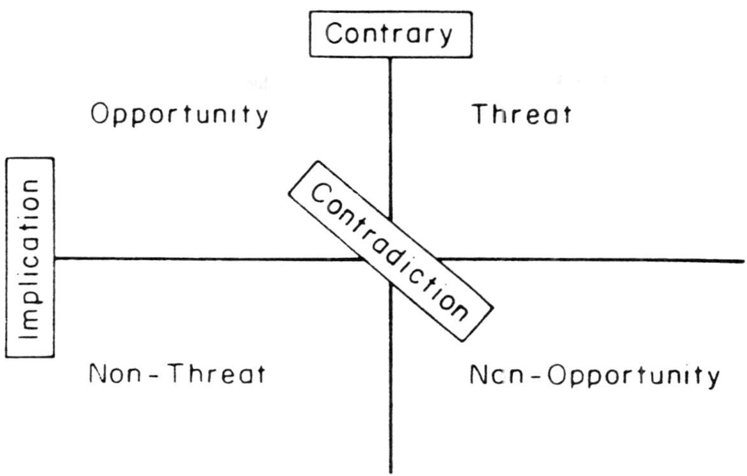

Figure 9. Mapping cognitive structure as a semiotic square

square, as shown in Figure 9. Using this analytic device, the reports of companies not involved in joint ventures show weak links to an environment characterized as threatening; their joint venture–active counterparts make many more references to an environment framed as providing opportunity.

STRENGTHS OF MAPS SHOWING COGNITIVE FRAMEWORKS

Maps that seek underlying cognitive structure are at odds, in their assumptions about cognition, with the assumptions that permeate almost all other mapping approaches. The underlying theory here suggests that the designation of important concepts, categorization of concepts, causal links and arguments are all highly influenced by previous experience and by routine. Thus, apparently "fresh" analysis and decision is structured by what worked in the past. The analyst wishing to understand and predict decision-making behavior must find a way to tap this underlying structure. Considerable experience with semiotics in particular is available to the strategy researcher attempting to map cognition from this perspective (Parmentier, 1994).

A second strength of these mapping methods is the alternative they provide for showing commonalities among maps. As Gottlieb's chapter in this book demonstrates, organizations share perceptual codes, and these appear to be an important source of coordinated action. Previous mapping methods, which look only at service vocabulary, may miss underlying commonalities that affect coordination or may emphasize surface similarities that are not supported by underlying commonalities (Winograd and Flores, 1987).

These mapping methods are, however, the most interpretive and difficult to replicate of all those described in this chapter. For example, semiotics, the method emphasized in this chapter, relies on an a priori value scheme, a foundation that may skew interpretations of the data. Outside observers worried that the data might have been fitted to a schema the researcher may already have had in mind can cite the very argument upon which this method is based: Attention and interpretation are influenced by deep, unconscious structures. If rigorously applied, however, the method of semiotics provides some check against this manipulation. Semiotics forces the researcher away from the *events* that are narrated toward *structures* of the narration (Greimas and Rastier, 1968). Developers of the method would argue that it is easier to subjectively frame overt, surface material, whereas narrative structures have a life of their own. The structures are not directly linked to the narrator or to the reader, but to the universe created in the narration (Gottlieb, 1989).

MAPPING RESEARCH ISSUES

So far this chapter has discussed a range of mapping techniques for illustrating and analyzing cognition. Though each approach has its own strengths and limitations, as a set they also pose some common choices for the researcher. This concluding section looks at some of these more generic research issues.

"Cognitive map" is a metaphor, and the analogy to a geographical map provides useful insights into what cognitive mapping involves. For our purposes, *cognitive mapping* can be thought of as akin to the science of cartography. The *territory* to be mapped involves organizationally relevant "mental relationships" held by one or more individuals; the *cognitive map* itself is "the representation on paper" that models, often graphically, particular features of the chosen territory (Langfield-Smith and Lewis, 1989).

Cognitive mapping can also be thought of as a form of content analysis, which Krippendorff defines somewhat enigmatically as "a research technique for making replicable and valid inferences from data to their context" (1980, p. 21). But cognitive mapping can be distinguished from more frequently used methods of content analysis that count and quantitatively analyze some chosen unit of text (Krippendorff, 1980; McCormack, 1982; Weber, 1983). In cognitive mapping, it is the *relationship* between cognitive elements that is being studied. Cognitive maps allow the reader to move back and forth between an understanding of the whole and its reduction and analysis by parts. The ultimate benefit of the cognitive map, in our view, is that it encourages this holistic synthesis of an actor's view of the world, and it is the emphasis on relationships that makes the mapping analogy particularly apt.

Once mapping has been chosen as a method, the researcher faces a number of important decisions. Consider a hiker and a pilot referring to maps of the same territory. The maps will necessarily show common topographical features. Still, the source of the information that is mapped, the scale and detail of maps, and most important, the features that are unique to each type of map, clearly depend on the needs of each map reader: the purpose for which each map was drawn.

In cognitive map making, researchers have to decide on the purpose a map will serve and the territory it will cover. Subsequent decisions include the specific source of information to be mapped (how are we to infer the contents of internal cognition?); the relevant features of the territory (what aspects of cognition do we wish to depict?); the appropriate range (how large and how detailed does our map need to be?); and the method of cartography that will represent mental relationships in a meaningful and reasonable isomorphic way. Each of these choices involves trade-offs, and raises concerns for the potential map maker. In our view, however, the possibility of graphically representing even a portion of the mental activity behind strategic and other organizational assessments makes the map making venture well worth pursuing by comparative researchers.

The Purpose of a Cognitive Map

Cognitive maps, as artifacts of human reasoning, can be used to study virtually any question raised by those who are interested in human activities, and virtually every social science field, from anthropology to sociology, has pursued cognitive studies relevant to administration or management. The question is not finding an appropriate subject for study, but focusing on the subjects for which cognitive maps provide the greatest insight.

A cognitive map can, for example, be interpreted as a direct product of cognitive processes, or as a tool to elucidate cognitive activity that does not necessarily have map-like properties. At the extreme, the choice is between treating the map as cognition itself or as an imperfect representation of cognitive processes that bear little resemblance to the map used by the researcher. Most researchers in management fields have not taken this distinction too seriously. As in this chapter, most assume that something very close to "real" cognition is being represented by the map. As more work is done, this assumption will be more carefully explored, particularly since the way cognitive maps can be used for intervention and prescription is affected by the extent to which they closely mirror cognitive processes.

Once a useful purpose for cognitive mapping has been defined, the researcher must specify the territory to be mapped. Perhaps the most basic decision in defining territory is the choice between a map that reflects individual cognition and a map that is interpreted as the shared perceptions of a group.

However, bias, lapses of memory, social norms about "desirable" answers and protection of sensitive strategic information all limit the use of such individual maps for understanding strategy. Working with data from multiple individuals may moderate these concerns, since each person's biases, forgetfulness, and images of the "proper" and the "sensitive" will be somewhat different.

The map maker who creates *one* map representing an organization or group faces cartographic challenges that those mapping individual cognition are less likely to encounter. If the map is a composite of maps from several individuals, it will be necessary to devise sensible decision rules for handling inevitable inconsistencies. Studies by Ford and Hegarty (1984), among others, demonstrate methods for aggregating cause maps, but much more work on aggregation methods needs to be done. Analytic techniques that specify the degree of similarity between each individual's map and an aggregate map would provide a measure of the strength or dominance of a "common" map. Successful measures of similarity would contribute to studies of group think, corporate culture and "industry wisdom." These methods are also needed to look for change in individual cognition over time.

Aggregation is complicated by the fact that some similarities are more important than others, and some differences are only marginally relevant. Furthermore, the implications of differences found among maps is not clear. Diversity among maps, and even inconsistency within a single map, could indicate the complicated interpretations that result from living in a complex, nonlinear world (Lincoln, 1985). Since strategy focuses on the messy problems, we must be concerned that mapping methods do not impose an unreflective artificial order on aggregate or individual maps.

Sources of Data

A third basic research decision of cognitive mapping involves the choice between interactively generating the data to be mapped and post hoc analysis of data generated for some other purposes. In each case it might be argued that the data relevant to the researchers' purpose could not be easily winnowed from direct observation or documents generated for other purposes. Researchers following the same line of reasoning have elicited cognitive maps from individuals and groups to clarify complex and ill-structured decision situations (Eden, Jones and Sims, 1981; Roos and Hall, 1980; Mason and Mitroff, 1981; Ramaprasad and Poon, 1985), sometimes using computer programs to help construct the map. Simulation and other experimental methods provide another generative source of data.

Interactive methods will almost always generate more detailed and comprehensive maps on specific topics of interest than methods based on the analysis of documents prepared with other purposes in mind. But interactively generating directly relevant data does have some potential disadvantages. Schwenk (1985) points out, for example, that individuals tend to impose order on recollected events, which may lead to overly rational maps and theories. Another problem involves the ease with which individuals see interconnections, once asked. At the extreme, maps drawn from interactivity-elicited data have a tendency to show "everything related to everything else" (Eden, Williams and Smithin, 1985). This is particularly a problem because previously unconsidered connections are quite likely to emerge during the data collection process. Interactive methods that generate new insights may be very productive for practitioners and consultants who wish to *improve* understanding, but may confound other research projects.

Post hoc analysis circumvents these disadvantages. It is usually more economical than interactive methods for looking at mental representations. Documentary data also are always available; a real attraction to exploratory work in particular is that it is possible to return to the text if additional concepts become theoretically interesting. This is less true with laboratory-based research and other generative methods, which tend to be more focused and capture cognition at a specific point in time under circumstances that are difficult to duplicate.

Work with documentary evidence also allows the researcher to compare mental representations from large numbers of people, facilitates the study of thought over long periods of time and helps the researcher understand the perceptions of decision-making entities that no longer exist. These advantages must be weighted against the insights and subtleties available to the researcher who interacts directly with the subject.

The major disadvantage of post hoc analysis is that the data usually come from communications to other organization stakeholders. The researcher must be concerned that the pattern that emerges from the data has more to do with the speaker's or writer's beliefs about effective persuasion and presentation of self than it does with underlying assessments of the subject being discussed.

Another disadvantage of post hoc analysis is that interpretation of extant materials can rarely be checked with the subject. Even if the informants are available, they compare their thoughts "now" against a map of "then."

Features Included on the Map

The most theoretically useful elements to map from an available set can only be specified in terms of the purpose of mapping activity, but almost all researchers will have to grapple with the amount of detail to include in the map. An important aim of graphic representation is systematically to reduce complexity to make cognition more apprehensible than it is in the original data, and there is clearly a limit to how many concepts can be depicted before the resulting map ceases to be illuminating. It is useful to remember, however, that a detailed map of downtown Boston, for example, is complex and hard to understand. In fact, geographic maps (and cognitive maps) are not just pictures worth a thousand words. They are, as Lynch (1960) long ago taught us, tools for understanding a complex and often confusing reality, and they have no intrinsic need to be instantly comprehensible. Advances in computer graphics do suggest promising directions for making complex representation more graphically understandable (Hall, 1992). It is possible to retain a very complex map in computer memory, displaying and comparing selected parts for analysis. Layering, use of color, three-dimensional representation and even animation are also tools that promise to increase the complexity of pictorial representations of cognitive maps while increasing comprehensibility (Clarke, 1995).

Conclusion

Over the millennia, the study of strategy has been informed by many disciplines, and cognitive science promises further to enrich this mix. I am especially enthusiastic about cognitive mapping as a methodological tool because it offers a way of accessing enormous, and relatively untapped, sources of data generated by organizations.

Our challenge now is to devise and apply methods that distill the many but elusive clues of cognition in some reasonably rigorous and replicable way. This chapter presents cognitive mapping as a technique that can meet these standards. I hope that it encourages others to add mapping to the portfolio of research methods they use.

Note

1. This chapter has been adapted by Rolland G. Paulston from A. Huff (ed.) (1990) *Mapping Strategic Thought* (pp. 11–52). Chichester, UK: John Wiley & Sons, Ltd. Used with permission.

REFERENCES

Anderson, J.R. (1983). *The architecture of cognition*. Cambridge, MA: Harvard University Press.

Argyris, C., and D.A. Schon. (1978). *Organizational learning: A theory of action perspective*. Reading, MA: Addison-Wesley.

Axelrod, R.M., ed. (1976). *The structure of decision: Cognitive maps of political elites*. Princeton, NJ: Princeton University Press.

Barber, T.X. (1996). "Restrictive versus open paradigms in comparative psychology." *American Psychologist, 51*(1), 58–59.

Barley, S.R. (1983). Semiotics and the study of occupational and organizational cultures. *Administrative Science Quarterly, 28*, 393–413.

Basu, A., and R. Blanning. (1994). Metagraphs: A tool for modeling decision support systems. *Management Science, 40*, 1579–1600.

Berelson, B. (1952). *Content analysis in communications research*. Glencoe, IL: Free Press.

Berlin, B., and P. Kay. (1969). *Basic color terms: Their universality and evolution*. Berkeley: University of California Press.

Bougon, M.G. (1983). Uncovering cognitive maps: The self-Q technique. In G. Morgan (ed.), *Beyond method* (173–188). Beverly Hills: Sage.

Bougon, M.G. (1992). Congregate cognitive maps: A unified dynamic theory of organization and strategy. *Journal of Management Studies, 29*(3), 369–389.

Brown, S.M. (1992). Cognitive mapping and repertory grids for qualitative survey research: Some comparative observations. *Journal of Management Studies, 29*(3), 287–307.

Bruner, J.S., J. Goodnow, and G. Austin. (1956). *A study of thinking*. New York: John Wiley and Sons.

Bryant, J. (1983). Hypermaps: A representation of perceptions in conflicts. *Omega, 11*, 575–586.

Calori, R., G. Johnson, and P. Sarnim. (1994). CEOs' cognitive maps and the scope of the organization. *Strategic Management Journal, 15*(2), 437–457.

Chen, K. (1979). Value oriented social decision analysis—enhancing mutual understanding to resolve public policy issues. *IEEE Transactions on Systems, Man, and on Cybernetics, 9*, 567–580.

Clarke, K. (1995). *Analytical and computer cartography*. New York: Prentice-Hall.

Cossette, P., and M. Audet. (1992). Mapping of an idiosyncratic schema. *Journal of Management Studies, 29*(3), 325–347.

Diffenbach, J. (1982). Influence diagrams for complex strategic issues. *Strategic Management Journal, 3*, 133–146.

Eden, C., and F. Ackermann. (1992). The analysis of cause maps. *Journal of Management Studies, 29*(3), 309–324.

Eden, C., S. Jones, and D. Sims. (1981). The intersubjectivity of issues and issues of intersubjectivity. *Journal of Management Studies, 18*, 37–47.

Eden, C., H. William, and T. Smithin. (1985). *Synthetic wisdom—designing a mixed mode modelling system for organizational decision-making*. Working Paper 01M/84, University of Bath, England.

Fahey, L., and V.K. Narayanan. (1986). Organizational beliefs and strategic adaptation. In J.A. Pearce and R.B. Robinson (eds.), *Best papers, Proceedings: Academy of management* (7–11). Briarcliff Manor, NY.

Finch, L., J. Landrey, D. Monarchi, and D. Tegarden. (1987). A knowledge acquisition methodology using cognitive mapping and information display boards. Proceedings of the Twentieth Annual Hawaii International Conference on Systems Sciences (470–477).

Fiol, C.M. (1990). Explaining strategic alliance in the chemical industry. In A.S. Huff (ed.), *Mapping strategic thought*. Chichester, UK: John Wiley and Sons.

Fiske, S.T., and S.E. Taylor. (1984). *Social cognition*. New York: Random House.

Ford, J., and H. Hegarty. (1984) Decision makers' beliefs about the causes and effects of structure: An exploratory study. *Academy of Management Journal, 27,* 271–291.

Golden, J.L., G.F. Berquist, and W.E. Coleman. (1976). *The rhetoric of Western thought.* Dubuque, IA: Kendall/Hunt.

Gottlieb, E. (1989). The discursive construction of knowledge. *International Journal of Qualitative Studies in Education, 2,* 131–144.

Greimas, A.J., and F. Rastier. (1968). The interaction of semiotic constraints. In *Yale French Studies: Game, play, and literature.* New Haven, CT: Eastern.

Hall, R.I. (1984). The natural logic of management policy making: Its implications for the survival of an organization. *Management Science, 308,* 905–927.

Hall, S.S. (1992). *Mapping the next millenium: The discovery of new geographies.* New York: Random House.

Hambrick, D.C. (1981). Strategic awareness within top management teams. *Strategic Management Journal, 2,* 263–279.

Hart, J.A. (1977). Cognitive maps of three Latin American policy makers. *World Politics, 30,* 115–150.

Hart, S., M. Boroush, G. Enk, and W. Hornik. (1985). Managing complexity through consensus mapping. *Academy of Management Review, 10,* 587–600.

Hodgkinson, G.P., and G. Johnson. (1987). *Exploring the mental models of competitive strategists: The case for a processual approach.* Paper presented at the Managerial Environments Conference, Boston, MA.

Huff, A.S. (1983). A rhetorical examination of strategic change. In L.R. Pondy et al. (eds.), *Organizational symbolism.* Greenwich, CT: JAI.

Huff, A.S. (1990). *Mapping strategic thought.* Chichester, UK: John Wiley and Sons.

Huff, A.S., and C.R. Schwenk. (1990). Bias and sensemaking in good times and bad. In A.S. Huff (ed.), *Mapping strategic thought.* Chichester, UK: John Wiley and Sons.

Isenberg, D.J. (1986). The structure and process of understanding: Implications for managerial action. In D. Gioia and H. Sims, Jr. (eds.), *The thinking organization.* San Francisco: Jossey-Bass.

Johnson-Laird, P. N. (1983). *Mental models.* Cambridge, MA: Harvard University Press.

Kelley, H.H. (1972). Casual schemata and the attribution process. In E.E. Jones, D.E. Kanouse, H.H. Kelley, R.E. Nisbett, S. Valins, and B. Weiner (eds.), *Attribution perceiving the cause of behavior.* Morristown, NJ: General Learning.

Kelly, G.A. (1955). *The psychology of personal constructs.* New York: Norton.

Krippendorff, K. (1980). *Content analysis: An introduction to its methodology.* Beverly Hills: Sage.

Langfield-Smith, K., and G. Lewis. (1989). *Mapping cognitive structures: A pilot study to develop a research method.* Working Paper no. 14. Melbourne: University of Melbourne, Graduate School of Management.

Levi, A., and P.E. Tetlock. (1980). A cognitive analysis of Japan's 1941 decision for war. *Journal of Conflict Resolution, 24(2),* 195–211.

Lincoln, Y. (1985). *Organization theory and inquiry: The paradigm revolution.* Beverly Hills: Sage.

Lynch, K. (1960). *The image of the city.* Cambridge, MA: MIT Press.

Martin, J. (1992). *Cultures in organizations: Three perspectives.* New York: Oxford University Press.

Mason, R.O., and I.I. Mitroff. (1981). *Challenging strategic planning assumptions.* New York: John Wiley and Sons.

Mason, R.O., and I.I. Mitroff. (1983). A teleological power-oriented theory of strategy. In R. Lamb (ed.), *Advances in strategic management.* Vol. 2 (31–41). Greenwich, CT: JAI.

McCormack, T. (1982). Content analysis: The social history of a method. In T. McCormack (ed.), *Culture, code, and content analysis* (143–178). Greenwich, CT: JAI.

Miller, G.A. (1956). The magical number seven, plus or minus two. *Psycoological Review, 63,* 81.

Mitroff, I.I., and R.O. Mason. (1980). Structuring ill-structured policy issues: Further explorations in a methodology for messy problems. *Strategic Management Journal, 1,* 331, 342.

Murphy, G.L., and D.L. Medlin. (1985). The role of theories in conceptual coherence. *Psychological Review, 92,* 289–316.

Nelson, K. (1987). Some evidence for the cognitive primacy of categorization and its functional basis. In P.N. Johnson Laird and P.C. Wason (eds.), *Thinking.* Cambridge, MA: Cambridge University Press.

Parmentier, R.J. (1994). *Signs in society: Studies in semiotic anthropology.* Bloomington: Indiana University Press.

Paulston, R. (1994). Comparative and international education: Paradigms and theories. In T. Husén and N. Postlethwaite (eds.), *The International Encyclopedia of Education.* Vol. 2. Oxford: Pergamon.

Porac, J.F., and H. Thomas. (1987). Knowing the competition: The mental models of retailing strategists. In G. Johnson (ed.), *Business strategy and retailing.* Chichester, UK: John Wiley and Sons.

Putnam, L.L., and P. Geist. (1985). Argument in bargaining: An analysis of the reasoning process. *Southern Speech Communication Journal, 6,* 225–245.

Ramaprasad, A., and E. Poon. (1985). A computerized interactive technique for mapping influence diagrams: MIND. *Strategic Management Journal, 6,* 377–392.

Roos, L.L., Jr., and R.I. Hall. (1980). Influence diagrams and organizational power. *Administrative Science Quarterly, 25,* 57–71.

Sapir, E. (1944). Grading: A study in semantics. *Philosophy of Science, 11,* 93–116.

Schank, R.C., and R.P. Abelson. (1975). *Scripts, plans, goals and understanding.* New York: Erlbaum.

Schwenk, C.R. (1985). The use of participant recollection in the modelling of organizational decision processes. *Academy of Management Review, 10,* 496–503.

Shaklee, H., and B. Fischhoff. (1982). Strategies of information search in causal analysis. *Memory and Cognition, 10(6),* 520–530.

Shrivastava, P., and G. Lin. (1984). *Alternative approaches to strategic analysis of environments.* Paper presented at the 4th Annual Strategic Management Society Conference, Philadelphia, PA.

Sowa, J.E. (1984). *Conceptual structures.* Reading, MA: Addison-Wesley.

Staw, B.M., P.I. McKechnie, and S.M. Puffer. (1983). The justification of organizational performance. *Administrative Science Quarterly, 28,* 582–600.

Stubbart, C.J., and A. Ramaprasad. (1985). An interpretive examination of a strategic decision maker's beliefs about the steel industry. Working paper.

Taylor S.E., and J. Crocker. (1981). Schematic bases of social information processing. In E.T. Higgins, C.P. Herman, and M.P. Zanna (eds.), *Social cognition: The Ontario symposium.* Vol. 1. Hillsdale, NJ: Erlbaum.

Thompkins, P.K. and G. Cheney. (1985). Communication and unobtrusive control in contemporary organizations. In R.D. McPhee and P.K. Thompkins (eds.), *Organizational communication.* Newbury Park, CA: Sage.

Toulmin, S.E. (1958). *The uses of argument.* Cambridge, MA: Cambridge University Press.

Walton, E.J. (1986). Managers' prototypes of financial firms. *Journal of Management Studies, 23(6),* 679–698.

Weber, R.W. (1983). Measurement models for content analysis. *Quality and Quantity, 17,* 127–149.

Weick, K.E. (1967). *The social psychology of organizing.* Reading, MA: Addison-Wesley.

Weick, K.E. and M. Bougnon (1986). Organizations as cognitive maps. In H.P. Sims and D.A. Gioia (eds.), *The thinking organization*. San Francisco: Jossey-Bass.

Whorf, B.L. (1956). Science and linguistics. In J. B. Carroll (ed.), *Language, thought and reality: Selected writings of Benjamin Lee Whorf*. Cambridge, MA: MIT Press.

Winograd, T., and F. Flores. (1987). *Understanding computers and cognition*. Reading, MA: Addison-Wesley.

Winston, P.H. (1984). *Artificial intelligence*. Reading, MA: Addison-Wesley.

Envisioning Spatial Metaphors from Wherever We Stand

Martin Liebman

Every viewpoint is a part of some picture, but not the whole picture. Only in the articulation of viewpoints can we understand anything about truth: that is, truth is a fundamentally interactional, social phenomenon.

Susan Leigh Star, "The Sociology of the Invisible."

The social map, as with any charted or sketched construction, sustains a number of components. This chapter first addresses three possible components: metaphors, incongruities and conceptual centers. The perspectives offered in several contemporary novels then illustrate the significance of these components to mapping particularly and academic research generally. Both discussions develop a unifying relationship between metaphors (considered here to result from the combining of two incongruities) and social maps as they spatially integrate incongruent, unrelated or dissimilar convictions, ideas, world views or whatever elements the mapper chooses to illustrate.

A social map presents its creator's perspectives of the world within the framework of chosen boundaries. Illustrating what stands in one place that is different from what stands next to it is one reason to create maps. A twofold metaphoric quality is attributed here to social maps. First, a social map defines a field of dissimilar entities or thoughts. The inhabitant icons bounded within a social map occupy territory in formerly unanticipated relationships so that the juxtapositions create innovative similarities—metaphors. Second, as it extends written research, a social map offers readers a new comparative level, a method of regarding similarities (defined by and within the map's constructed boundaries) formerly considered dissimilar.

The social map's importance derives from the increasingly numerous socially and culturally incongruent beliefs and convictions arising in contemporary postmodern society, particularly academic theories and intellectual discourse communities. However, social maps ultimately illustrate people whose world views begin from a place that is their center, their social horizon's nucleus. While the postmodern teems with a criticism and censure of centers, individuals seem to retain the notion that they are central in their environment, that they are the center of a particular universe. Contemporary novels whose authors recognize and work with the ideas of social space and the metaphors associated with mapping, support this centering perspective. In a brief analysis of four randomly selected authors, I show how characters, place and space play on mapped fields of intellect, discourse and self-discovery.

THE SOCIAL MAP IS AN INTERACTIVE METAPHOR COMPARING PERSPECTIVES

Metaphors clearly create a similarity rather than find a preexisting similarity (Black, 1962, p. 37). Combining previously unrelated ideas, they yield a previously unrecognized similarity. In terms of comparative research, we might consider the research domain and the resultant research paper. For example, is there a similarity between schooling on a distant peninsula and the research article written following extensive studies of that education system? Of course, no similarity exists between that education system and the written paper. One is not the other. Yet a cognitive interaction joins the distant peninsula's people, the researcher's article and the researcher's readers. An interaction further obtains when the author elaborates the words with a map. Geographically, such a map situates or pinpoints the distant peninsula, focusing the reader's geographic perceptions. The map potentially yields (at least to a literate and knowledgeable reader) cultural, ethnic, religious and other distinguishing characteristics joining or separating the people of the distant peninsula and the reader.

This geographic focus and the cultural incongruities synthesize to create an interaction among persons, although no discernible similarities exist between the participants in this interactive potpourri, as the reader creates metaphors linking unlikely elements. This cognitive link conforms to Mary Hesse's (1980) interaction view of metaphor. Hesse describes the interaction view as linking a primary and secondary referent (pp. 111–12), while Max Black considers the metaphor's "focus" and "frame" (1963, p. 28). Warren A. Shibles (1971) notes that Black's descriptive words, *focus* and *frame*, form a mixed metaphor. (I will make the point momentarily that the social map introduces a tertiary metaphoric referent, designating the social map as the metaphor's *face,* further mixing metaphors.)

Noting that a metaphor consists of two referents, Hesse suggests that the primary referent (focus) transfers descriptive traits to the secondary referent (frame), establishing a unique interpretation of the secondary referent's descriptive words and introducing an essence or vision the primary referent did not previously convey. The reverse is also true, as the primary referent influences how the audience understands the secondary referent. Previously held notions conditioning both the focus and the frame interact, creating a fresh perspective of similarities rather than maintaining previously held differences. The dissimilarities now drive the similarities; the metaphor's context changes the audience's perceptions so neither referent is independent in the audience's cognitive vision.

Hesse submits as examples the metaphors "man is a wolf" and "hell is a lake of ice." Each metaphor creates a novel vision of the primary referents, man and hell, by associating new models, new ways of seeing the descriptive properties of previously unrelated yet familiar objects. Metaphors depend entirely on the language community's common understanding and the agreement regarding meanings attached to both the primary and the secondary referents. Metaphors arise and sustain through "literal" or precise and unexaggerated language (Hesse, 1980, p. 111). Cultures that are unaware of the physical antecedents of the words *wolf, hell* or *ice* would have difficulty understanding the relationships. Even then, as Hesse notes, the metaphor's creator presupposes the words' referential commonalities among the language community for which the metaphor is intended.

Take as metaphors the sentences "Man is lupine" and "Hell is a lough of solid water." These ideas may have comparative implications for some persons. However, the readers may consider the metaphoric quality strained or botched. These two strained metaphors lack what a humorist calls "punch." Lupine, lough and solid water do not strike the reader or hearer with the force of the more common, vision-enhancing words *wolf, lake* and *ice.*

Evoking Metaphor's Face

The focus and frame associations correspond to the research investigation and the written findings. The grounding in a common language makes the knowledge transfer possible. As metaphor, the subject of the research (the primary referent or focus) is the written article (the secondary referent or frame). The researcher creates an original similarity where no preexisting similarity existed. The secondary referent does not *become* the first referent; that is, the written referents are not literally identical to the subject of research just as man is not literally identical to a wolf and hell is not literally identical to a lake of ice. Concerning this question of identity, Hesse concludes "there should be a patent falsehood or even absurdity in taking the conjunction literally" (1980, p. 113).

In his list of assumptions, physicist John Barrow (1988) maintains the world is rational because people accept that "A" is not equal to "not A." There is a parallel assumption that the statement "A is A" is accepted as an identity. "A is B," however, rather than being false, may instead be considered an interactive metaphor when the audience understands the use of language and agrees to the metaphoric conjunction. In any instance, an audience should recognize the inherent absurdity of stating "A is B" and the falsehood of that statement in any literal sense.

Falling into a category noted earlier as the tertiary referent or the "face" of the metaphor, social maps further extend the idea that the reader cannot interpret metaphoric conjunctions literally. The researcher studies a social system (the primary referent or focus), offers observations in a written article (the secondary referent or frame) and illustrates either a portion or all of the research findings in a social map (the tertiary referent or the face), asking readers to accept that "A is B is C" with the understanding that there is an inherent falsehood and absurdity in believing that A is literally B or C, or that B is literally C. This three-part metaphor system charts out in this manner:

A	B	C
Focus	Frame	Face
Primary Referent	Secondary Reference	Tertiary Referent
System Studied	Published Findings	Social Map

Figure 1

Viewing the social map as the tertiary referent or face of the metaphor raises the map to the level of a word or phrase, specifically a noun and all its grammatical functions. Consequently, we may either include a map in a text as an appositive, or set a map alongside explanatory text with neither a number nor caption. (Edward R. Tufte's books, [1983, 1990] are resources enabling fresh and novel thinking regarding visual and textual associations.)

THE SOCIAL MAP'S CONTENT AND NATURE

Mappers find, identify, measure and record incongruities, projecting concurrently identifiable features and specific differences that would otherwise be inappropriate, unsympathetic or incompatible. The vocabulary of social incongruity is extensive: bizarre, bohemian, counterculture, deviant, different, dissenting, dissimilar, diverse, incompatible, revisionist, and varied are but a few words that can identify incongruous mapped references. Anyone desiring to locate an accommodating social paradigm traverses the map, a wanderer in search of space where shared convictions flourish, congruity prevails, the icon is pleasing to the mind's eye and the surrounding terrain is hospitable. Only incongruities that erode space would remain were it not for shared convictions, congruity, pleasing icons and hospitable terrain.

Social maps display ideas, narratives or voices, but they also locate people who occupy multiple spaces because of their many pursuits and self-expectations. As mathematicians define space as "the set of all points" (cf. Burrill, et al., 1995), we may reason that social space is the set of all persons. The comprehensive social map (an inconceivable construct) shows a point for each person's many roles. To date, however, less ambitious but far more practical social maps leave to each person the process of identifying the self within the map's parameters.

Excavating the Map's Terrain

Rolland Paulston's original social map and Val Rust's revision of that map (each shown in other chapters) share common elements. Rust identifies several such elements, one being "a metaphorical utility" drawing attention to the connections and interactions of the intellectual discourse communities each map identifies. Common also, according to Rust, are some morphological characteristics dealing with the mapping process and map symbols, and some interactive criteria dealing with the selection and location of the intellectual communities presented on these maps. These morphological characteristics and interactive criteria share certain similarities with the real and conceptual incongruities previously noted.

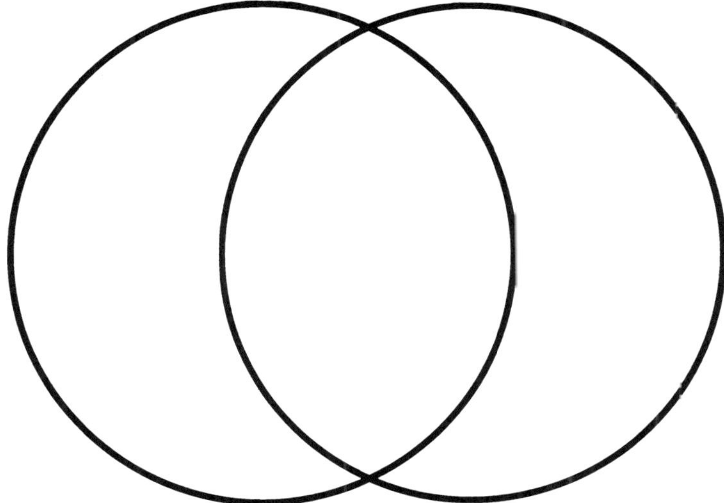

Figure 2

The most obvious element common to both Paulston's and Rust's respective maps is the impression of a (possibly unconsciously placed but seemingly) universal middle, a social centrality beset by many antiuniversalist communities. Although each mapped community is self-central while spatially distanced and ideologically detached from other communities, the mappers' world visions include specific core locations.

Paulston's map illustrates this core through the overlapping or interlocking of two circles, as portrayed in my Figure 2. We easily identify these circles with the mapped earth hemispheres illustrated in elementary geography texts. The intellectual discourse communities Paulston identifies could then as well be analogous to autonomous but determined crustal plates set adrift on the magma of social unrest. Working within a spatially similar domain, Rust elects to open the terrain of the social field, removing the intellectual discourse communities from the restraining hemispheres while retaining Paulston's axial coordinates. As in Figure 3, this axis draws the viewer's attention to a universal nucleus with cross hairs. Whether the entities represented on the map navigate its terrain or stop to build, the cross hair on Rust's map represents either a mariner's compass or a surveyor's sextant.

These observations deal with a visual reaction to social maps, with the power these maps have to stir both emotion and imagination. The working dynamics most active in these maps are those already noted: their facil-

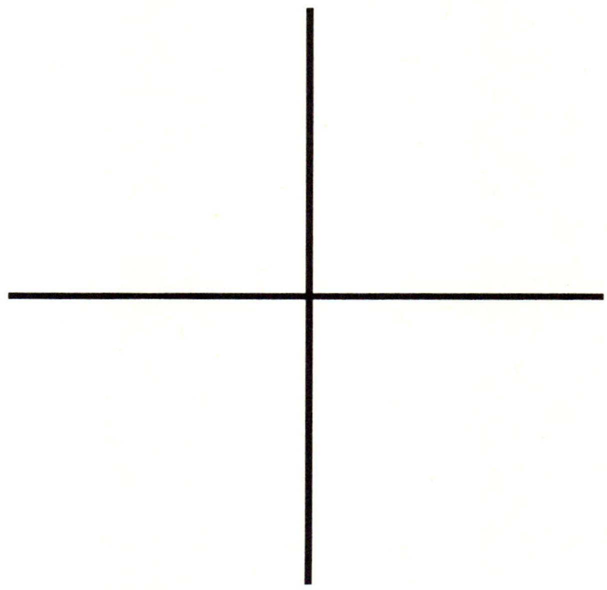

Figure 3

ity to focus attention from cores or centers, and the real and conceptual in-congruities they animate. Mapped centers are the easier of the reactions to deal with so I will address them first.

A few of the questions to propose at this point come from Maxine Greene (1994), who submits the core concerns of epistemology studies: "What is the connection . . . between knower and the known? What does location or vantage point have to do with the processes or results of inquiry? Is neutrality or lack of bias ever conceivable? Is disembodied inquiry, or inquiry devoid of prejudgment, possible or desirable?" (p. 425). The centrality I find conspicuous in the two social maps identified above, the limitations imposed by two encompassing circles and directed cross hairs, echo a contingent reliance on empirical science, or what Greene calls "mainstream science . . . viewed . . . through the lenses of positivism" (p. 426). Can the postmodern world of diverse rubrics offer space to center-positioning maps?

The connection between the knower and the known is a line, not straight, not even necessarily unbroken, twisting between the knower and the knowable. This line affects what Greene calls "the processes or results of inquiry" so that "neutrality or lack of bias" is in extreme doubt. It seems unlikely that inquiry can be disembodied when the knower is corporeal and feeling. Perhaps disembodied inquiry will evolve from intercomputer com-munication without the annoyance of human contribution. Until then, some part of the human mapper will appear in social maps.

Discounting the immediate possibility of a world dominated by intercomputer-based inquiry and a flat social terrain, there is one explana-tion why disembodied inquiry remains improbable. Simply stated, try as we may to expand our views of the universe, we still, as relativity theory ar-gues, see all of space from where we stand—the point in the whole of space that is the self.

Is There a Human Fixation with Centers?

Possibly no visual stimuli have impressed us more in recent decades than those transmitted from outer space. Although the images of planets and gal-axies depict objects too distant and too conceptually remote to be more than curiosities, one image served as a perspective-shattering catalyst. The first live transmission of "Earthrise" seen from the moon's surface, gave us a new perspective on what had been the center of man's existence throughout his-tory; yet we barely realized at the time the social implications of viewing the center as a horizon.

The impact of this new perspective was then and yet today remains shrouded in confusing prepositions. For so long had people looked "up" at the moon that when men finally stood on the moon and looked at Earth, they noted that they were looking "down." Astronauts and reporters alike were unable successfully to conclude a new idea of what makes up our relationship with the dimensions of space. No prepositions seem accurately to suggest a viewer's relationship to spacebound objects. Up and down seem incorrect; across, for the location of Earth viewed from the moon, shows a relationship between the two bodies based on polar juxtapositions that fails to consider how the two spheres would appear from a triangulated apex some half-million miles away. (This is well-illustrated in the movie *Star Trek: The Wrath of Khan*. Khan's physical defeat, at the hand of the space-seasoned James T. Kirk, results from Khan's spatially limited twentieth-century vision: He cannot manipulate his commandeered starship or think of movement in the terms demanded by travel in the three dimensions of space.) While we know what we saw in those pictures of Earth transmitted from the moon, we do not know our relationship to the perspective. By avoiding the question, we retain some feeling of security, for to extend our horizons too far leaves us vulnerable to too many encroachments and subsequent loss of individual meaning.

The interminable search for meaning, according to astrophysicist Trinh Xuan Thuan (1993), is a demonstration of the human desire to find some perspective usually tied to a centrality that can be considered an essential element of individual security and social certainty. Thuan writes, "From the very beginning, human beings have attempted to dispel their mortal dread of infinite space by structuring the world around them in ways that make it seem a more familiar place" (p. 14). Thuan summarizes the history of humans' endeavors to center their world, chronicling the progression of discovery beginning with the pre-Neanderthal world of spirits, followed by the development of mythologies and gods, Plato's geocentric cosmology, Aristotle's dual cosmos of imperfection and perfection, Aquinas's personification of God as Watcher/Mover, Copernicus's heliocentrism, Galileo's heresy, Newton's determinism and Darwin's eviction of Eden from Adam and Eve. Humans have never maintained preeminence in the total system, but through each generation of inquiry and discovery, individuals viewed everything both near and far from a personal locus circumscribed by universal activity.

World-structuring projects attempted to cultivate or manipulate the human environment, creating individual and social external buffers extending toward the horizons. Individuals considered themselves secure at the buffer's center, when everything within the buffer was familiar. However, familiarity exhibits a dual edge; it may suppose either intimacy or intrusion. Familiarity creates security, it dispels a "mortal dread of infinite space." Familiar places convey security. The unknown, on the other hand, is familiarity's intrusive edge; it is a haunting, element we want not to intrude on our space in a familiar way. Each of us wants a core, a center, that is comfortably congenial. (It is there, as noted later, that Hamlet was secure in his dreams, and Einstein sought sanctuary from the exigencies of even those who would be closest to him.)

Thuan's book, a brief analysis of the universe for the amateur, is a metaphor for centralities in both words and pictures. For example, he describes a red giant as a star consisting of successive onionskin layers of elements, the core of the star composed entirely of hydrogen atoms. The hydrogen atom, at the grade-school level, consists of one central proton and one circling electron. He uses phrases such as "the cores of galaxies," "spiral counterparts" "galactic nucleus," "period of rotation," and "implosion of the core." Thuan's map of the physical universe consists of many distinct centers of stars, solar systems and galaxies. This perspective, too, is the basis of Paulston's original social map. The map depicts many distinct centers, theories or communities of commonly held ideas moving in some manner about a vague center.

We are not at the center of the universe, as was made evident in live coverage of "Earthrise" from the moon's surface, although we have not the slightest notion of how to address our relationship with those parts of the universe that are not Earth. While we are not centers of the universe, while the universe does not revolve around us, while we decry centers as modernist, positivist notions that hold no relevance in this postmodern era, Paulston's and Rust's mappings of paradigms and theories extend a core mapping motif into the postmodern era. In the following section I consider why these mappers map the contemporary social environment using tools that are reminiscent of centering devices.

We Are Who We Are; If We Change, We Remain Who We Are
While individuals are distinctive, why they are characteristically different is singularly typical. Allow me to reflect on who we are as individuals by considering who I am.

I am the center. The panorama unique to me focuses the universe as a ring circumscribing my centeredness. The reality of my centeredness is bounded and augmented by the conscious reception of my senses. My sensory uniqueness results from the geometric situating of stimuli to my receptors. Although experiences and choices guide my individuality and uniqueness, these experiences obtain from where I am, so that who I am is subject to the sum of experiences and available choices found within the horizons of my particular space and time. As I move across the planet's surface and suffer the indignity of time, the spatial center that is the relationship of me with the horizon develops an individuality derived from sensory reception, thus creating a self that is central to all else. My life experience is one of growth, an expansion of myself not to transcend the bounds of my centrality, but to extend the bounds of my centrality.

The origins of a social movement, a cultural theory or an ideology derive from a maneuvering of some similar (although not necessarily similarly obtained) individual world views toward a center. A gathering of persons to a center creates a social core where the panorama unique from that center focuses the universe as a circumscribing ring, although the social movement's center may be both internally and externally discerned as an aggregate of views. The individuals who make up such a potpourri may emerge less meaningful while the core of the social movement, cultural theory or ideology emerges as more significant than its autonomous members. The group, to authenticate its social leverage, declares the proprietorship of a center; it asserts as its purpose the desire to endow humanity with its own created significance, developed and best understood from its ideological core, to proclaim its thesis from that core, to expand humanity's social horizon and tame the demons haunting just beyond the frontier. The core presents itself as the universal metaphor. (If the reader does not define "self" as the center, then our disagreement is not about the existence of such centers, but where we place our respective centers. Despite whose eyes we choose to locate where we are, we each find ourselves somewhere in the middle of it all.)

The desire to "extend the bounds of my centrality" is expressed in a variety of ways. Shakespeare's laconic style condenses all I write above into some mere score words: "I could be bounded in a nutshell and count myself a king of infinite space, were it not that I have bad dreams" (*Hamlet*, 2. 2. 258–260). Bad dreams may follow an injustice precipitated by someone within the dreamer's immediate horizon. Hamlet's dreams define his perspective of the injustice his father and mother suffer at the hand of the dead king's brother. For Hamlet, the heart of a nutshell would suffice his need for space, define his fundamental horizon, save the necessity to avenge his father's death. Hamlet's commitment to the will of his father's ghost compels him to venture beyond the rim of the shell, beyond the bounds of what is familiar to him. This is a journey we may all dread: to move beyond the known to diminish whatever demons encroach our space.

Albert Einstein (1982), the wizard of an infinite empirical universe, represents his personal space as a tangible element. "I am truly a 'lone traveler' and have never belonged to my country, my home, my friends, or even my immediate family, with my whole heart; in the face of all these ties, I have never lost a sense of distance and a need for solitude—feelings which increase with the years" (p. 9). Read without additional context, these words characterize Einstein (according to one reader of an earlier draft of this chapter) as a misanthrope. Considering this "lone traveler" statement as a part of Einstein's idea that "empty space [is] a special case of ponderable matter" (p. 74) reveals him as a man who recognized that solitude is a prerequisite to opening one's personal space to the humor of the muses, sufficient reason for each of us to covet a particular, individual space. Our perspectives, our centers, our horizons, produce the social incongruities dictating the necessity for personal buffers to protect us from the suspicion and paranoia that veil human society.

The Mapping Metaphor and Literature

The fiction writer's world thrives on visual images and plays on words. Fiction writers map landscapes, space-scapes, hope-scapes and hopeless-scapes paved with roads that have curious beginnings and bizarre endings. They create unconventional icons and peculiar names and offer these with the prosaic idea that their readers will suspend real-world beliefs and find these strange creations all too unpretentious. For the price of six dollars per paperback, these writers offer to guide us to a distant peninsula of mind. Because these writers, too, are people, we may find the populations of these novel peninsulas often operate with the same imbalance of emphasis on space and time as does the world their readers consider real.

Geographic and mapping metaphors flourish in contemporary literature, although I admit I am more responsive to these images now than I was several years ago, and for that reason it may only seem that literature has taken a turn toward the idea of human space. If space has had some essence in literature, it seems time has received a greater measure of consideration. The reasons may be too complex to consider here (where space is restricted even if the core idea is space). Literature's anguishing interest in the theme of time is based on a reality stated in *As You Like It* (2.7.20): "And so, from hour to hour, we ripe and ripe, And then from hour to hour we rot and rot; And thereby hangs a tale." This view of time is a metaphor for maturation and subsequent corruption on life's vine; time, at first the companion to growth and learning, is soon the consuming enemy of all life. For centuries writers and intellectuals consumed themselves in the subject.

However, in *The Sun Also Rises*, Hemingway focuses on the spatial liability of the bullring, where life is tenuous and space more than time differentiates the participants who live and die there. "In bull-fighting they speak of the terrain of the bull and the terrain of the bullfighter. As long as a bullfighter stays in his own terrain he is comparatively safe. Each time he enters into the terrain of the bull he is in great danger" (p. 213). In his non-fiction work, *Death in the Afternoon* (1960), Hemingway explains the terrain *(terrena)* or social space created by man and animal in the bullring: ". . . in the broadest technical sense the terrain of the bull is called that ground between the point where he is standing and the center of the ring; that of the bullfighter is the ground between where he is standing and the barrera" (p. 455). Note that the bullfighter chooses to enter the bull's terrain, a determined action, the product of his will. Space, on the other hand, is a naturally occurring phenomenon to both the bull and the bullfighter. That the space of the bullring is enclosed, that it is circular, that man and bull occupy it, may determine what space each allocates to the other, how much space each demands of the other to say, "I feel safe here and at this distance." Were the two in a large open field the distances alone would change, not their attitudes toward the other. To both (the personified) bull and bullfighter, space is important; it is the comfortable margin between their respective centers and the horizon defining and separating a threat from security. The bullring, particularly when we consider the number of actors involved beyond the bullfighter and the bull, is a complex, continually unfolding social map. More recent writers, too, work in the motif of similarly complex social maps.

Hamlet, Einstein and Hemingway each illustrate examples of insulated being, islands circumscribed by individual subsistence and endurance. This life-is-an-island mapping theme predominates the characters and stories next discussed.

Four Contemporary Novels

Each of these novels, distinctive for their authors' diverse backgrounds and subject matter, offers a unique sense of geography evoking curious territories and boundaries. Only Cynthia Kadohata's novel, *The Floating World*, uses a geographic attribute in the title, although some might consider Robert Silverberg's title, *The Face of the Waters*, as also geographically representative. These two titles also convey spatial images. The titles of the other two novels, Vlady Kociancich's *The Last Days of William Shakespeare* and Scott Bradfield's *The History of Luminous Motion*, express linear time. All four novels offer insights that further our ideas of distinct cultures and places. An interesting analogy, however, is that the two books with titles conveying spatial images hearken back to *As You Like It*'s ripening of life; the two books with titles referencing linear time hearken back to *As You Like It*'s onset of social rot. That space is life and time is decay may be closer to the contemporary view of being than the anxiety with time and disregard of space that overwhelms more seasoned literatures.

As a study of the contemporary human condition, Silverberg's novel, *The Face of the Waters*, extends deep into the present concern for our perspectives of or from space. For the purposes of this chapter, however, Silverberg extends the vision of the social map as an ever-changing and challenging creation.

Liquidity and the Social Map

All maps eventually change, some due to new discoveries or refined measurements, others due to political change and geographic or ideologic shifts. Social maps should be no exception; in fact, change could be not only the rule but the consequence of social mapping. People whose concerns include social position, the perspectives others have of them and how they fit into society, may wish to alter their world view and thus shift their mapped location. Similarly, others may never concern themselves with how some persons see them as they continue to find contentment in deeply held personal truths. How may we visualize these dynamics when developing social maps? Science fictionist Robert Silverberg offers a stimulating example of the vigorous style that I imagine possible for social mapping.

In his 1991 novel, *The Face of the Waters*, Silverberg creates the water-world Hydros, populated by an amphibious race the humans call Dwellers. A physically descriptive slang, Gillies, finds its way into some of the more eccentric characters' conversations. Dwellers live on free-floating islands they construct of fibrous materials harvested from the ocean floor. Ocean currents propel these islands about the surface of the planet. Hydros is an Earth penal colony where humans are interred for life, a sentence that carries over to ensuing generations because no technology permits the landing of interstellar craft on the planet's liquid surface. As the story begins, a small human population has coexisted on some islands with the Dwellers for one hundred and fifty years. Interisland trade links the human cottage industries. The Dwellers require no commerce and take no interest in human habits.

Although the islands float on Hydros's surface, human ingenuity finds a way to chart the planet's fluid surface to permit accurate interisland sailing, a need culturally unfamiliar to the amphibious, ecologically connected Dwellers. Lawler, the physician to the human community on Sorve Island, is at the dockside shack of Delagard, the owner and captain of an interisland ship. Delagard shows Lawler

. . . a laminated plastic globe about 60 centimeters in diameter, made
of dozens of individual strips of varying colors fitted together by some
master craftsman's hand. From within it came the ticking sound of a
clockwork mechanism. . . .

Circular purple medallions the size of a thumbnail were moving slowly up and down along the chart, some 30 or 40 of them, perhaps even more, driven by the mechanism within. Most went in a
straight line, heading from one pole toward the other, but occasionally one would glide almost imperceptibly into an adjacent longitu-
dinal strip, the way an actual island might wander a little to the east
or to the west while riding the main current carrying it toward the
pole. Lawler marveled at the thing's ingenuity.

Delagard said, "You know how to read one of these? These here
are the islands. This is Home Sea. This island here is Sorve."

A little purple blotch, making its slow way upward near the
equator of the globe against the green background of the strip on
which it was traveling: an insignificant speck, a bit of moving color,
nothing more. Very small to be so dear, Lawler thought.

"The whole world is shown here," [Delagard continues] "at least
as we understand it to be. These are the inhabited islands, in purple—
inhabited by Humans." (p. 78)

It is interesting that the globe Delagard shows Lawler depicts differently those
islands inhabited by Dwellers and humans and those inhabited only by
Dwellers. It makes commercial sense that the human navigators would have
no reason to travel to Dweller islands where no opportunity exists for trade,
but this does not obscure the fact that this globe is both a geographic map
and a social map of Hydros. The opening sentence of Silverberg's last para-
graph quoted above identifies the social map: "The whole world is shown
here, *at least as we understand it to be*." This narrative, too, furthers an
understanding of the differences separating my view of the social map and
Delegard's globe of Hydros from Borges's (1970) Aleph: the Aleph reveals
what the world *is*, not what the person absorbed in viewing the Aleph understands the world *to be*. Most interesting, though, is how the globe/map
of Hydros illustrates the fluid properties I associate with the social map,
properties explored by Japanese-American novelist Cynthia Kadohata.

The World Is Islands Surrounded by Dry Land
In her 1989 novel, *The Floating World*, Kadohata describes her childhood
ukiyo (the book's Japanese title). She writes that she and her family abbre-

viated their perceptions of life in the United States, their world view, to a
single notion: "we were stable, traveling through an unstable world" (p. 3).
The instabilities Kadohata observed consisted of encounters with the non-
Japanese society her family experienced while traveling through the United
States as her parents searched for work. Their individual, familial and Japa-
nese-community stabilities could not be entirely isolated from the external
instabilities. Yet, as she matured, Kadohata began to speculate that new
people who float into her world might reveal fascinating ideas, as the stranger
who asked of her, "Where do you locate your expertise in this world?" (p.
10). It was much later when Kadohata realized this geographic metaphor
asked, "Who are you?"

This is a query not heard on Silverberg's Hydros, where Dwellers
never question the presence of humans, and most humans do not attempt
to understand the Dwellers. Had the humans asked about life on Hydros,
their journey to "the Face of the Waters" where, in some manner not fully
described by Silverberg, they become incorporated into the planet's
bioculture, would not be so difficult. What the humans do not come to realize in their century and a half on Hydros and what Kadohata does not realize until she matures beyond the confines of the Japanese community is
that, as Margaret Archer (1989) explains concerning the organization of all
prevailing consciousness, "all things capable of being grasped, deciphered,
understood or known by someone" (p. 104) shape the cultural system.

Kadohata's is also a world where a definitive form does not confine
ideas, where every idea passing for fact can dissolve into fiction, so that as
one world floats away another world takes its place; the resulting world is
one of ongoing fictions where Kadohata experiences a dichotomy between
being and feeling: "I was free. But I didn't feel free" (p. 30). It is a world
where the teenage Kadohata senses security and safety when her parents root
their family in a community of shared self-perceptions and shared ideas. The
influences patterning all the individual stabilities combine to create a larger
communal stability, an island in an otherwise unstable world filled with fictions that too often float into Kadohata's personal world. Kadohata recalls
the counsel of her grandmother on the subject of racial fictions:

> *Hakujin* don't know when a smile is an insult . . . *Hakujin* were white
> people . . . if you hated white people, they would just hate you back,
> and nothing would change in the world; and if you didn't hate them
> after the way they treated you, you would end up hating yourself, and
> nothing would change that way, either. So it was no good to hate them,
> and it was no good not to hate them. So nothing changed. (p. 9)

Outside the Japanese-American communal stability is a universe of un-friendly floating worlds in the early stages of what Mike Davis (1992) describes as the *Blade Runner* scenario . . . the fusion of individual cultures into a demonic polyglottism ominous with unresolved hostilities" (p. 82). Young Kadohata finds only social depravity beyond her community's social boundaries, boundaries seemingly created by a caring adult community offering Kadohata stabilities and securities not forthcoming from the outside. "Neglect, I think, was a quality I associated with the world"(p. 97).

Kadohata's awareness parallels that of Vlady Kociancich's protagonist in his 1991 book, *The Last Days of William Shakespeare*. Kociancich describes a wall of family providing stabilities through assimilated like ideas and similar needs. Family and community space is a map filled with persons of diverse yet symbiotic functions. Kociancich and Kadohata find that ideas they presumed to be facts often become fictions in an unstable world. While Kadohata writes, "no idea had definite form, every fact could dissolve into a fiction" (p. 35), Kociancich notes that even the fact of self-identity becomes fiction when viewed historically. "When one looks back in time . . . every fact becomes a fiction, every person a character, every place a scene on a stage . . . I call up my memories and see myself, from the outside" (p. 67). Kociancich observes that age and aptitude oblige persons at all levels to assume some social authority; authority, however, often camouflages individual instabilities so that "each person, within the individual constraints of office and wall, constructs his own little house, his temple . . . [self-] idolatry sustains our desire to survive the disillusionments and dangers of this world" (p. 104).

Kociancich's protagonist, "the prizewinning writer Santiago Bonday, 'the Master,' as he is known" (p. 2), a novelist of on-again, off-again, almost forgotten and even questionable literary repute, chooses to live far outside the city, separate from what he calls the "dangers of this world." His seclusion offers little security, however; certainly less in his old age than the security young Kadohata's community provides. Bonday creates a personal, anchored, lonely island surrounded by a world of floating islands where he is an authority only to himself, while what he possesses, including his own being, is often appropriated to serve the self-idolatry of authorities made superior by power. Young Kadohata recognizes that happiness and stability flourish in an isolated community; Kociancich's protagonist, choosing isolation from any community outside himself, suffocates within the bounds of his unsociable boundaries.

Scott Bradfield's eight-year-old protagonist, Phillip, in the 1989 novel *The History of Luminous Motion*, shares a similar world view to that offered by Kociancich's theme of suffocating on one's own floating island. Phillip exists in *ukiyo*, but one lacking the larger support community that embraces young Kadohata. Traveling contemporary Southern California with his mother, Margaret, behind the wheel of a deteriorated station wagon, Phillip sees a world consumed by incessant movement, a world where he and his mother maintain neither course nor mission. Yet this floating tribe, a wired, dysfunctional mother and a coping son, derive stability from the feeling of completeness Phillip perceives while enveloped in his isolation. Philip feels that his completeness with his mother is so solid that he describes their being as "geographical weight and mass" and his mother as "a place, a voice, a state of repose" (p. 3). This is a framing of certainty, security, belonging without conditioned membership. It is a self-actuated and self-contained world of assurance.

Phillip's world sustains only while he can rely on his mother; that is, when she is self-reliant in the way she has chosen to maintain her self-reliance, primarily through theft and prostitution. When she changes, becomes "lost in endless dialogues between herself and her own reflection" (p. 21), Phillip enters a smaller world, drawing from himself certainty, security and belonging. Assuming the role of self-provider, he finds the world outside himself is "filled with sharp things, things that banged and brushed . . . things that crowded and pressed . . ." a world where Phillip sees himself "growing more and more immaterial" (p. 21). When his mother again becomes central to his certainty, security and belonging, his world softens; he becomes, again, "just a passenger, and like all passengers, fundamentally unconcerned with landscape and plot, enveloped only by the simple movement of it all, the cumulate graph of those coherent points where [he and his mother] ate, slept . . . and awaited movement again" (p. 43).

Standing Inside the Map

Although the major characters in these four novels interact differently with their worlds, the obstacles they encounter all spring from the same place, the individual centers and buffers we each seek to protect us from the unknown. On Hydros, the buffer is the ecological ignorance separating humans from the reality of the water planet's indigenous biocomplex; Cynthia Kadohata's buffer is her family and Japanese-American community; Santiago Bonday's buffer is his estate in the countryside; Philip's buffer is his mother. Mapped conventionally, we might imagine these protagonists' buffers as the doughnut drawing in Figure 4.

The center circle is the individual; the larger circle is the system of

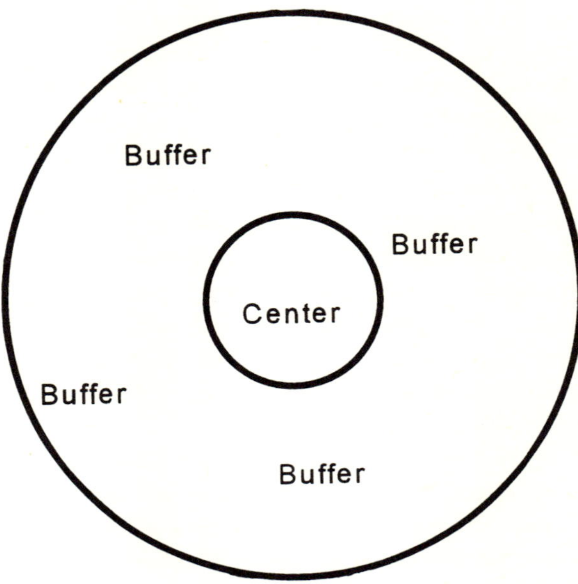

Figure 4

buffers. Whatever these protagonists do not want in their world, the unknown, exists beyond the boundaries of the larger circle. From this perspective we could as well mistake an individual for the hole in a doughnut. As the four authors cited above make clear, individuals are more than two-dimensional; the buffer does not provide total protection. How is it possible that people find their lives occasionally disrupted? From where do the disruptions and incongruities we can combine to create new metaphors, new world views, arise?

The buffers are not as solid as the two-dimensional doughnut image implies. Kadohata's buffers, for example, include many other people. Although she envisions herself as the center of this group, each individual also holds the self-perception of a center. In addition, each person surrounding Kadohata moves about in a three-dimensional plane underscored by a two-dimensional surface. Imagine, then, each person in Kadohata's buffer as a marble on a flat surface (Figure 5).

Movement on the surface opens and closes spaces around Kadohata's center. The outer edge of the buffer expands and contracts. Some marbles move so far from her center that they become inconsequential buffers. In this marbles-on-a-flat-surface map we see that the buffer does not provide total protection from the outside. How might this map of Kadohata's buffers better indicate the intrusive elements she encounters?

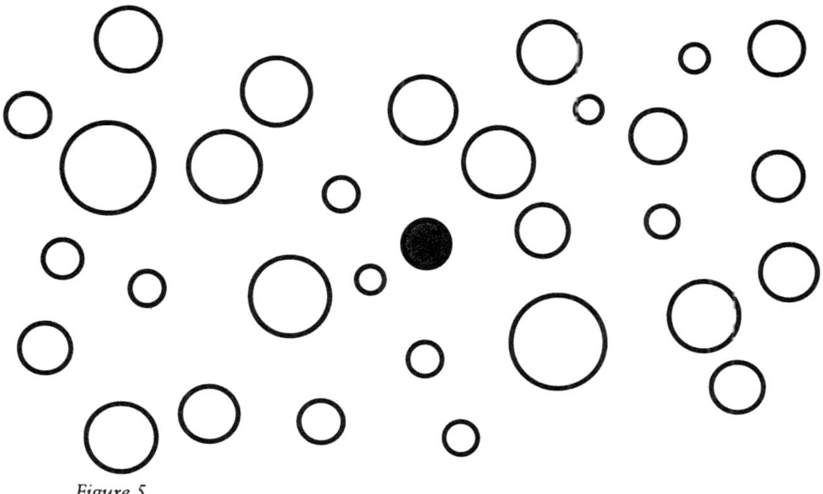

Figure 5

A David Povilaitis drawing in Rudy Rucker's 1984 book, *The Fourth Dimension*, offers insight on how haunting disruptions gain entry. Rucker's discussion of physical dimensions includes a description of the one-dimensional world in Edwin Abbott Abbott's nineteenth-century satirical novel, *Flatland*. Rucker suggests sliding colored patterns on the surface of a soap bubble represent the residents of Flatland's two-dimensional space. A cutaway section of the Flatland map is the mathematical definition of a line: a continuous extent of length, straight or curved, without breadth or thickness." This cut-away is impossible to illustrate for any line drawn here would, of course, have thickness.

In *Flatland*, one resident, "A Square," discovers the third dimension; he not only slides around the world but can pass through to the surfaces; that is, he can move just above or just below Flatland in a dimension previously unknown. With this discovery, A Square now occupies one of three separate planes. A new Flatland map as in Figure 6 now offers a different way of seeing A Square's new relationship to Flatland's physical environment as it takes us on a journey with A Square progressing freely "through" (rather than "around" and "in") Flatland, taking up positions on both the "top" and "bottom" of the map and partially embedded in the map as he enters from the "top," passes through, and exits onto the "bottom" (Figure 6).

Flatland suggests an explanation why incongruities seem to appear from nowhere, why buffers as suggested in the marble model for Kadohata's *ukiyo* do not isolate individuals in groups from outside influences. The next map in Figure 7 is a side or "cut-away" view of the Kadohata marble model

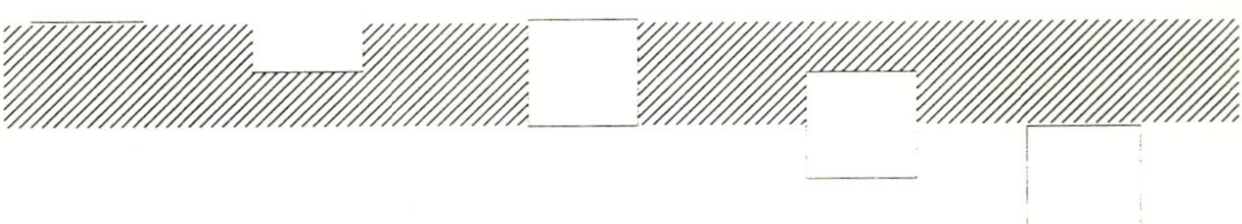

suggested earlier. Here the solid circle resting on top again represents Kadohata. Her family/community buffers consist of the varied circles also resting on top of the flat surface.

Embedded within the surface are other circles not of her family/community. These embedded circles are incongruities. They represent outside influences her family and community attempt to protect her from. These embedded circles are the disruptions that may literally "spring up" at any moment.

Similar to A Square and the other inhabitants of Abbott's *Flatland*, we tend to move in two dimensions; the dissimilarity is that we travel on the surface rather than between surfaces (the tops of our heads and the soles of our feet do not form the measure of our world's depth), and that the surface we navigate is three-dimensional. Factors including our physical height and thickness, visual perceptions and freedom of movement in all directions give us third-dimensional movement not dissimilar to that discovered by A Square. How may we apply this discovery of what is hidden beneath the map of Kadohata's community buffer to the social maps offered by Paulston and Rust?

Discovering the Third Dimension

Like the astronauts' view of Earth from the moon, social maps offer perspectives that tend to look down on rather than look around in the social

world. The two variations of macro mapping of paradigms and theories offered by Paulston and Rust are similar to the first map of Kadohata's floating world, where viewers look down on the marbles representing family and community members rather than at a cross section of the map, one that, as shown, offers a perspective suggesting a different way of seeing the world. The "flat map," the physical paper, appears little different from the "soap bubble" surface populated by Flatland's inhabitants. The ideas and theories inhabiting social maps are each a created *ukiyo* floating on the surface just as do the colors on a soap bubble or the inhabitants of Flatland. While we expect movement and each move is mappable, the flat map offers very little perspective. It is, to use another familiar metaphor, a contour map, but one offering no hint of depth. How could we represent the social map otherwise? What would be the application of another perspective?

Just as we can offer a perspective from a "cut-away" edge of the Kadohata marble map, we may similarly view Paulston's and Rust's respective maps. Working with Paulston's map, consider the possibility of viewing the map in cross section, cut away where indicated by a dotted line. If we stand in Figure 8 at the point marked "x" and look eastward, Paulston's map may appear as shown in Figure 9: a world of both direction and dimension.

This map also offers a hypothesis applicable to concerns regarding mapping and its capacity for discovering, revealing and placing what Susan Star (1991) calls "previously hidden narratives." We can make the map reader aware of unheard voices, represented here by the blank circles em-

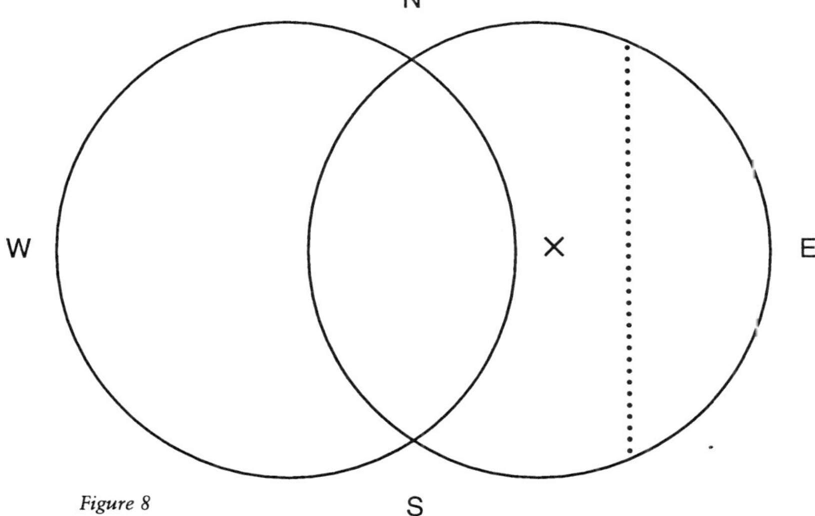

Figure 8

LEGEND
DEPENDENCY
CONFLICT THEORY
MODERNIZATION
HISTORICAL MATERIALIST
NEOFUNCTIONALIST
HUMAN CAPITAL
HIDDEN NARRATIVE

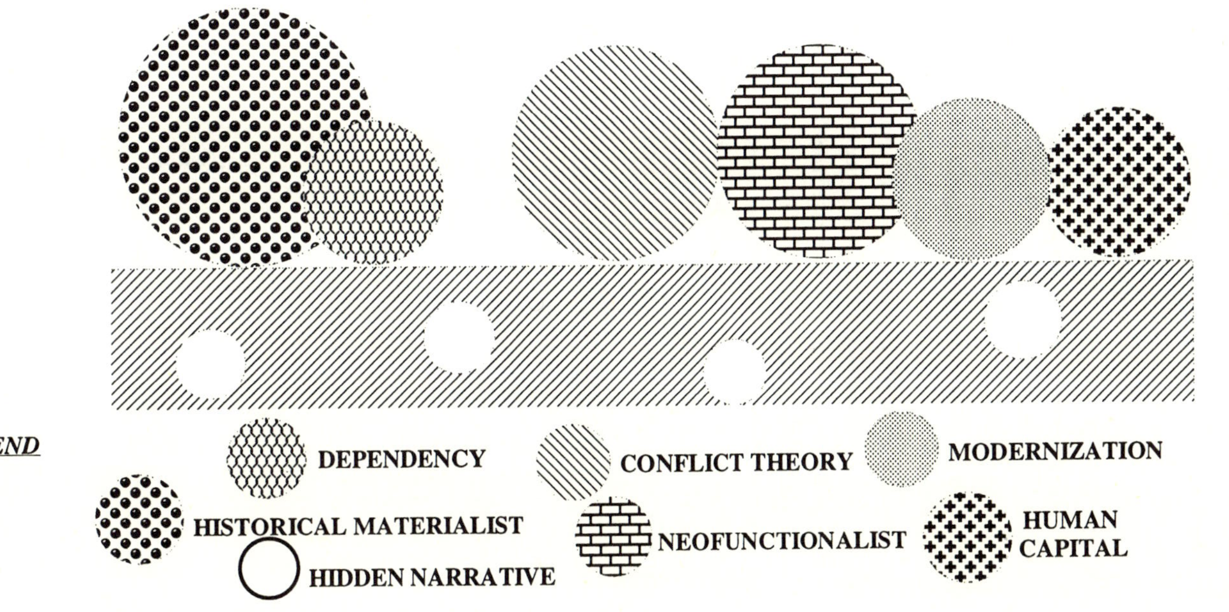

Figure 9

bedded under the surface of the map. Viewing the map this way is similar to standing on the north wall of the Grand Canyon, gazing across at people standing on the far wall while being aware that there are hidden stories under the surface, embedded in the stone of the canyon's wall just under where these people we see are standing. These hidden narratives await not discovery, but a recognition that places them on the map, that seems to make them "spring up" and take their places among the developing, moving and growing theories already placed within the social map's parameters.[1]

CONCLUSIONS

One objective of research is to discover the unknown, what is different, and to make it familiar to an audience. I offer that research of this nature creates a metaphor and that the social mapping project initiated by Rolland Paulston offers researchers a method of providing their research articles a "face," extending the metaphor to a new, third level.

This has been a too-brief overview of metaphor and the style's method of dealing with incongruity. It seems evident that metaphor calls attention to incongruity, but also immediately shows the commonality of traits not before evident in the association of incongruous ideas. Individuals and groups, however, do not seek incongruent relations and associations. They do not wander the social terrain hoping they will find animosities toward their person or their ideas. What they seek is congruity. They yearn for and aspire to accord, agreement, compatibility and harmony. The problems piqued by disassociative allusions and the resulting incongruity are well documented in the four novels presented in the second half of this chapter.

Like young Cynthia Kadohata, most people find congruity in family and associated community. Others, like the human community on Silverberg's Hydros, must endure some form of social reorganization before they subdue personal or person-directed animosities. Vlady Kociancich's protagonist, Santiago Bonday, does neither; he floats through life taking whatever the surrounding society decrees to unload on him. For eight-year-old Phillip in Bradfield's novel, every association except that with his mother is or becomes an obstacle. Life on the road as a dysfunctional (as viewed by some other persons) family member is his preferred life style; it is only on the road with his mother behind the wheel that he experiences congruity with the world. In these novel fantasies there remains the underlying realization that fiction never falls far from what we otherwise understand to be real life.

At times we cannot discern the incongruity of the real and the conceptual, possibly because incongruity is merely the pursuing of congruity. That is why we map yin and yang as a conceptual whole, as a circle containing Borges's Aleph of being. That is why each of the randomly selected words associated with social incongruity—bizarre, bohemian, counterculture, deviant, different, dissenting, dissimilar, diverse, incompatible, revisionist and varied—has an opposite word signifying social congruity—customary, resident, establishment, normal, alike, adaptable, similar, correspondent, dependable, continuous and imitative. As revealed in the simple maps offered here, incongruity is often, conceptually, just below our feet, just as A Square finds the third dimension just above and below the surface he shares with his fellow Flatlanders. The task left to mappers is to duplicate A Square's discovery, to find what yet remains hidden below the flatland of social maps and bring these discoveries to the surface.

NOTE

1. Editor's note: See an earlier discussion of this idea in Deleuze and Guattari (1987, p. 480) where they reference Paul Virilio's work on smooth and striated space— i.e., on how "the sea constitutes a smooth space with 'a fleet in being,' and how a vertical smooth space of aerial and stratospheric domination springs up" (1975, p. 93).

REFERENCES

Abbott, E.A. (1993). *Flatland: A romance of many dimensions.* 1884. Reprint, New York: Barnes and Noble.

Archer, M.S. (1989). *Culture and agency: The place of culture in social theory.* Cambridge, England: Cambridge University Press.

Barrow, J.D. (1988). *The world within the world.* Oxford: Clarendon.

Beardsley, M.C. (1958). *Aesthetics.* New York: Harcourt, Brace.

Black, M. (1962). *Models and metaphors: Studies in language and philosophy.* Ithaca, NY: Cornell University Press.

Borges, J.L. (1970). *The aleph and other stories, 1933–1969.* Edited and translated by N.T. DiGiovanni. New York: Dutton.

Bradfield, Scott. (1989). *The history of luminous motion.* New York: Knopf.

Burrill, G., J.J. Cummins, T.D. Kanold, and L.E. Yunker. (1995). *Geometry: application and connnections.* Lake Forest, IL: Glencoe.

Davis, Mike. (1992). *City of quartz: Excavating the future in Los Angeles.* New York: Vintage.

Deleuze, G. and F. Guttar. (1987). *A thousand plateaus: Capitalism and schizophrenia.* Translated by B. Massumi, Minneapolis: University of Minnesota Press.

Einstein, A. (1982). *Ideas and opinions.* New York: Crown.

Greene, M. (1994). Epistemology and educational research: The influence of recent approaches to knowledge. In L. Darling-Hammond (ed.), *Review of research in education* 20. Washington, DC: American Educational Research Association.

Hemingway, E. (1960). *Death in the afternoon.* New York: Charles Scribner's Sons.

Hemingway, E. (1986). *The sun also rises.* 1926. Reprint, New York: Scribner.

Hesse, M. (1980). *Revolutions and reconstructions in the philosophy of science.* Bloomington: Indiana University Press.

Kadohata, C. (1989). *The floating world.* New York: Viking.

Kociancich, V. (1991). *The last days of William Shakespeare.* Translated by Margaret Jull Costa. New York: William Morrow.

Paulston, R. (1994). Comparative and international education: Paradigms and theories. In T. Husén and T.N. Postlethwaite (eds.), *The International Encyclope-*

dia of Education (923–933). Vol. 2. 2nd ed. Oxford: Pergamon.

Richards, I.A. (1948). *The philosophy of rhetoric.* New York: Oxford University Press.

Rucker, R. (1984). *The fourth dimension: A guided tour of the higher universes.* Boston: Houghton Mifflin.

Shibles, W.A. (1971). *An analysis of metaphor in the light of W.M. Urban's theories.* The Hague: Mouton.

Silverberg, R. (1991). *The face of the waters.* New York: Bantam.

Star, S.L. (1991). The sociology of the invisible. In D.R. Maines (ed.), *Social organization and social process: Essays in honor of Anselm Strauss.* New York: Aldine de Gruyter.

Thuan, T.X. (1993). *The birth of the universe: The big bang and beyond.* New York: Harry N. Abrams.

Tufte, E.R. (1983). *The visual display of quantitative information.* Cheshire, CT: Graphics Press.

Tufte, E.R. (1990). *Envisioning information.* Cheshire, CT: Graphics Press.

Virilio, P. (1975). *L' insecurité du territoire.* Paris: Stock.

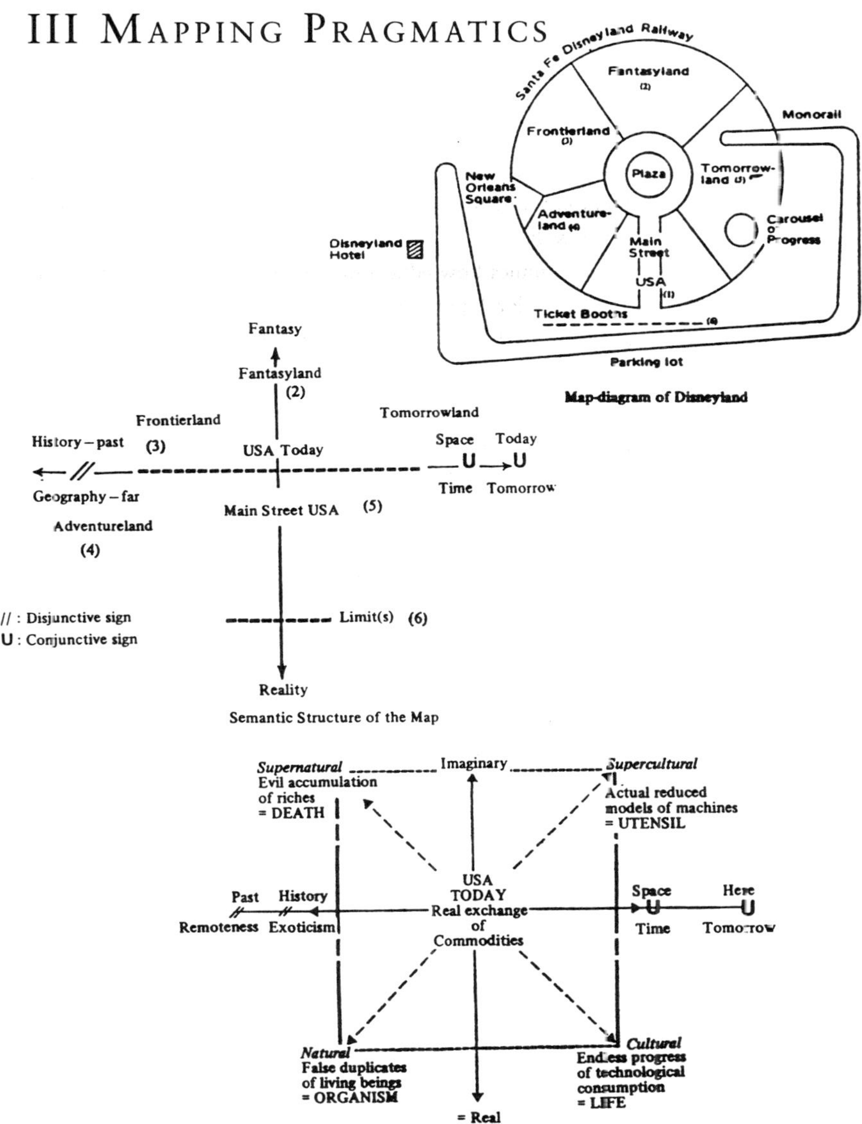

Map-diagram of Disneyland

Semantic Structure of the Map

Semantic structure of the ideological representation in Disneyland

To support his contention that a degenerate utopia is ideology mapped into the form of a myth, Louis Marin (1990) illustrates structuralist mapping practice in creating meaning out of space. Here, his three-part mapping of the original Disneyland reveals the interplay—and "deep structure"—of mapped geographical, semiotic, and ideological space. Marin argues that "... by acting out Disney's utopia, the visitor realizes the ideology of America's dominant groups as a mythic founding narrative for their own society" (p. 241). See his *Utopics: The semiological play of textual spaces*. Atlantic Highlands, N.J.: Humanities Press International. © Éditions de Minuit, Paris. Reprinted with permission.

Introduction

In her chapter, "Mapping Gendered Spaces in Third World Educational Interventions," Nelly P. Stromquist contends that social cartography is seen in some circles as an expression of postmodern thought. It can, however, also be construed as a visual representation that borrows concepts and metaphors from social geography, a field that also operates in a modern and, at times, critical tradition as well.

Stromquist examines how educational interventions to advance the conditions of women in developing countries are being conceived and implemented in substantially different ways depending on the actors that support those interventions. Key institutions in such efforts are national governments, nongovernmental organizations (NGOs), international development agencies and persons (mostly women) in academia. In her chapter, Stromquist identifies features that characterize each type of actor in terms of their social class origins, their positionality in the social context, the educational and social terrain addressed in their interventions and the instances of boundary-crossing that are emerging. Particular attention is given to the distinctions between academic feminism and popular feminism and how these approaches impact the objectives, content and pedagogic strategies of their educational interventions. These distinctions are conceptualized as differential spaces with networks, time referents, and boundaries.

Education designed to remove negative gender stereotypes and to create new gender meanings in society touches many axes, ranging from cultural norms to the educational system's capacity to implement innovations. Though social cartography allows graphic representations, to date these tend to be uni- and bi-dimensional. Stromquist's chapter explores how an essentialist social mapping methodology can be utilized both to enhance the understanding of feminism as a new social movement, and to invite others to present their own views in this shifting physical and conceptual space.

Esther E. Gottlieb, in her essay, "Mapping the Utopia of Professionalism: The First Carnegie International Survey of the Academic Profession," examines how the Carnegie Foundation's International Survey constructed the academic profession as a utopian concept, existing everywhere yet nowhere in particular—as the world professoriate. Here, professors are seen as citizens of the world where professionalism represents their condition. Gottlieb examines in some detail how the questionnaire was progressively de-discursified; that is, how its markers of discourse indicating a discursive situation were erased so that the instrument finally came to sound scientific. Questions were addressed to professors in the world at large, but actually nowhere in particular. Nevertheless, each country was free to add specific local questions that transformed their questionnaires from "the international version" to "the Australian questionnaire" or "the Dutch questionnaire." The examination of these local questions reveals local concerns such as unionism or class and ethnic origin, whereas the international version relies on professionalism as a global ideology that is not connected to location but to "the profession," an ideological place that does not exist in time or space, but only in the world.

This creates the professoriate as a utopia of worldwide professionalism. No wonder, then, that *The Chronicle of Higher Education*'s reception of the survey involved showing the results on a world map. Although only the countries in the survey are shaded-in, the whole world is present. This map is significant, she argues, because it supplies the image that the categories "worldwide" and "global" depend on.

Christopher Mausolff suggests in his paper, "Postmodernism and Participation in International Rural Development Projects: Textual and Contextual Considerations," that Norman Uphoff's *Learning from Gal Oya: Possibilities for Participatory Development and Post-Newtonian Social Science*, alerts us to the possibility of many valid ways of seeing participation in rural development projects. This essay builds on Uphoff's work by continuing to explore the implications of multiple framing perspectives for participatory practice. The case for methodological pluralism is explored by mapping alternative perspectives in the literature on participation. The observation that each of the complementary perspectives represented in the literature (positivism, pragmatism, hermeneutics and critical theory) continues to provide significant practical insights supports the postmodern critique against totalizing grand theories.

Another reason that a postmodern sensibility may be relevant to development managers is that it is in indigenous and other often isolated communities that the metanarratives of modernity are especially likely to make invisible what Liebman and Paulston refer to as "the mininarrative knowledge claims of cultural clusters." It may be that local conceptions of needs and desired ways of meeting these needs are locally constructed in small groups and hence vary considerably across projects and regions. This proposition is evaluated in seven different rural development communities in Honduras using a technique called participatory self-evaluation. The results of these evaluations have been summarized into social maps of the communities' project goals. The resulting maps show considerable variety in the values expressed by the community groups. This outcome, Mausolff contends, suggests a uniquely postmodern justification of participatory development:

that beneficiary participation helps to make projects more responsive to the local narratives of community groups.

Christine Fox's chapter, "Listening to the Other: Mapping Intercultural Communication in Postcolonial Educational Consultancies," explains that intercultural communication situations between consultants and indigenous educators are gradually changing from a colonial or neocolonial style of interaction to a more authentic dialogue. Her chapter addresses the concept of authenticity in such relationships and examines possibilities for mapping the multiple ways in which interlocutors interpret meaning. She first examines some of the dilemmas created by unequal power relations in intercultural communication, and traces some historical consequences. Rather than seek explanation in a linear, time-bound model, connections and interconnections of social interaction are viewed spatially and mapped as fields. Secondly, she examines the language of educational consultancy in action. A number of communicative "filtering" devices are identified and discussed. They seem to form the basis of an emerging, identifiable intercultural discourse. Fox also looks at some examples of intercultural dialogue where the interlocutors attempt to reach understanding, but where certain blocks to successful communicative interaction are identified. The usefulness of social cartography to describe and critique these enabling (or blocking) filters is then discussed.

JoVictoria Nicholson-Goodman, in her chapter, "A Ludic Approach to Mapping Environmental Education Discourse," states that questions of what environmental education (EE) should aspire to teach have engendered a flurry of dialogue in recent decades. Today, an ever expanding EE discourse explores new ways of seeing the relationship between humans and the rest of the natural world, humans and science-and-technology communities, humans as social beings in enclaves variously competing with each other for resources or attempting to work together to protect an increasingly ravaged planet. The implications of the questions raised spread themselves over a broad range of disciplinary fields, even leading to the creation of new fields of inquiry, particularly in the natural and social sciences. The result is a wildfire of ideas and applications for educators to consider as they prepare to educate youth to cope with "realities" of the future. Friction between varying truth-and-value choices forms the combustible core, with political, philosophical and sociocultural knowledge claims feeding the controversy. Within this scenario, educators are expected to choose programs and applications that can assist their students to make "wise" choices for the future.

In her chapter, Nicholson-Goodman argues that the methodology of social mapping, as it makes visible sometimes bewildering relationships between old and new ways of seeing within the field of EE discourse, may serve both as an aid to clarifying truth-and-value choices implicit in the discourse and as a means of orienting advocates and practitioners alike to the range of perspectives that currently compose its open, and rapidly expanding boundaries. This cartography of ideas also serves as a ludic approach to truth-and-value conflicts. It is a "playful" form of resistance to any totalizing discourse that would seek to silence all others. This chapter portrays the process of social cartography with reference to a range of pronouncements that variously revolve around or direct attention to the intertextual field of EE discourse.

Martin Liebman's "Social Mapping: The Art of Representing Intellectual Perception" suggests that social mapping offers comparative education and the social sciences a facility similar to that mapped geography conveys: a graphic that evokes personal spatial presumptions informing insight, comfort, and possibly relief when an identifiable landmark, icon or name on the map permits the mind to claim "You are here," whether the map reveals a network of perspectives or the floor plan of a building. Social mapping, clearly a member of the larger, long-recognized cognitive mapping model family, extends cognitive mapping beyond its earlier concern for "knowing" geographic location or for the arranging of information. Merging an explicative method with the higher human abstractions and predilections, social mapping illustrates genuinely cognitive constructs.

The innovative maps reviewed in Liebman's chapter exemplify the possibilities social mapping extends to comparative education and other research fields where interests focus on exploring and understanding the interaction of ideas, people, and social movements. Varying significantly in their origins and purposes, ranging from ingenious to romantic to analytic, these maps permit us to see the mappers' conceptions and unique regard of the world, as well as how details and features of knowledge interrelate from each mapper's unique vision. Thus, perceived abstraction and vision attain the status of research through cognitive mapping practice and analysis.

Mapping Gendered Spaces in Third World Educational Interventions

Nelly P. Stromquist

Knowledge can be remapped, reterritorialized, and decentered in the wider interests of rewriting the borders and coordinates of an oppositional cultural politics.

Henry Giroux, "Democracy and the Discourse of Cultural Difference"

Being at the heart of feminism—one of the strongest social movements in the twentieth century—gender is a highly contested concept, with various individual and institutional actors struggling to define it in ways suitable to their interests. Accordingly, it may be illuminating to explore a spatial approach to the study of the condition and status of women in society and to identify how diverse actors seek to carve up particular cognitive spaces to maintain and defend their view of reality.

Since the 1975 Mexico meeting celebrating the International Women's Year and the subsequent world conferences on women in Copenhagen (1980) and Nairobi (1985), as well as the multiple national and regional nongovernmental organizations (NGOs) and governmental meetings on gender issues, attention to the conditions of women in developing society and the design of policies and projects to advance women's conditions have become increasingly institutionalized. Major conceptualizations of gender now range from "women in development" (WID) to "gender and development" (GAD) to "empowerment."[1] These conceptualizations offer different perceptions of the problems facing women and thus of the solutions needed. It is beyond the scope of this chapter to deal extensively with these concepts. Suffice it to say that WID, introduced by the United States Agency for International Development (USAID) in the early 1970s, is characterized by its focus on the role and status of women in society, while GAD (the preferred concept among several international agencies since the 1930s) introduces the need to treat women in more relational terms. It compares women and men in terms of the productive and reproductive roles they fulfill, and is more sensitive to the social and cultural contexts affecting the relations between men and women. Empowerment (a concept introduced by third world women, primarily Indian feminists and women of color, around 1984) refers to the imperative that women become agents—through knowledge, skills and confidence-building—of their own transformation from the very beginning of any change effort (hooks, 1984). Aid agencies uphold WID and GAD approaches, while women-run NGOs operate more within empowerment frameworks.

Educational interventions[2] to advance the conditions of women in developing countries, as is the case for interventions in other sectors of the economy and society, are being conceived and implemented in substantially different ways depending on the actors that support those interventions. Key institutions in such efforts are national governments, NGOs, international development agencies and individuals (mostly women) in academic settings.

This chapter examines work on gender issues utilizing mapping as both a metaphor and an analytical tool. As an analytical tool, mapping borrows some fundamental notions from the field of geography: namely place, space and location (Hanson, 1992). Here I proceed by focusing on institutionally occupied spaces, with particular features regarding terrain and distance. These institutional spaces, occupied by key actors, create particular dynamics linked to knowledge production, borderwork and boundary-crossing. Attention is centered on five key actors: international aid agencies, national governments, nongovernmental organizations (both international and domestic) and feminist academics. Mapping entities, their conceptual distances, and physical connections is done with the purpose of elucidating positions, identifying the relations between these entities and exploring the convergence and divergence of their ideas. Mapping, more than other types of intellectual enterprise, tends to reflect the perspective of its producer, as the mapping exercise is built on a close point of reference, usually the researcher's own experience or perspective.

Paulston and Liebman assert that mapping efforts "contain some part of that person's knowledge and understanding of the social system" (1994, p. 223). My position, therefore, should be made clear. I function full-time in academia in the field of international education; my special emphases are gender issues and policy. I do regular consulting with multilateral and bilateral aid agencies, and maintain a frequent and close contact with feminists in NGOs in developing countries and in academia here and abroad. At one point in my professional life, I spent four years working for an international aid agency. The mapping I propose results to a great degree from my experience and trajectory. Although my assertions about the strength of the various actors and their definitions of gender issues are probably not far from "reality," more contestable are my views about the degree of contact the actors keep with each other.

Actors active in the definition of and response to gender as a social marker operate both within and outside governmental structures, and therefore have different objectives, functions and networks. In considering the various actors in educational development work, I discuss their features in terms of their position in the sociopolitical context of their respective countries, the educational and social terrain addressed in their educational interventions, and the instances of boundary-crossing that may be emerging. Particular attention will be given to the distinction between academic feminism (emanating from university settings) and popular feminism (growing from the claims made by poor women) and how these approaches affect the objectives, content and pedagogic strategies of educational interventions. While academic feminism comes in multiple variants (contrast, for instance, my position with that of Lather—see her chapter in this volume), it is still accurate to differentiate it from representations that predominate in other settings, particularly international bureaucracies.

Postmodernism and Mapping

Some provisos may be in order before proceeding to mapping gendered spaces. Social cartography is seen in some circles as an expression of postmodern thought in that it facilitates the creation of smaller and more contextualized narratives, thus enabling the emergence of new voices and perceptions. While it is true that, in looking at spaces occupied by diverse entities, their perspectives are more readily recognized than if we were to employ a metanarrative with high levels of aggregation, mapping does not have to be mutually exclusive with the use of a particular set of normative principles that are to be applied to entire societies and even the world at large. For instance, maps can present multiple views regarding democratic rule and we can also map the values of what we might consider democratic rule to encompass.

Postmodernism has made significant contributions to our understanding of knowledge. It has brought forth the importance of multiple perspectives and multivocality, and the fact that the "grand" narratives or general truths we so trusted are really imbued with power and privilege and are often the extension of this power and privilege. Anyon observes that postmodernism has shown us "the importance of the local, the validity of deconstruction, and the centrality of discourse" (1994, p. 118).

I cannot avoid presenting here my personal position. There are some aspects of postmodern feminism I find enlightening: its sensitivity to the multiple voices of "women," the recognition of their multiple perspectives, the questioning of binary categories (white/nonwhite, rich/poor, straight/gay, woman/man). But I am, as are others (Di Stefano, 1990; Anyon, 1994), uncomfortable with the notion that diversity in human nature is so great that we should deal only with highly contextualized categories and avoid the use of "totalizing" and yet narrow categories. Totalizing categories do exist in society and are often used by dominant groups for the purposes of exclusion. To combat this oppression, one must create counterhegemony or alternative views (for an expanded view of this, see the chapter by Bartolovich in this volume). This process inevitably begins by using these categories in order to question them subsequently. It is ironic that in order to seek a degendered society—to make gender no longer serve as a powerful and arbitrary social marker—we have to proceed first by showing how men and women relate to each other. One of the most important topics in feminism today is the question of domestic violence. To address domestic violence, of which wife-beating is the most salient component, one must not only talk about men and women as binary categories, but also give more attention to heterosexual relations.

The denial of all universals is intellectually disarming and politically naive. The most important and most used concepts in the social sciences are by definition encompassing. They embody substantial differences but at the same time have the virtue of offering a simple and ingenious stroke of thought that captures the many distinctions under an effective conceptual umbrella. What would we do without concepts such as democracy? the family? industrialization? the West? poverty? These concepts have not been dropped despite the heterogeneity they comprise. Feminism, even if subject to substantial variation in meaning, is also entitled to function. In postmodernism the concept of difference reigns, but I prefer to work with "different similarities."

Endorsing the validity of multiple voices could be expanded to say that everybody is right. A normative and emancipatory view of society would object to this, arguing that there are discrete, more desirable views of society than others. One negative, certainly unintended, effect of postmodernism is that it fosters the guilt of finding something "wrong," inasmuch as the identification of something unfair or unequal presupposes that there is a universal standard to which all should aspire. To be sure, any encompassing concept of women will conceal different levels of experience and oppression. While nuanced accounts are needed, we should not lose sight of the fact that there are categories we must use. Challenges to the "universal" woman are present in analyses that seek the intersection of class, race, gender and culture. These more complex analyses strengthen rather than weaken the gender construct.

Linking education to a project of social emancipation presupposes knowledge of certain facts and possession of certain visions about gender in society. Not everything can be negotiable or open to different interpretations. In response to a comment that she was trying to present "black feminist thought" as a uniform phenomenon, Collins presented a rejoinder that is equally appropriate to my argument: "The use of difference taken to its logical conclusion leaves us with a group of politically unorganized, unique 'individuals' and firmly entrenches us in liberal democratic politics that allegedly protect the 'individual's' right to be free" (1992, p. 518).

Social mapping can be an effective exercise within a modern, "rationalistic" perspective. By this, I do not adopt the position that considers modernization as the only path to development, but rather one that assumes the existence of some objective facts and inequalities that must be addressed. Within these parameters, I construe social cartography as a visual representation that borrows concepts and metaphors from social geography but operates in a modern, albeit critical, tradition.

A non-postmodern approach does not have to suppress agency. As Anyon argues, "useful theory is that which leads to successful political activity" (1994, p. 129). She further adds that this theory needs to acknowledge the complex narratives that connect larger social structures and daily life, and to seek what sociologists have long called middle-range theories—those that connect local activity to widespread social constraints.

Analytical Considerations

It is a truism that all social interaction takes place in space. But it is not space per se. Shilling, a sociologist, declares that space is no longer seen as an environment in which interaction takes place, but is itself deeply implicated in the production of individual and collective identities and social inequalities (1991). Space simultaneously structures and is structured by individuals.

Social spheres are constituted by different institutional actors. These actors, because of the spaces they occupy and the interests they represent, produce a particular understanding of the world. It is not that their perception happens to be different; rather, the interests they represent lead them to engage in differential perceptions. In the area of international development, this means that each actor has access to and manages a particular set of concepts and perspectives, and propounds a particular understanding of national development and of the forces that cause it

To the metaphor of mapping, I want to add three others—the metaphors of border, borderwork, and border-crossing. These are important because they invite and often challenge us to delineate our positionality as well as that of others. The spaces occupied by the various actors and the relations they keep with each other create specific constellations. Communications tend to be more open among institutions or people who share the same definition of a situation; it is also the case that by engaging in frequent contact with each other, institutions come to share certain meanings.

Borders exist to protect us from the challenge of others as well as to preserve our identities; borders can function as lines of separation but also lines of cooperation, when we cross them, as in this book, to share or gain meaning. Boundaries may be activated as lines of separation by recognition of irreconcilable opposing perspectives in other social actors. Thorne (1993) distinguishes between borderwork—those interactions that strengthen boundaries—and border-crossing—those actions that move us into the other's sphere or space. Borderwork maintains boundaries by activating the distinctions that separate entities or categories of people; in contrast, border-crossing creates opportunities for developing shared values and views. Boundaries may be more fluid in times of analytical or strategic convergence. We become willing to cross borders also when we stand to gain something by doing so. As Thorne (1993) observes, instances of border-crossing are important from a change perspective, because it is in this space that the renegotiation of difference takes place.

The intellectual spaces occupied by institutions also vary according to power. Institutions with large political and financial resources occupy larger cognitive spaces in the sense that it is easier for them to convey their perspectives across a wider social network than it is for small institutions or groups. It is also clear that border-crossings usually occur among the weaker rather than stronger institutions. But even when boundary-crossing occurs, it is often limited among the social actors to getting information that will confirm the differences among the social actors.

Let us examine now the four selected social actors (international development agencies, national governments, women-run NGOs and feminist academics) in the area of gender in educational development. Figure 1 identifies how I see the various actors, and locate them in relation to one another.

INTERNATIONAL DEVELOPMENT AGENCIES

These institutions occupy a terrain characterized mostly by "doing": the design and implementation of projects and programs are their main activities. These activities are predicated on frequent oral and written exchanges be-

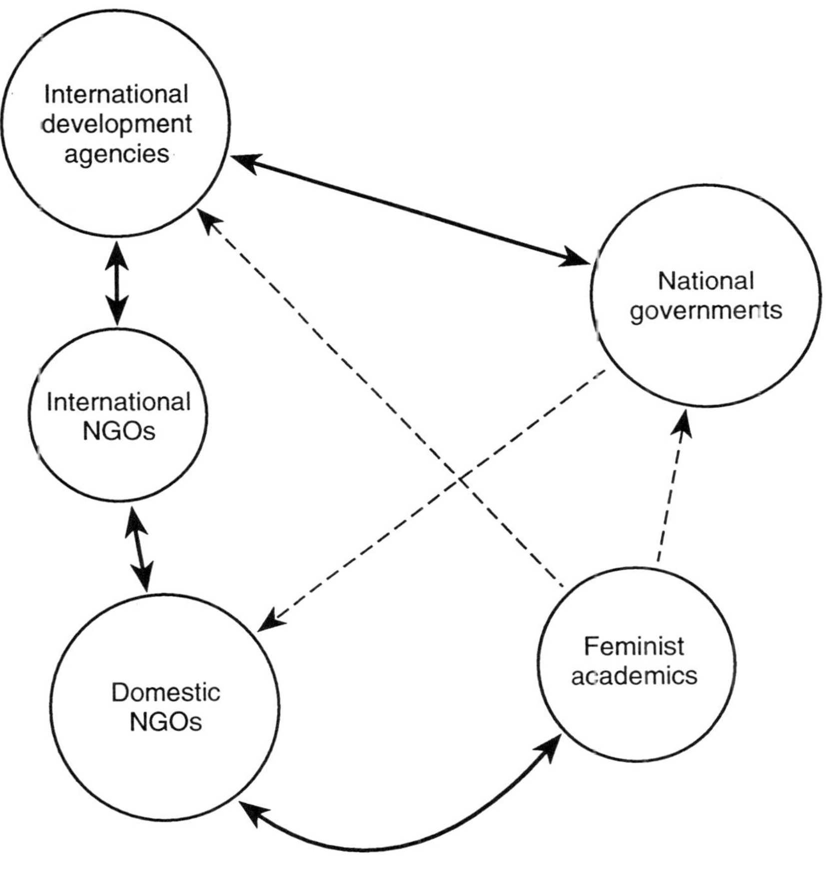

LEGEND: _______ denotes strong border crossing
-------- denotes sporadic border crossing

Figure 1. A map of key actors and border-crossing in international gender work

tween agency staff and their counterparts—staff in ministries of developing countries. This modality of work, in addition to the agencies' linkages with capitalistic and patriarchal structures, leads them to see solutions in highly fragmented forms, usually by governmental sector (e.g., education, health, agriculture, labor).

International agencies comprise an array of multilateral and bilateral development agencies; of course, there are important variations among them. As a whole, they see national development primarily as economic growth, despite new definitions such as the human development index advocated by the United Nations Development Program (UNDP). Gender issues are seen as a collection of random inequalities caused by inefficient decisions. In the view of many agencies, inefficient decisions in the area of education tend to be a segmented array of "erroneous" preferences and practices by parents, employers and schools. The position of international development agencies is essentially atheoretical; their main line is that of "integrating" women into development efforts. Several international agencies are now more sensitive to issues associated with social justice such as democracy (Stromquist, 1994), yet continue to conceptualize it in terms of formal, institutional and elections-derived power—ways that bypass recognition of other forms of women's emergent political power. For instance, USAID is currently enacting an initiative to work on democracy entitled "Women and Local Development Programs." This effort considers the participation of women in both national and municipal elections and governance, but avoids discussing or examining democracy within the household.

International agencies have considered women only when they are very poor, rural or marginally urban. A strong example of this is reflected in the World Bank's policy paper on gender, entitled *Enhancing Women's Participation in Economic Development*. Although the foreword of this document makes reference to social justice, it soon sees the gender problem as one of "reducing poverty." Attention to women is justified as "investments" likely to promote economic growth, promote efficiency, reduce poverty, help future generations and promote sustainable development (World Bank, 1994, pp. 22–28). The reasons behind women's low status and how this situation benefits men are not addressed. Thus, most attention given to women treats them as dependent "vulnerable groups" whose basic needs are confined to the domestic sphere.

In the area of education, agencies tend to assume that demand and supply of schooling are forces easily malleable, that educational decisions are based primarily on knowledge of the rate of return to schooling for boys and girls and that preferences stated by individuals are essentially the result of personal choice (as opposed to being the consequence of oppressive socialization experiences). Within international aid agencies, the role of economics itself as the outcome of power differentials is not considered (Kardam, 1991).

The largest multilaterals, such as the World Bank and UNDP, influence the way gender is defined in development work. Smaller but specialized agencies such as the United Nations Development Fund for Women (UNIFEM) and the United Nations International Research and Training Institute for the Advancement of Women (INSTRAW) exist exclusively to address women's issues in the third world, but their resources are extremely small (about $6 million in 1994 for both these agencies together). Among the bilaterals there is substantial variety in their dealing with gender issues, though one can observe an improvement over time in terms of their creation of organizational units and procedures to address women's issues (Stromquist, 1986, 1994). In a few agencies, this improvement is also reflected in that they now consider women's autonomy and empowerment, but even in these agencies there is still a marked disparity between the discourse in their official documents and the work they actually support.

International development agencies do not see the need for women's emancipation or liberation or for combating ideological forces that sustain women's subordination. Therefore, programs supported by international aid agencies, including agencies highly committed to the advancement of women, such as the United Nations Children's Fund (UNICEF) do not work, as a rule, with the ideological content of women's conditions. At most, they offer women training for specific tasks based on their traditional role; that is, being responsible for reproduction and for the care of home and children (IBAM/UNICEF, 1991, p. 71). Some exceptions exist among donor agencies. The Scandinavian and Dutch agencies are notable for their firm proactive stand on gender issues and for affirming the concept of autonomy. Their work is valuable and often transformative. Unfortunately, the support they can provide is very small compared to the needs in developing countries.

Key agencies such as World Bank and the International Monetary Fund (IMF) are not sufficiently sensitive to the social impacts of structural adjustment programs (SAPs). And while the World Bank has become more sensitive to the impact of SAPs on women, several observers note that this attention is limited to women's role as reproducers. UNICEF is very sensitive to SAPs (as demonstrated by its various writings on the topic) but, unlike the World Bank and IMF, it does not have access to ministries of finance or economic planning, which make final decisions on national budgets after the enforcement of SAPs (Chinery-Hess et al., 1989).

Institutionalized spaces not only develop their own discourse but purposefully avoid others. Within international aid agencies some words have a particularly taboo status. These words, interestingly, are central to feminist analysis: patriarchy, oppression, exploitation, ideology, subordination, power differences. Discourse used within official discourse today refers to economic competitiveness, decentralization, vulnerable groups, safety nets and sustainable development. Women are mentioned, but the social relations of gender are not. Although primary attention is given to mothers in their social reproduction roles, there is also some consideration today, given the widespread economic crisis, of their productive functions. In this case, attention to women is justified as an efficient strategy to increase output and maximize resources. Moser (1989) notes that the most popular approach to women among agencies emphasizes efficiency, defined as the increase of women in the labor force under the "assumption that economic participation increases women's status and is associated with equity" (p. 1813). In both cases, when agency attention is paid to women's role as reproducers or to women's role as producers,[3] there tends to be an objectivization of women.

International development agencies have as their main interlocutor the recipient states. While there is much contact with governmental agencies, they do not talk to other important actors of the civil society. Since the discourse, problem definition, and solutions regarding the condition of women in national bureaucracies resembles or is influenced by the international aid agencies, programs and projects evidence a high degree of homogeneity.

Inasmuch as gender issues are usually defined as problems faced by poor women, the activities set in motion seek to enable women to provide better services to their families; thus the offering of skills in nutrition, health, hygiene and family planning. International aid agencies, like the national governments they deal with, do not examine—or rather avoid—the tensions between production and reproduction. The fact that ideological norms create psychological and material obstacles to women are reduced to marginal adjustments to suit "women's needs," such as offering programs at the times women are able to attend (after having fulfilled their conventional household manager tasks). Through educational assistance, aid agencies generally seek to "integrate" women into society by giving them greater levels of education in order to increase their productivity in the marketplace and their efficiency in the home.

Aid agencies conduct sophisticated work, as is evident in the graph presented below (Figure 2), showing the connection of education to health and farm productivity. It will be observed that this configuration allows no

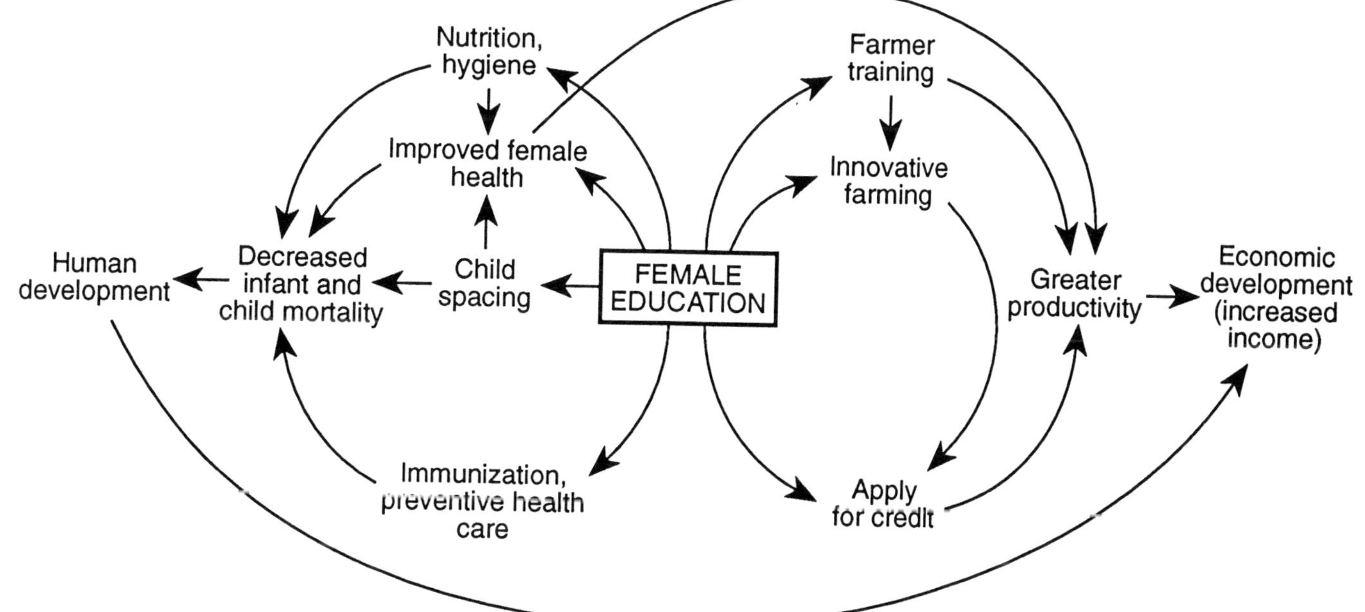

Figure 2. The functional linkages between female education, human development and economic development; an example of a modernization perspective typically held by international development agencies regarding gender and educational development

room for gender ideologies and chooses to see decreased infant and child mortality as a key indicator for human development, assuming that having fewer children and being more certain that the children will live will somewhat decrease the subordination of women in the home and community. It does not consider a woman's health other than as it relates to her being a mother. It sees the problem in agriculture as one of training and credit, not one of land reform. In all, a highly technocratic scheme for women's advancement is promulgated.

Donor agencies have the money and thus the influence. They do not engage in border-crossing to apprise themselves of different or opposing positions. While they engage in much border-crossing with national governments, they cross boundaries to visit NGOs only where there is a need to rely on them as implementers of selected projects, as NGOs tend to have a comparative advantage in terms of outreach to and acceptance by marginal groups. Some multilateral organizations, mostly UNICEF and UNIFEM, maintain vigorous contact with NGOs. This border-crossing can be explained in terms of their explicit mandate to work with women and the recognition that some very creative and efficient work occurs in NGOs.

Major international initiatives such as Education for All (held in Jomtien, Norway in 1990), the *Social Summit for Development* (held in Copenhagen in 1995), the *Fourth World Conference on Women* (held in Beijing in 1995), have been and will be realized with the support of the United Nations, which in turn depends mainly on major industrialized countries for contributions of international aid. These events are interesting manifestations of symbolic border-crossing, moments at which governmental delegations meet with NGO representatives and try to respond to their demands. I consider these events symbolic border-crossing because while there is an increasing tendency among developing nations to accept the petitions and recommendations presented by the NGOs, there continues to be a very weak implementation of these requests.

National Governments

Developing countries present great heterogeneity in respect to each other. As Sengupta remarks, "a typical developing country is a combination of different sectors at different levels of development with different types of problems" (1993, p. 9). Despite variation in levels of industrialization, rigidity of cultural norms and the nature of women's status, states in the developing regions are characterized by a definition of gender that calls for protection of low-income women against disease and, to a lesser degree, poverty. Development policies of states in the third world today emphasize the promotion of export-oriented industries and give scant consideration to subsistence agriculture and the informal sector of the economy, both of which comprise large numbers of women. Again, programs offered to these women only rarely seek to provide them with a critical understanding of their reality or to enable them to move into a higher socioeconomic level via access to land, resources and capital.

Much of the national governments' policies toward women have been shaped by declarations endorsed at the world-level meetings sponsored by the UN. The UN, it must be noted, has been a target of feminists from the North, through whose pressure this organization enacted the International Women's Year (1970) and, subsequently, convened the various world conferences on women (Mexico, Copenhagen and Nairobi). Over the past decade, many of the national governments—in response to agreements made at the Nairobi conference at the end of the UN Decade for Women (1985)—have established machineries to address women's issues, usually known as WID Units. Except for those in the Caribbean and a few in Asia and Latin America, WID Units have been weak in promoting change, partly as a consequence of limited funds but also partly as a function of the absence of personnel trained in gender issues in governmental ministries and related agencies. WID Units have operated mostly with the notion that women should be "integrated in development."

The boundary-crossing of national governments reciprocates that of international aid agencies, and their discourse closely resembles that of these agencies. There is limited dialogue between civil servants and women working in either NGOs or women's groups. A survey of fifty-eight national machineries revealed that these exchanges are very infrequent (Stromquist, 1996). They occur mainly when the government wishes to count on access to hard-to-reach populations or wishes to engage in practices that are labor intensive and can be done more cheaply outside the government.

National governments' work on women's education emphasizes mostly schooling, and, in particular, primary education. The main educational problem affecting girls is defined as one of access. In countries where enrollment levels are high, the problem is identified as one of completion. Governmental initiatives seek to improve access and permanence in the system through the creation of more schools, the limited provision of scholarships or an array of incremental adjustment measures such as those dealing with class schedules, the provision of more women teachers or the setting of lower scores for junior high and senior high school admission (the latter a feature exclusive to African countries).

Governments operate on a narrow definition of women's problems in society and also avoid discussion of power imbalances. Their work on adult women covers small numbers and concentrates on increasing women's efficiency in running the household and raising children. There has been an increase in the number of educational projects that address the productive roles of women, especially through income-generating projects; again, these are limited in number and usually funded at levels that do not permit reasonable success. Moser (1989), after reviewing the approaches to women by international aid agencies and the governments they deal with, concludes that these efforts address practical rather than strategic needs of women (this is discussed further below).

Some of the literature on social change argues that it is more accurate to think of the state as a contested terrain, where opposing demands are resolved and where new understandings can be imposed. Our mapping exercise in Figure 1 indicates limited contact between feminists and the state and between NGOs and the state in developing countries. The full extent of contestation in terms of shaping policies and receiving national resources is yet to be realized.

WOMEN-RUN NGOS

Within the set of actors mapped in Figure 1, I consider both NGOs in the developed countries (usually called international NGOs) and those in the developing countries (the domestic NGOs). It has become clear that the strongest advocates for women's advancement are women themselves. The number of women-run NGOs has increased exponentially in the past decade. In Latin America, where women NGOs are particularly active, historian Francisca Miller considers that feminism has become a "primary vehicle for social criticism," both in liberation movements and in redemocratization efforts in Brazil, Argentina, Chile, Paraguay and Uruguay (1991, p. 187). One informed account estimates that by 1992 there were some 12,000 women NGOs in Chile. In Argentina, which had a relatively late feminist development, the number is estimated at about 200.

The terrain of NGOs is also characterized by doing: workshops, advice on immediate problems, mobilization, advocacy. NGOs seek to respond to critical problems of women's everyday lives and focus preferentially on low-income urban and rural women. On the other hand, many of them work with concepts such as patriarchy, which is usually defined as a hierarchical system that considers women inferior to men and thus exploits and oppresses women. Miller notes that feminists in Latin America frequently use the word "vindication" (*reivindicacion* in Spanish), thereby claiming not merely new rights or rights equal to men, as those demanded by US or European feminists, but

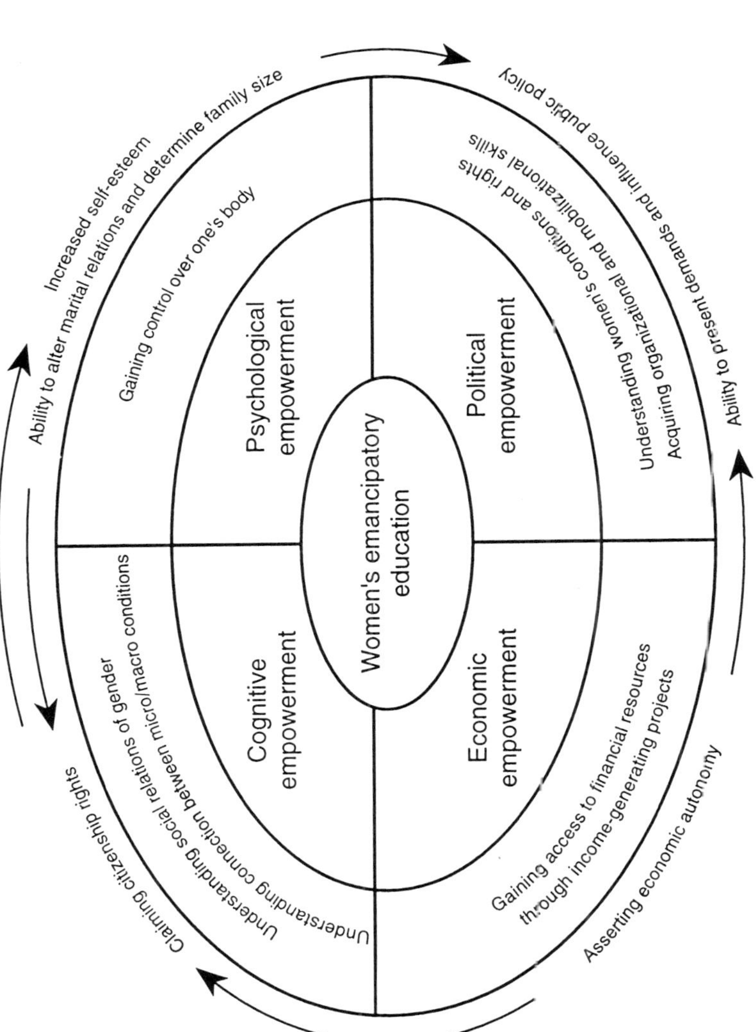

Figure 3. A conceptualization of education from an empowerment perspective

also inherent rights (1991). The discourse of women-centered NGOs makes reference to patriarchy, mobilization, civil society, social movements, power bases, social change/transformation and structures of domination. Figure 3 maps the conceptualization of empowerment in the design of education for women; contrast this with the formulation presented in Figure 2.

In the past the boundary-crossing between international and domestic NGOs tended to center on the provision of funds from the North to the South, with women of the South generally being seen as the "Other" (Samarasinghe, 1994). Recently, there is a greater sharing of cognitive spaces. Feminism in developing countries is much more linked to poverty and survival issues than in the North. This can be seen in the top two NGO priorities discussed at the World Conference on Women in Beijing in September 1995: (1) elimination of conflict, such as domestic violence, violence in the streets, and civil wars; (2) access to basic needs, such as clean water, food, housing and health.[4] But there has been a gradual recognition that some problems are as critical in the North and in the South. For instance, spouse violence, which has been a persistent major issue among third world feminists, has been identified as the top gender issue in the US for 1995.

In developing countries, a new form of feminism is emerging: popular feminism. This movement is increasingly targeting the state for many of its demands. While poor women are concerned with the improvement of neighborhood and families (food, sewage, transport, water, day-care services, schools, health posts), this concern is expanding into improvement of women's status, women's rights and violence in the family (Schild, 1994; Barrig, 1989). NGOs, unlike national governments, are much more likely to engage in compound interventions that consider themes such as domestic violence, women's control over their own bodies, literacy, sex education and legal rights in close association with one another. They also seek to promote gender awareness and empowerment. Many women NGOs see the importance of organizing and mobilizing women to change; many also tend to work explicitly on the transformation of society. Their actions take place under conditions of hardship, yet they engage in significant aspects of social reconstruction. The empowerment emphasis is clearly exemplified in the Self-Employed Women's Association (SEWA) of India, which has even developed its own banking system. In contrast to the perspective of national governments, women NGOs defend the right to treat women as citizens, individuals with their own worth, independent of their capacity as mothers and wives. Several observers maintain that feminism, by paying attention to social justice and representative democracy, is forging a new conception of citizenship and politics, powerfully reflected in the slogan "Democracy in the home and in the country," which was effectively used by feminists in the movement for return to democracy in Chile.[5]

Much of the financial support for domestic NGOs comes through the international NGOs. With the current attacks on the welfare state taking place in many industrialized countries, forces sympathetic to the plight of women in the third world are now having to fight for their own rights as well. In the US, this is manifested in the defense of women's right to abortions, which from a feminist perspective is seen as part of the fundamental right of women to control their lives.

Since popular feminism centers on the conditions of poor women in society, it offers clear instances of the intersection between gender and class. Popular feminism incorporates social class de facto by involving feminists from middle social classes in coordinating capacities within women NGOs.[6] Poor women in survival activities such as soup kitchens, vegetable gardens and wholesale buying activities have also joined hands with feminist work done by middle-class and professional women.

It has been noted that women NGOs are placing increasing demands on the state, which suggest greater border-crossing by NGOs than by governmental bureaucrats. A key factor in this recent behavior is the feminist concern regarding the impact of structural adjustment programs (SAPs) on women, particularly on the increase in the labor of poor women. Women-run NGOs reject the pre-eminence of the market as the primary mechanism for resource allocation, they seek people-centered sustainable development, they demand access to productive land and they want to empower people to strengthen their own capacities. These demands on behalf of women, which go well beyond the proposals coming from governments, are reflected in the amendments to the *Platform for Action* (many of which were approved by governments in Beijing) made by the Women's Caucus, an umbrella group gathering various international women's networks from some fifty-two countries (The Women's Caucus, 1994). Women NGOs are also demanding that governments reject World Bank and International Monetary Fund (IMF) prescriptions for economic recovery and that they refuse to pay the external debt that so drains the resources of developing nations.

While government-NGO interaction has increased, there is still considerable distance between NGO and state activity, compared to the contacts that NGOs have with each other and with academic feminists. An exceptional event in this connection is President Chamorro's invitation to feminists in Nicaragua (Grimones, 1992) to negotiate with the Norwegian mission for international cooperation—a decision that may, however, have been prompted by Norwegian leverage. In a way, the spaces occupied by women through their NGOs have had to be on the margins to enable women to define themselves and strengthen their organizations while avoiding dealing with hostile institutions.

International women's gatherings such as those in Mexico, Copenhagen, Nairobi and Cairo in 1994, have contributed to the creation of numerous cross-national informal networks. At the 1995 meeting in Beijing, with an attendance of some 25,000 members, NGOs provided an exceptional space for learning and exchanging views.

NGOs are emerging as powerful actors in gender work in developing countries. Most of their efforts have centered on adult women. Yet schooling is receiving more attention, as NGOs push for new boundaries in terms of curriculum content, methodologies of instruction and purpose of education, in favor of an education that is liberating and multidisciplinary.

These groups operate through networks that combine other domestic and international NGOs. They get their information through newsletters and journals on women; two worldwide feminist media networks, represented by the International Tribune Center (based in New York) and ISIS International (based in Rome with a regional office in Chile), play a crucial role in the dissemination of feminist information. Women NGOs also manage to send representatives to national, regional and international meetings, which often include among the participants women from some of the most disadvantaged social classes, such as the various indigenous and black ethnic groups. In 1992, the Latin American and Caribbean region held its sixth feminist meeting. The first Asian Indigenous Women's Conference took place in 1993. Ready access to cyberspace is enabling them to communicate in unhampered and constant ways. Today there is a proliferation of feminist and women's journals and networks. The year 1994 saw the sixth international fair of feminist books, in Australia. Since English is the international language of scientific exchange, many smaller NGOs do not benefit from the expanded literature. On the other hand, strong regional networks are being produced and substantial exchange occurs through informal means. With increased access to fax and e-mail, some of the national NGO groups enjoy a fast and vibrant degree of communication with one another.

NGO networks operate somewhat distant from feminist academic spaces. As a result, some of the rich empirical work of activists has not been properly recognized in academia nor has it always been helped by the academics. The Women's Caucus, an international NGO, reflects the perspective of the NGOs regarding their allies. They are identified as including "women leaders, grassroots activists, analytic thinkers and writers" (The Women's Caucus, 1994, p. 35).

In discussing this group, I consider not only the university-trained women working in universities, but also those working in research centers and institutions, a phenomenon quite important in developing countries, especially in those that saw their universities decimated during dictatorial regimes such as in certain Latin American countries. The reader will note that these actors are identified as persons rather than by their location in institutions. This is due to the fact that gender work in universities can be located in women's studies programs, in progressive departments, and among individuals, but it is not influential enough to shape entire institutions. Feminism exists only in pockets within universities. While feminist intellectuals working within autonomous research institutions tend to face a friendlier environment in those settings, they also represent only a small element of the professional staff.

The academic world is a terrain of much reading, writing and talking, a place where knowledge is both consumed and produced. Activities revolve around programs of study, courses, students, research. Academics inhabit a world of writing, engaging largely in written production, based either on primary work or on synthesis and critique of others' writing.

The distance that academia places itself from the action may lead to naive or impractical assertions. The positive effect of distance, however, is that it permits reflection and thus facilitates the production of systematic knowledge and theoretical work. Academia's autonomy—whether in industrialized or developing countries—enables women to produce research that can be critical of policies and actions by state and international aid agencies. Ironically, feminist research is still questioned in academia, as some peers—notably those in the "hard" sciences—see feminist work as unnecessary or nonscientific. The discourse in feminist work in academia includes gender ideologies, patriarchy, the gender/class/race interaction, postmodern feminism (an example of which is Lather's chapter in this volume), nonsexist education, the feminization of poverty and the impact of globalization on the sexual division of labor.

The contribution of feminist academics to work in gender and development has been consequential. Academic voices were the first to be critical of WID strategies, noting that integration into a capitalist mode of production would result in a greater exploitation of women in their productive and reproductive roles (Beneria and Sen, 1981; Sen and Grown, 1985). Academics such as Molyneux (1985) have introduced the concepts of "practical and strategic needs," bringing attention to the contrast between efforts to satisfy immediate basic needs and work that attempts to produce macrostructural changes. Other feminists (Schmukler, 1994; Stromquist, 1992b) have noted that practical needs are not to be underestimated because they can serve as the stepping stones to more transcendental work. Schrijvers (1991) has made important contributions through her detailed discussion of the concept of "autonomy" and Stromquist (1992a), synthesizing the work of various Latin American women NGOs, has addressed the concept of "empowerment." Feminist academics have also used their skills to examine the limits of policies and to separate rhetoric from actual accomplishments (see, for instance, Feijoo, 1992). It is the work of academics, moreover, that has shown how macroeconomic policy, particularly that represented in SAPs, does not take into account human costs but rather market costs (for a more detailed examination of this work, see UNICEF, 1987). A characteristic of feminist academic efforts has been that of borderwork, as they have repeatedly questioned the assumptions made by governments and the solutions attempted by them.

Feminist academics engage in much border-crossing, particularly those who seek to keep abreast of developments regarding gender issues. Important forms of border-crossing with international aid agencies and governments occur through consultancy work, where the internal space of otherwise close institutions may be revealed, if only momentarily. Academics do border-crossing also by investigating the agencies' work, something that is less feasible (for recent research studies of agency work in women and basic education, see Stromquist, 1986, 1994). For many an academic, border-crossing into NGO spaces is a source of inspiration and optimism. Through informal networks of friends and colleagues working in NGOs, through access to their documents, newsletters and reports, and ultimately by visiting the NGOs, knowledge is gained about the realities of the intersection of gender and class, the intersection of gender and race, the feasibility of empowerment programs and feminist work under adverse economic and political conditions.

Regarding education, feminist work has also initiated a critique of existing liberatory popular education in Latin America. Rosero charges that popular education still does not explicitly address gender issues; that is, it does not consider fully production outside and inside the home, and the contradictions of multiple demands upon women. Nor does it strengthen women's organizations so that they become sound interlocutors in decision making and policy making at the global level (Rosero, 1993, p. 80).

Reflections on Mapping as a Methodology

Mapping, by forcing us to think in terms of discrete entities occupying specific spaces, makes us aware of the spaces we inhabit and the positions we take relative to others. In this way, maps help to capture the intersection between our personal and public worlds. They make transparent our dif-

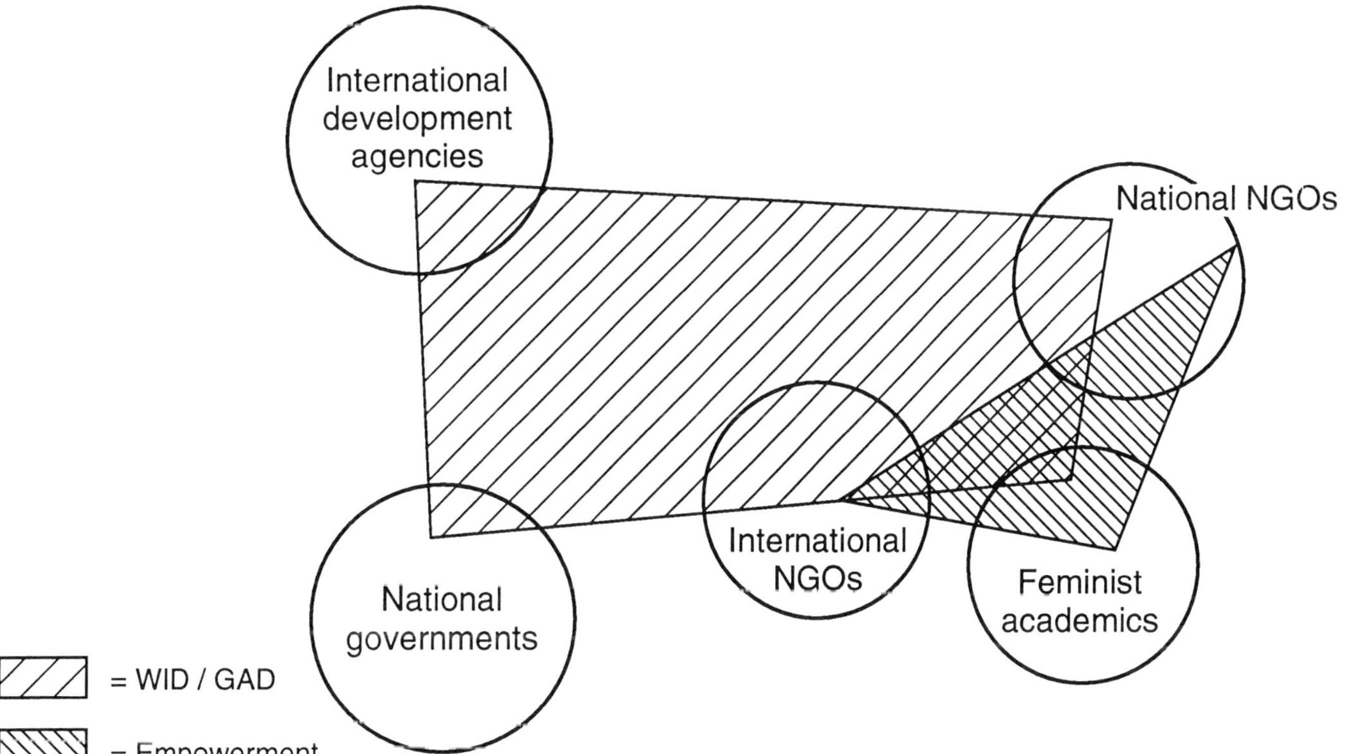

Figure 4. *Mapping conceptual proximity and remoteness among key actors in gender issues*

ferences and invite discussion of how differences emerge and are maintained. Social cartography enables us to identify discrete networks and the degree of contact between them. Figure 4, for instance, attempts to present a visual representation of how the gender cognitive space, constituted by WID/GAD versus empowerment approaches, is shared by the selected actors.

Through the examination of border-crossing we gain insights into the degree of convergence being shaped among actors that keep close contact with each other. Mapping is useful to identify moments at which these spaces are crossed, how they are crossed, or why they are seldom crossed. At a more fundamental level, mapping can facilitate the identification of sources antagonistic and favorable to a particular social movement.

Maps, however, cannot show multiple dimensions in a given space and thus, while enabling us to perceive location, cognitive space and distance, they cannot be easily used to reflect border-crossing and borderwork. This chapter has explored how a mapping methodology can be utilized to enhance the understanding of feminism as a new social movement. Mapping can be used to invite others to present their own views of this physical and conceptual space (Paulston and Liebman, 1994). Thus, mapping may also be used for remapping and for achieving increasingly accurate understandings of our changing social world.

Conclusions

Gender work occurs in terrains that engage in distinct discourse and alliances. Affinities between international aid agencies and national governments lead them into a discrete constellation. Likewise, similarities between feminist work in NGOs and in academic settings create a second constellation, even though much weaker than the first. The constellations do not emerge because the respective sets of actors hold competing knowledge claims, as has been observed in other social mapping exercises; rather, the actors hold different political understandings of a problem that involves substantial power ramifications—that of the condition of women in society.

NGOs have emerged as new spaces for collective action and their strength is increasing. Their border-crossing is greater among each other than it is with other actors. In this case, border-crossing is not occurring to renegotiate differences, as Thorne detected in her study of gender in schools (1993), but rather to develop consensus across the various groups in the women's movement. The fluidity of border-crossing among NGOs is fostering a congruence of definitions and positions that is likely to benefit the feminist movement. Another important feature of this border-crossing on gender work is that it goes beyond and overrides the nation-state and is truly transnational in nature. In being so, it signals the opening of a new world space with changing political frontiers and potentials.

Border-crossing also means entering a different cultural space and a different cognitive space. This is the kind of activity feminist academics and a few NGOs do with the state and some international aid agencies. On the other hand, borderwork—necessary to maintain one's sense of identity—is limited mostly to feminist academics. This is an area where much more work needs to be done in order to defend the inroads made so far and to politicize an area that must be seen from a political perspective.

NOTES

1. For a distinction between WID and GAD, see Moser (1989).

2. I am using the term intervention to denote a self-conscious action with specific objectives in mind that, in seeking to create change, alters the regular features (however temporarily) of a given situation.

3. I say "some attention" advisedly, because the amounts of funds devoted to productive projects for women are frequently extremely small.

4. In contrast, the Beijing agenda identified by multinational agencies and national governments (and being shaped through UN-coordinated experts and regional meetings) is much more diffuse. It includes poverty, education, health, violence against women, effects of armed conflict, economic structure and policies, inequality of men and women in decision making, gender equality, women's human rights, media and the environment.

5. For a discussion of the concept of autonomy among women NGOs in Latin America, see the book edited by Barrig and Wehkamp (1994).

6. Another instance of the growing alliance between feminists of distinct social classes is the creation of the National Women's Network in May 1990 in Peru. This network represents a very heterogeneous group, comprising women in mothers' clubs, popular dining halls, mothers in the Milk Glass Program, labor unions, peasant federations, political parties, the Catholic Church and universities.

REFERENCES

Anyon, J. (1994). The retreat of Marxism and socialist feminism: Postmodern and poststructural theories in education. *Curriculum Inquiry,* 24(2), 115–133.

Barrig, M. (1989). The difficult equilibrium between bread and roses: Women's organizations and the transition from dictatorship to democracy in Peru. In J. Jaquette (ed.), *The women's movement in Latin America* (62–81). Boston: Unwin Heyman.

Barrig, M., and A. Wehkamp, (eds.). (1994). *Sin morir en el intento: Experiencias de planificacion de genero en el desarrollo.* The Hague: NOVIB.

Beneria, L., and G. Sen. (1981). Accumulation, reproduction, and women's role in economic development: Boserup revisited. *Signs,* 7(2), 279–298.

Chinery-Hess, H., et al. (1989). *Engendering adjustment for the 1990s.* London: Commonwealth Secretariat.

Collins, P. (1992). Reply. *Gender and society* 6(3), 517–519.

Di Stefano, C. (1990). Dilemmas of difference: Feminism, modernity, and postmodernism. In L. Nicholson (ed.), *Feminism and postmodernism* (63–82). New York: Routledge.

Feijoo, M.C. (1992). *Women confronting crisis: Two case studies in the greater Buenos Aires area.* Paper presented at the conference on Learning from Latin America:

Women's Struggles for Livelihood. Los Angeles: University of California at Los Angeles.

Giroux, H. (1991). Democracy and the discourse of cultural difference: Towards a politics of border pedagogy. *Journal of Sociology of Education, 12*(4), 501–519.

Grimones, M. (1992). *Women in power in Nicaragua.* Paper presented at the conference on Learning from Latin America: Women's Struggles for Livelihood. Los Angeles: University of California at Los Angeles.

Hanson, S. (1992). Geography and feminism: Worlds in collision? *Annals of the Association of American Geographers, 82*(4), 569–586.

hooks, b. (1984). *Feminist theory: From margin to center.* Boston: South End Press.

IBAM/UNICEF. (1991). *Mulher e politicas publicas.* Rio de Janeiro: Instituto Brasileiro de Administracao Municipal.

Kardam, N. (1991). *Bringing women in: Women's issues in international development programs.* Boulder, CO: Lynne Rienner Publishers.

Miller, F. (1991). *Latin American women and the search for social justice.* Hanover, NH: University Press of New England.

Molyneux, M. (1985). Mobilization without emancipation? Women's interests, state and revolution in Nicaragua. *Feminist Studies, 11*(2), 227–254.

Moser, C. (1989). Gender planning in the third world: Meeting practical and strategic gender needs. *World Development, 17*(11), 1799–1825.

Paulston, R., and M. Liebman. (1994). An invitation to postmodern social cartography. *Comparative Education Review, 38*(2), 215–232.

Rosero, R. (1993). Challenges to the democratization of popular education from the perspective of gender. *Convergence, 26*(1), 73–81.

Samarasinghe, V. (1994). The place of the WID discourse in global feminist analysis: The potential for a reverse flow. In G. Young and B. Dickerson (eds.), *Color, class, and country: Experiences of gender* (218–231). London: Zed.

Schild, V. (1994). Recasting 'popular' movements: Gender and political learning in neighborhood organizations in Chile. *Latin American Perspectives, 21*(2), 59–80.

Schmukler, B. (1994). Maternidad y ciudadania Femenina. In C. Talamante, F. Salinas, and M. de Lourdes Valenzuela (eds.), *Repensar y politizar la maternidad: Un reto de fin de milenio* (51–58). Mexico DF: Grupo de Educacion Popular con Mujeres.

Schrijvers, J. (1991). *Women's autonomy: From research to policy.* Amsterdam: Institute for Development Research, University of Amsterdam. Mimeographed.

Sen, G., and K. Grown. (1985). *Development, crises, and alternative visions: Third world women's perspectives.* Stavanger, Norway: Media-Redaksjonen.

Sengupta, A. (1993). *Aid and development policy in the 1990s.* Helsinki: United Nations University World Institute for Development Economics Research.

Shilling, C. (1991). Social space, gender inequalities and educational differentials. *British Journal of Sociology of Education, 12*(1), 23–44.

Stromquist, N. (1986). *Empowering women through knowledge: Policies and practices in international cooperation in basic education.* Report prepared for UNICEF. Stanford: Stanford University.

Stromquist, N. (1992a). Education for the empowerment of women: Two Latin American experiences. In V. D'Oyley, A. Blunt, and R. Barhnardt (eds.), *Education and development: Lessons learned from the third world* (263–282). Vancouver: Detselig.

Stromquist, N. (1992b). Women's literacy and the quest for empowerment. In J. Claessen and L. van Wesemael-Smit (eds.), *Reading the word and the world: Literacy and education from a gender perspective* (50–70). Oegstgeest, the Netherlands: Vrouwenberaad Ontwikkelingssamenwerking.

Stromquist, N. (1994). *Gender and basic education in international development edu-*

cation. New York: UNICEF.

Stromquist, N. (March 1996). *Joining the state: Addressing gender issues within governmental structures*. Mimeographed. Paper presented at the annual CIES conference, Williamsburg, VA.

Thorne, B. (1993). *Gender play: Girls and boys in school*. New Brunswick, NJ: Rutgers University Press.

UNICEF. (1987). *The invisible adjustment: Poor women and the economic crisis*. Santiago, Chile: The Americas and The Caribbean Regional Office.

The Women's Caucus. (1994). *Draft amendments to the WSSD draft declaration and draft programme of action*. New York: Women's Environment and Development Organization. Mimeographed.

World Bank. (1994). *Enhancing women's participation in economic development*. Washington, DC: World Bank.

Yates, L. (1993). Feminism and Australian state policy: Some questions for the 1990s. In M. Arnot and K. Weiler (eds.), *Feminism and social justice in education: International perspectives* (167–185). London: Falmer.

Mapping the Utopia of Professionalism

The First Carnegie International Survey of the Academic Profession

Esther E. Gottlieb

> *A Map of the world that does not include Utopia is not worth even glancing at, for it leaves out the one country at which Humanity is always landing.*
>
> Oscar Wilde

Preface

> *The geographic lines dividing the world of scholarship are becoming blurred. A global community of academic interests is emerging. Scholars everywhere, while maintaining national distinctions, acknowledge common concerns— not just intellectually but professionally as well. And in the century ahead, three critical issues (student access, governance, and the teaching/research/ service mission) will influence profoundly the shape and vitality of higher learning all around the world.*
>
> E.L. Boyer, P.G. Altbach and M.J. Whitelaw,
> *The Academic Profession: An International Perspective*

The *Chronicle of Higher Education*'s report of the results of the Carnegie International Survey of the Academic Profession included showing the numerical results of one question from the Carnegie questionnaire on a world map. Although only the fourteen countries that participated in the survey are shaded-in on the map, the whole world is present (see Figure 1). This map is significant, because it supplies the image that the categories "worldwide" and "global" mentioned in the epigraph above depend on.

As Lynch observed,

> [an] image is valuable not only in this immediate sense in which it acts as a map for the direction of movement; in a broader sense it can serve as a general frame of reference within which the individual can

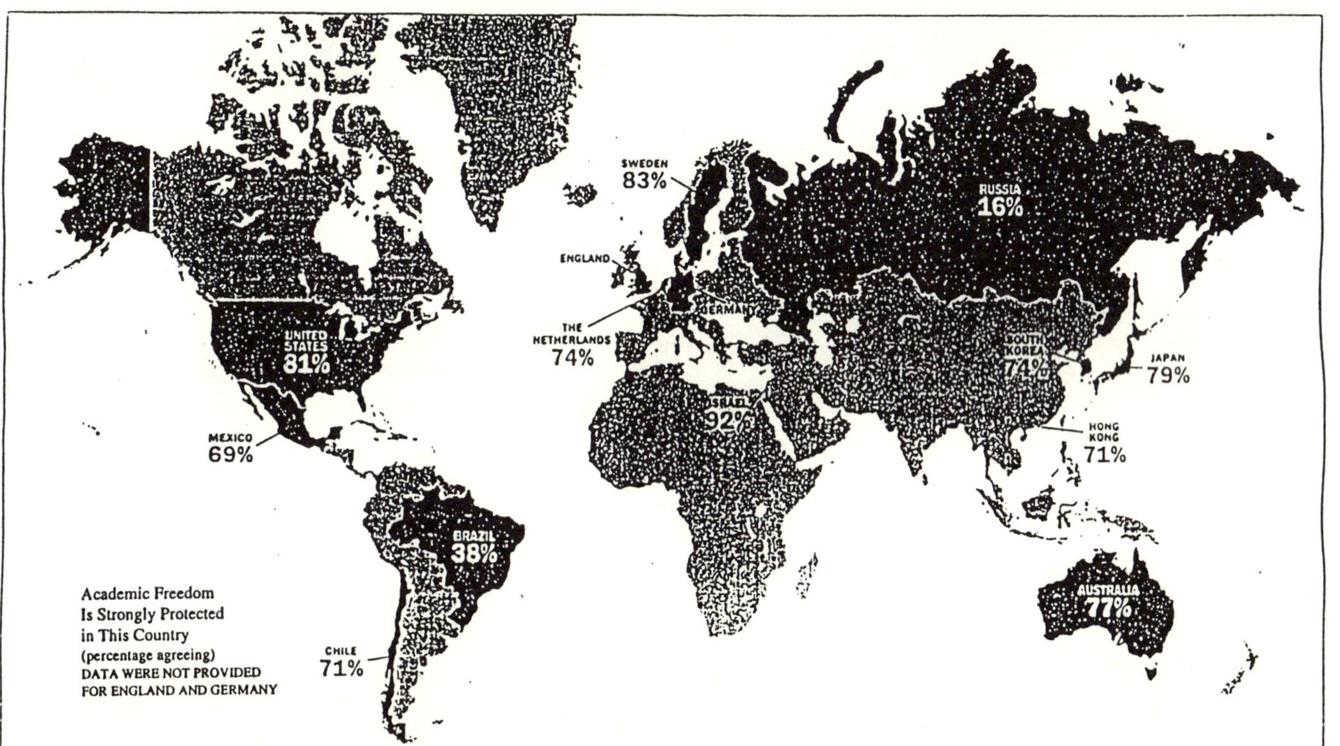

Figure 1. Map extract of questionnaire results from the First Carnegie International Survey of the Academic Profession. From C.J. Mooney (1994). The shared concerns of scholars. Chronicle of Higher Education, June 22. Map by John Gragasin. © 1994, The Carnegie Foundation for the Advancement of Teaching. Reprinted with permission.

act, or to which he can attach his knowledge. In this way it is like a body of belief, or a set of social customs: it is an organizer of facts and possibilities. (1960, pp. 125–126)

In our case a map of the world is the organizer, the frame of reference for results of the International Survey, as is evident from coverage of the survey in the print media. "Scholars across the globe may work under vastly different conditions, but they have much in common": This is the opening sentence of the report on the Carnegie survey in *The Chronicle of Higher Education* (Mooney, 1994). *Higher Education and National Affairs* (ACE) entitled its article "U.S. Higher Education Concerns Are Global, Survey Shows"; similarly, *The Washington Post*'s subtitle reads, "Study Gives Voice to Professors' Global Discontent," and refers to "professors around the world" who "rued the fact that despite their knowledge they were not among their countries' most influential opinion leaders" (Jorden, 1994). According to the *New York Times* report on the survey, "Today professors all around the world support the idea of wider engagement in real life issues" (Honan, 1994). These examples attest to the success of the Carnegie Foundation in convincing its audience that the International Survey represents global perspectives.

How are such qualifiers as "the academy worldwide" and "faculty around the world" (Boyer, Altbach and Whitelaw, 1994, pp. 1, 14) sustained? Do they depend on how many countries participated in an international survey? "International" according to Webster means "between or among nations"; that is, involving more than one nation, whereas "worldwide" or "global" mean "relating to the earth as a whole" or "extending throughout the world." How did these terms become conflated? Does the representation of the world by an international survey depend on how many countries participated, in which particular countries the data was collected or that at least one country from each continent participated? From the above examples we know that the answer does not lie in coverage or number; rather it is a matter of rhetoric—the techniques of persuasion. How does the International Survey construct the image of a "worldwide" and "global" study? Just using "international" and "around the world" interchangeably in reporting the results does not by itself effect a discursive conversion of one term into the other.

Since it is true that this preface to the reader was written after the fact I could at this point outline the process of uncovering how the image of the "world professoriate" was produced, but then it would turn into a deconstruction of my own construction, which is another paper for another time. As my title declares, I have to show below how the Carnegie study constructs the professoriate as the utopia of worldwide professionalism, and I do this by deconstructing the survey instrument and its various versions.

In order to proceed in good faith, I need to open with a statement of my own "subject position." I admit to having "native" knowledge of the subject. Being one of the principal investigators of the survey in Israel makes me an active participant in the Carnegie survey process, rather than an outside observer. Yet I alone am responsible for this version of the story. As Richard Harvey Brown (1987) has argued, if we took rhetoric seriously as a context-giving tradition, "we would become more aware that there are alternative ways of truth telling, and that we are therefore responsible for the form we use to tell our truths" (p. 3).

The methodology used in this study overlaps with that used currently in other social sciences investigating discursive construction of social reality (Bazerman and Paradis, 1991; Schwichtenberg, 1993), disciplinary ways of knowing and knowledge producing (Gottlieb, 1989; Messer-Davidow, Shumway and Sylvan, 1993; Smoodin, 1994) and ways of representation in spatial/ geographical terms (Shohat and Stam, 1994). This means adhering explicitly to a constructivist epistemology (Goodman, 1978; Rorty, 1982) whose consequences for education are that "objects" such as educational policies, plans and questionnaires are understood to be constructed, not "given" or "found," and the issue of how such objects are constructed, in particular the attention to the form in which discourse is constructed, becomes crucial.[1]

Such an explicit undertaking will depart from previous attempts to engage educational comparativists in methodological debates (e.g., Adams, 1990; Masemann, 1990; Rust, 1991; Welch, 1991; Paulston and Liebman, 1994) by actually using interdisciplinary traffic between disciplines such as art, architecture, literature, history, anthropology, cultural studies and social geography to analyze a specific practice in education. Doing this will also mean giving in to postmodernism's "seductive example of extremity," and endorsing its effect of "radicalizing, consolidating, and promoting alternatives for practice in the ongoing internal critiques of disciplinary traditions" (Marcus, 1994, p. 385). The attention to the "literary" or "rhetorical" features of high-profile planning and advisory documents or carefully crafted "international" questionnaires in no way undermines their scholarly credibility and status. Running the International Survey through a discursive and textual analysis has nothing to do with the intentions or actions of the actors involved.

In 1991 the Carnegie Foundation for the Advancement of Teaching initiated its International Survey of the Academic Profession. For several decades, Carnegie has surveyed US faculty in institutions of higher education, publishing the results, financing research, sponsoring commissions and publishing reports and manifestos. The survey under study here is the first international survey of the academic profession undertaken by a US foundation. The data were collected in fifteen participating countries, namely Australia, Brazil, Chile, West Germany, Egypt, South Korea, Hong Kong, Israel, Japan, Mexico, the Netherlands, Russia, Sweden, the UK and the US. Each country had a director or team who worked directly with Carnegie. Carnegie provided some funding for the collection of data, but additional local funding was needed in most cases. The survey resulted in the collection of 19, 290 questionnaires. Respondents were a sample from each country's teaching faculty at institutions of higher education.

The directors were invited to two conferences in Princeton, New Jersey. In the first conference, directors presented a background paper on higher education in their countries (June 17–19, 1991), and worked on the survey instrument and data collection procedures. By the time of the second conference, most countries had already collected the data, and directors presented preliminary statistical analysis of the results, as well as negotiated the final format of the whole data set (April 7–9, 1993). Carnegie handled the data analysis and supplied each country with detailed tabulations of their raw data as well as preliminary aggregated means for all countries on the seventy-two questions (over 200 items). The directors were supplied with copies of the data from the fourteen countries collected to date (Egypt's data unavailable at press time). This data set has not yet been made public. On June 20, 1994, the highlights of the results were officially released by *Carnegie News*, followed by the publication of a special report by Boyer, Altbach and Whitelaw (1994). Carnegie will also publish a book with case studies, with a chapter on each country presenting the results and analysis by the director(s) of the survey. This volume is edited by Philip Altbach (in press), the consultant of the survey.

Constructing the Questionnaire

The questionnaire was negotiated between the respective national study directors of the twelve original participating countries (Brazil, Chile, West Germany, Egypt, South Korea, Israel, Japan, Mexico, the Netherlands, Russia, Sweden, the UK and the US; Australia and Hong Kong joined later) and the Carnegie Foundation. Initially Carnegie prepared a list of 142 items on fourteen aspects of faculty work life,[2] compiled from a number of sources: 1) the Carnegie National Survey of US faculty, 1989; 2) a 1989 survey of university academics in the United Kingdom; and 3) items suggested by the country director(s).

This list of items was circulated to all involved. This version underwent consolidation into four parts under twelve different subheadings (two fewer than the original list) to become the first draft of a questionnaire.[3] In June 1991, the directors met in Princeton for the first conference, which focused on refining the draft of the proposed survey questionnaire, yet conflicting views of what should be included and what excluded remained unresolved. One of the methodological questions involved what percentage of the questionnaire would have to be the same across all countries in order to keep this a comparative study of the professoriate. The strongest opposition to the first (May 22, 1991) and the second drafts (October 22, 1991) came from the European countries, who urged dropping questions that were too "US-oriented," and adding items that in their opinion better reflected the local context.[4] Carnegie did not give up its idea of a scientific instrument, and continued to insist that "asking the same questions will make this study more comparable." This prolonged the process of finalizing the international version for almost another year of negotiations. The results were that the (then) West Germany, Netherlands, and UK questionnaires kept at least 56% of the international version intact in their local versions. (South Korea, meanwhile, went ahead and collected 1000 full questionnaires using the nonfinal version (October 22, 1991), and repeated the process later using the final version; it would be interesting to compare the two sets of data.)

The final twelve-page international questionnaire included over 200 questions covering a variety of topics, under six subheadings.[5] In addition to demographic data such as gender, age, degrees, specialization and academic preparation, it included questions on academic career patterns, work loads and allocation of time, attitudes toward teaching, research, service and governance, and a range of higher education and society issues, as well as international dimensions of academic life. As the special report on the study notes:

> A core of the questions was used in every country. However, research directors could omit questions from their own country's survey instrument if the topic was not applicable or not relevant in their country. In some instances, countries added questions to reflect specific national circumstances. (Boyer, Altbach and Whitelaw, 1994, pp. 25–26)[6]

The questionnaire was translated into Arabic, Dutch, German, Hebrew, Japanese, Korean, Portuguese, Russian, and Spanish. Australia, Hong Kong, Sweden, the UK and US used an English version.

Deconstructing the Questionnaire

While the questionnaire we used is based on The Carnegie Foundation's original format for surveys of college and university faculty in the United States, it was significantly modified to reflect the international context of the new study and to focus on new topics identified by the international research group. The very process of designing the questionnaire was itself a revealing exercise, as differences in priorities of the professoriate, and even the meaning of basic concepts, were discovered, debated, and resolved.

E.L. Boyer, P.G. Altbach and M.J. Whitelaw,
The Academic Profession: An International Perspective

What are these different priorities and new concepts, discovered in the process of constructing the international version, that are alluded to in the above text? What is the difference between the preliminary and final versions of the questionnaire, and how were differences between local and international contexts resolved?

Comparing the list of topics from the inventory of items from multiple sources (March 14, 1991) to the first draft of the questionnaire (May 22, 1991) shows consolidation of the fourteen topics into four areas. These four areas are reworded in the final international version as follows:

Who Are the Academics————————>	Personal Inventory
Academics' Work————————————>	Working Conditions
The Academic Career————————>	Professional Activity
Perspectives on Policy————————>	International Dimensions
	of Academic Life
	Higher Education and Society

Implicit objections to these headings can be reconstructed from the changes in the countries' survey versions. For example, the Dutch questionnaire replaced the very first section, "Personal Inventory," with "Your Career Path to the Present and Current Position." Only at the very end of the questionnaire do the Dutch ask (differently) the very first question in the International version:

What is your gender?————> I am: Female___ Male___

Here the Dutch version re-introduces a personal pronoun ("I am")
missing from the international version. The personal pronouns provide, according to Benveniste (1971), basis for expressing subjectivity in language.
Other classes of linguistic items that share the same status as *I* are:

> the indicators of *deixis* demonstratives, adverbs, and adjectives,
> which organize the spatial and temporal relationships around the
> "subject" taken as referent: "this, here, now," and their numerous
> correlatives, "that, yesterday, last year, tomorrow," etc. They have
> in common the feature of being defined only with respect to the
> instance of discourse in which they occur, that is, in dependence
> upon the *I* which is proclaimed in the discourse. (Benveniste, 1971,
> p. 226)

In fact, this is the clue to a key feature of language use in the final version
of the international questionnaire generally. The final international version strives to "de-discursify" the language of the questionnaire; that is,
it systematically deletes the *I* and *you* markers of "discourse" (in
Benveniste's sense), those linguistic forms indicating "person" (p. 225).
It erases markers of personal engagement or positioning in the discourse
(such as the first-person pronoun restored in the Dutch version). An examination of changes in the language of the items from one version to
another reveals this "de-discursification" process. In the original collection of items (March 14, 1991) from past US and UK surveys and items
suggested by country directors, the questions admit expressive personal
language such as:

I get appropriate recognition for the work I do.

I often find myself frustrated by the bureaucratic requirements at this institution.

My job is a source of considerable personal strain.

I find it difficult to work efficiently in my office during normal working hours.

I feel under pressure to do more research than I would like to do.

I feel that my work is hampered by lack of adequate technology.

I am pleased and proud to be a faculty member at this institution.

By contrast, the final version retains very few first-person expressions
(e.g., *If I had to do it again, I would not become an academic*). Even statements that made it into the first draft (May 22, 1991) on a list of items un-

der *How do you feel about each of the following statements?* are changed in the final version from *How do you feel?* to *How would you rate?* or *How would you evaluate?* Personal subjective feelings and attitudes are replaced by "objective" measures: *rating, evaluating.* "A language without the expression of person cannot be imagined," according to Benveniste, yet it can "happen that under certain circumstances, these 'pronouns' are deliberately omitted" (1971, p. 225).

Personal markers and discrete instances by which discourse provokes the emergence of subjectivity are removed from the final version of the Carnegie international questionnaire. The linguistic tools for achieving this move are varied; for example, although an original item from the list (March 14, 1991), *My job is a source of considerable personal strain,* did survive in the final version, it appears in a list of five *agree—disagree* items, prefaced by *Turning to your discipline and to your own career, we would like your opinion on each of the following statements.* In the final version of this question, most markers of place and time (indicators of *deixis),* which, according to Benveniste, organize the spatial and temporal possibility of subjectivity, have also been removed. In the final version the question seeks to connect the individual only to his professional associations *(your discipline, your career),* but not to any particular kind of institution (university, college; public, private), or to any specific system of higher education ("mass" vs. "elite" system, highly unionized or not). These usages in the international version serve to underline the value of the avoided forms of subjectivity, which are conspicuous by their absence.

Another change involves removing indicators of opinion with a positive or negative evaluation. For example:

Original item:	Final version (one item in a list):
I feel my work is often hampered by lack of adequate technology such as desktop computer, software, and audiovisual equipment.	*At this institution, how would you evaluate the following facilities, resources, or personnel you need to support your work?* *Computer facilities* *Excellent—Good—Fair—Poor*

Evaluative words—"hampered," "hindered," "lack of," "inadequate"— have all been changed to value-neutral terms—"the availability of," "the number of," "the amount of"—as in the following example comparing the same question from the second draft (May 22, 1991) and the final version.

<table>
<tr><td>

Second draft (October 22, 1991) question 42:

Please indicate the degree to which the following circumstances have hindered you from carrying out academic research in the past few years:

   a) Lack of adequate research funds

   b) Poor laboratory facilities

   c) Inadequate library support

   d) Heavy teaching load that restricts available time

   e) Too many distracting administrative assignments

   f) Too many other professional activities

   g) Political demands that prefer certain topics to other

</td><td>

Final draft question 54:

Please indicate the way your academic research is influenced by the following circumstances:

   a) The availability of research funding

   b) Facilities and resources for research here

   c) The number of courses I am assigned to teach

   d) The kinds of courses I am assigned to teach

   e) The number of students enrolled in my classes

   f) The quality of students available as research assistants

   g) The amount of student advising I do

   h) My non-academic professional activities

   i) My administrative work

</td></tr>
</table>

As can be seen, not only are the evaluative words replaced but also the reference to time: "in the past few years" is missing from the final version. The reference to "political demands" is replaced by "non-academic professional activities," again connecting the individual to his profession and not to politics, which are a local phenomenon. Although the changes are minimal, their direction is clear: toward restraining the linguistic apparatus that reveals subjectivity and specificity of both time and place. What takes its place are forms appropriate to the expression of objectivity, a high priority in the "scientific, universal" language style of quantitative research.

In constructing the international version, most markers of personal and temporal/spatial specificity were systematically removed, leaving few opportunities for the respondent to project or interpolate local conditions or produce subjective meanings. Whereas it was in the interest of the

Carnegie Foundation to produce a questionnaire from which all traces of the "local" and "national" were removed, as appropriate to an international study, it was by contrast in the interest of the country directors to restore some of the opportunities for expression of the local that the international survey avoids.

Conflicts between the Carnegie Foundation, initiators of the survey and the survey directors from some of the countries focused on questions inappropriate to the local context. Inappropriate questions included, for example, asking about income in a country where professors are unionized and their salaries are fixed by a scale. Conversely, some questions that had been dropped from the final international version as too nation-specific, such as the question on "union membership," were restored in specific national questionnaires, specifically, in this case, in the Australian questionnaire. Thus, while the directors sought to contextualize the questionnaire so as to reflect their country-specific conditions more accurately, the Carnegie strove for "one size fits all" formulations, toward objective (placeless) linguistic forms.

What is the international version trying to accomplish by decontextualizing the questionnaire? It seeks to avoid opportunities for introducing personal experience, local institutional structures and the local historical development of higher education. Avoiding linguistic forms appropriate to the expression of subjectivity in the exercise of discourse, the questionnaire strives for a neutral, universalizing "scientific" language, something like what Barthes (1968) called "zero-degree writing."

The Utopia of Professionalism

The questionnaire represents the production of an instrument to conduct research about the academic profession; the deconstruction of this representation uncovers the ideology controlling it. Ideology is the presentation of the imaginary relationship between people and their real condition of existence (see Marin 1984, p. 239, following Althusser). This is exactly what the international version achieved by progressively "de-discursifying" its language, erasing from it the traces of real local conditions: removing the local, it left only the universal, in this case the universalizing construct of professionalism. "Professionalism" is precisely an imaginary relationship of people to their real conditions of existence; that is, an ideology.[7]

Professionalism is a global ideology connected to no particular place but rather to an ideal. It connects individuals to the power of the "free professions" independent of the client or the employer, whether the company or the state (see Gottlieb and Cornbleth, 1989, p. 4). This imaginary relationship connects the individual not to a particular institution of higher education with its own history, in a particular system of education with its own history, in a specific location on the globe, but to his fellow professionals anywhere and everywhere—to "the profession." In the end, after several versions and erasure of the coordinates that place the subject in an instance of discourse, the Carnegie survey questionnaire achieved a universal, scientific, placeless style appropriate to an international instrument for collecting data in more than one country. Its questions are addressed to professors in "the world" at large, but actually nowhere in particular.

Where is the world at large? Such a no-place, according to Marin (1984), is utopia, an ideological place, a stage for ideological representation where ideology is put into play (p. 239). The questionnaire is utopic insofar as it reveals a plurality of hidden spaces whose incongruity with any place or country in particular lets us examine the space of professionalism as an ideology.

As we recall, each country could add specific local questions, which indeed six countries did (but by doing so they transformed their questionnaires from the "international" version to the "Australian questionnaire," the "Dutch questionnaire," etc). The examination of these local questions reveals local concerns such as unionism, class and ethnic origin and multiculturalism, whereas the international version relies on professionalism as a global ideology that is not connected to location but to the profession, an ideological place that does not exist in time or space, but only in "the world." The international version thus utopifies the professoriate, freeing it from social or geographical context, and elevating it to the level of reflecting conditions and attitudes worldwide.

The Carnegie report of survey results (Boyer, Altbach and Whitelaw, 1994) explicitly develops the global spatial reference:

> Our goal was to learn more about the condition of the professoriate from a larger perspective and, in the process, define priorities that could strengthen the academy worldwide. It would be especially interesting, we believe, to explore why faculty around the world define this as an intellectually creative time. (pp. 1, 14)

Such globalizing language paves the way for *The Chronicle of Higher Education*'s transformation of the results of the Carnegie survey into geographic inscription on a world map (Figure 1). This is an essential transformation of discourse into a spatial representation. Although showing only the results of one question, and suffering from misinformation (only Russia participated in the survey but the whole area of the former Soviet Union is indicated by shading), the *Chronicle*'s map is significant because it attests to the success of the ideological operation undertaken by the international questionnaire. First, by shading in the whole area of the countries that participated, the *Chronicle*'s cartographer persuades us that the whole country was covered by the survey; for example, because the US participated in the survey, Alaska is conspicuously shaded in, even though no single academic from the state of Alaska may have answered the questionnaire. Secondly, by shading the countries that participated in the survey, including the ones that did not provide data for this particular question, the map carries out another operation: The map persuades us that all the continents are "covered" since there is at least one country from each continent in the survey. Since the whole world was covered by the survey, we can speak confidently of what "the professoriate worldwide" thinks. This map is the visual utopia of professionalism.

MAPPING UTOPIA AND MAPS AS UTOPIA

The International Survey has constructed the academic profession as a utopian concept, existing everywhere yet nowhere in particular—as the world professoriate. "Professoriate" is a term used very seldom in a local specific context; professors usually identify themselves by their national, disciplinary or institutional affiliation—"a British professor," a "biology professor" or "a professor at Tel Aviv University." The International Survey successfully creates this nonspatial place where professors have become members of the professoriate and citizens of the world, where professionalism represents their relationship to their conditions in the world. This no-place does not mean the unreal or the imaginary; rather it signifies the place of the neutralization of difference and objectification of reality.

The most obvious coordinates of the construction of the professoriate as utopia are shown in Figure 2. Figure 2 "maps" the route that the original national surveys took. Each individual national survey is "discursive" (contextualized, "personalized") in referring to particular national structures, values, geographies, etc. These national surveys underwent an ideological screening process (indicated by the solid-line arrows in Figure 2) whereby the national surveys were assimilated to the international survey, and national particularism was reconfigured as the scientific, universal, and worldwide. This utopia of professionalism, as presented by the international survey, was in turn recontextualized in national terms (indicated by the dotted arrows) by some countries more than others (particularly Australia, West Germany, the Netherlands, and the UK, as shown in Figure 2).

The attention to the linguistic form in which the discourse of the in-

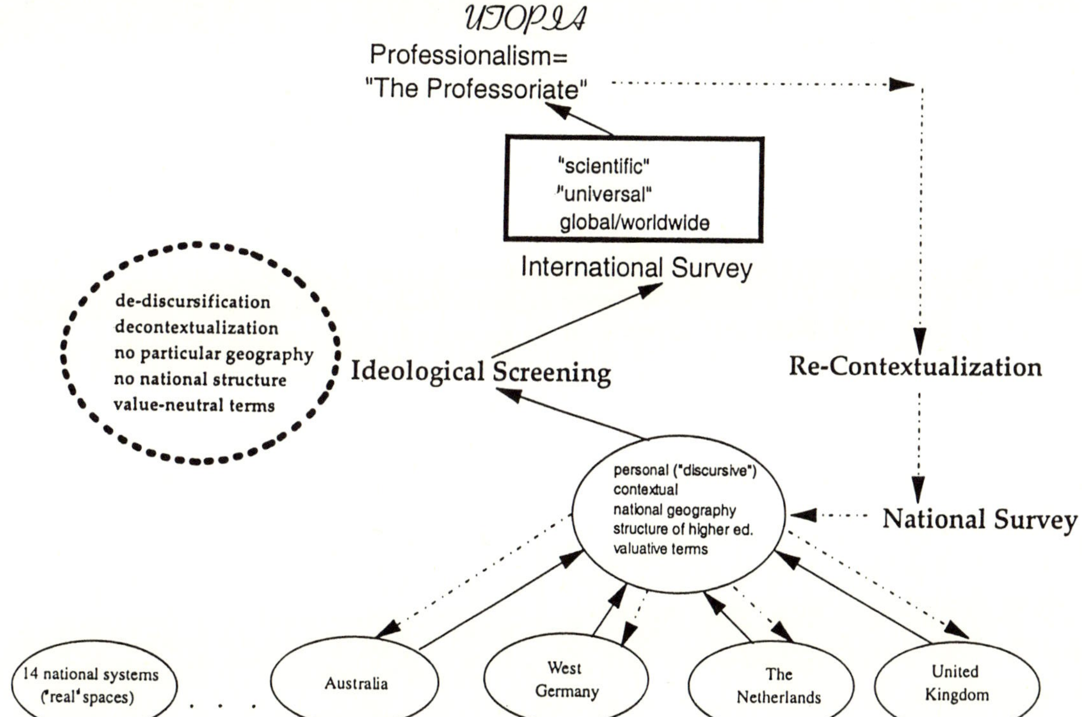

Figure 2. The schema of the construction of the professoriate as the utopia of worldwide professionalism

ternational questionnaire was constructed has uncovered the ideological screening that the national surveys and collection of items underwent. Guided by the ideology of professionalism this screening created professionalism as a utopian place. This place might be like the "Terra Incognita" marked in the first woodcut world map of 1482, where according to Turnbull, "Greek aesthetic sensibility dictated that there should be a southern body of land to balance that in the north. Australia was thus invented through the power of the map before it was 'discovered'" (1991, p. 41). The terra incognita of the professoriate was invented as much as discovered through a scientific instrument (i.e., the international questionnaire). Setting out to discover the professoriate worldwide entailed objectification of the real spaces professors occupy in their national contexts, much like the complex operation of a map, the art of inscribing and tying together places in a surface through networks of names and signs.

As Marin observed,

> The map is originally a net of itineraries and a system of potential routings all present at the same time, co-present. These paths are the opposite of a trip and its surprises and events. They are also the reverse side of a narrative unfolding in surprise and expectation, of a story limited by its character's viewpoints. With a map and its surface presentation, the viewpoint is no longer affected by surprise or expectation of the unusual. The gaze is everywhere present, and all points of view are the viewpoint, similar to Leibniz's God. All points of view are negated by its ubiquitous vision, everywhere present for everyone and every detail. . . . Vertically situated with reference to the map, the dominating gaze is in complete possession of all places. It is itself not a part of their system, but rather at the transcendent center organizing them into a system so as to render the elements interchangeable. (1984, p. 264)

By projecting the results of the International Survey onto a flat analogic model of the world (see Figure 1), *The Chronicle of Higher Education* constructs the professoriate as a kind of "Leibniz's God," present everywhere. In contrast with reporting the results narratively, picturing them on a world map universalizes the International Survey. In other words, the utopic operation of the International Survey manifests itself in the relationship between the surveyor's gaze and the representation of this reality (the results collected by the survey). This map is nothing less than the visual utopia of the professoriate (Figure 2).

This chapter has offered an explicit deconstruction of a discursive practice in the field of education. By using "discourse" both in the restricted Foucaultian sense (as in my past work; see Gottlieb, 1987), and in a more general sense as used by scholars of language, the analysis in this chapter seeks to combine heterogeneous theories of cultural practice in a more complex synthesis. The analysis proposed here has the potential to serve as a model for discourse analysis in other practical investigations in social cartography.[8]

1. See McHale (1992), *Constructing postmodernism,* especially pp. 1–3.
2. List of topics, international survey of faculty (March 14, 1991):

1. Individual Characteristics (Demographics)	8. Report of Scholarly Productivity
2. Training and Career Record	9. Department Autonomy
3. Principal Activities of Academics	10. Feelings about Work, Colleagues, and the Institution
4. Connections among Academics	
5. Political, Social, and Economic Perspectives of Academics	11. Rewards of the Academic Profession
	12. Perspectives on Professional Development and Career Opportunities
6. Institutional Supports and Constraints	13. Assessment of Faculty Performance
7. Teaching and Learning	14. Perspectives on Contributions to Society

3. List of topics, International Survey of the Academic Profession, 1991–93. First draft for a questionnaire (May 22, 1991):

PART I: WHO ARE THE ACADEMICS?	PART III: THE ACADEMIC CAREER
Individual Characteristics	Rewards and Advancement
Training and Experience	Attitudes Towards Work, Colleagues, and the Institution
Academic's Activities	
Professional Affiliation	Career Opportunities
PART II: ACADEMICS' WORK	PART IV: PERSPECTIVES ON POLICY
Scholarly Productivity	Higher Education in Society
Teaching and Learning	National and International Policy Issues
Governance	

4. For example, compare the two versions of the following question:

October 22, 1991 version:

What is your current employment status at this institution?

1) *Full-time, permanent* 2) *Full-time, contract* 3) *Part-time* 4) *Other*

In the final version, this question has a number of additional options:

What are the characteristics of your current employment?

a) *Working time*: 1) *Full-time* 2) *Part-time*

b) *Nature of employment*: 1) *Tenured* 2) *Indefinite contract without full ten-ure* 3) *Contract for fixed period* 4) *Other contract (please specify)*

5. List of topics, International Survey of the Academic Profession, 1991–1993. Final International Version:

Personal Inventory	Governance
Working Conditions	International Dimensions of Academic Life
Professional Activity	Higher Education and Society
Teaching, Research, Service	

6. Australia added twenty-six questions in addition to using all of the international version. West Germany omitted some questions and added thirty-six. Japan used all of the international version, in addition to "closing" most of the open questions and adding nine questions. The Netherlands omitted many questions (though not more than 44%, which was the maximum change) and added twenty-nine. The UK omitted a few, changed the wordings on many and added ten. The US used the international version and added six questions on multiculturalism. Israel, South Korea and the South American countries used all the items in the international version. Yet the translation, which was checked for accuracy and comparability in meaning using a process of back-translation, shows, as expected, that some phrasings and meanings are not identical; these questions were dropped from the Carnegie comparative analysis.

7. See also Larson's discussion of how the disparate occupational categories that we call professions (e.g., professors in higher education) are essentially brought together by ideology (1977, pp. 220–25).

8. I would like to acknowledge Cindi Carpenter at the West Virginia University Department of Academic Computing for her help with the graphics, and Philip Altbach for his continuing support for offspring studies from the Carnegie International Survey.

REFERENCES

Adams, D. (1990). Analysis without theory is incomplete. *Comparative Education Review, 34*(2), 381–385.

Altbach, P.G. (1991). Trends in comparative education. *Comparative Education Review, 35*(3), 491–507.

Altbach, P.G., ed. (in press). *The international academic profession: Portraits from 15 countries.* Princeton, NJ: Carnegie Foundation for the Advancement of Teaching.

Barthes, R. (1968). *Writing degree zero.* Translated by A. Levers and C. Smith. New York: Hill & Wang.

Bazerman, C., and J. Paradis, eds. (1991). *Textual dynamics of the professions: Historical and contemporary studies of writing in professional communities.* Madison: University of Wisconsin Press.

Benveniste, E. (1971). *Problems in general linguistics.* Translated by M.E. Meek. Coral Gables, FL: University of Miami Press.

Boyer, E.L., P.G. Altbach, and M.J. Whitelaw. (1994). *The academic profession: An international perspective.* Princeton, NJ: Carnegie Foundation for the Advancement of Teaching.

Brown, R.H. (1987). *Society as text: Essays on rhetoric, reason, and reality*. Chicago: University of Chicago Press.

Goodman, N. (1978). *Ways of worldmaking*. Indianapolis, IN: Hackett.

Gottlieb, E.E. (1987). *Development education: Discourse in relation to paradigms and knowledge*. Ph.D. dissertation, University of Pittsburgh.

Gottlieb, E.E. (1989). The discursive construction of knowledge: The case of radical education discourse. *Quantitative Studies in Education*, 2(2), 131–144.

Gottlieb, E.E., and C. Cornbleth. (1989). The professionalization of tomorrow's teachers: An analysis of teacher-education reform rhetoric. *Journal of Education for Teaching*, 15(1), 3–15.

Honan, W.H. (1994). Professors in 14–nation study say their ideas are ignored. *New York Times*, 20 June.

Jorden, M. (1994). Respect is dwindling in the hallowed halls: Study gives voice to professors' global discontent. *Washington Post*, 20 June.

Larson, M.S. (1977). *The rise of professionalism: A sociological analysis*. Berkeley: University of California Press.

Lynch, K. (1960). *The image of the city*. Cambridge, MA: MIT Press.

Marcus, G.E. (1994). On ideologies of reflexivity in contemporary efforts to remake the human sciences. *Poetics Today*, 15(3), 383–404.

Marin, L. (1984). *Utopics: Spatial play*. Translated by R.A. Volirath. Highlands, NJ: Humanities Press.

Masemann, V. (1990). Ways of knowing: Implications for comparative education. *Comparative Education Review*, 34(4), 465–473.

McHale, B. (1992). *Constructing postmodernism*. London: Routledge.

Messer-Davidow, E., D. Shumway, and D.J. Sylvan, eds. (1993). *Knowledges: Historical and critical studies in disciplinarity*. Charlottesville: University Press of Virginia.

Mooney, C.J. (1994). The shared concerns of scholars: Survey finds professors in 13 countries and Hong Kong have much in common. *Chronicle of Higher Education*, 22 June.

Paulston, R., and M. Liebman. (1994). An invitation to postmodern social cartography. *Comparative Education Review*, 38(2), 215–252.

Rorty, R. (1982). *Consequences of pragmatism: Essays 1972–1980*. Minneapolis: University of Minnesota Press.

Rust, V.D. (1991). Postmodernism and its comparative education implications. *Comparative Education Review*, 35(4), 610–626.

Schwichtenberg, C.M., ed. (1993). *The Madonna connection: Representational politics, subcultural identities, and cultural theory*. Boulder, CO: Westview.

Shohat, E., and R. Stam. (1994). *Unthinking Eurocentrism: Multiculturalism and the media*. London: Routledge.

Smoodin, E., ed. (1994). *Disney discourse: Producing the magic kingdom*. New York: Routledge.

Turnbull, D. ([1989] 1993). *Maps are territories; Science is an atlas*. Chicago: University of Chicago Press.

U.S. higher education concerns are global, survey shows. (1994). *Higher Education and National Affairs* (ACE); 27 June.

Welch, A.R. (1991). Knowledge and legitimation in comparative education. *Comparative Education Review*, 35(3), 508–531.

Postmodernism and Participation in International Rural Development Projects

Textual and Contextual Considerations[1]

Christopher Mausolff

My name is Pastor Jiménez. When I was 29 years old, I learned how to read and write thanks to my organization, the CNTC (Central Nacional de Trabajadores del Campo). Also, I learned about the reality of our national situation and I have come to understand the situation in which we peasants live in Honduras. I am ready to keep on advancing in the literacy group of my community.

Testimonial from a literacy manual of the
Honduran peasant movement organization, the CNTC

Introduction

Norman Uphoff's *Learning from Gal Oya: Possibilities for Participatory Development and Post-Newtonian Social Science* (1992b), alerts us to the possibility of many valid ways of seeing participation in rural development projects. Uphoff demonstrates that beneficiary participation in the Gal Oya project in Sri Lanka can be viewed through the lenses of chaos theory, existentialism, phenomenology and Einstein's theory of relativity. With these new framing choices, he argues, come new insights for improving participatory practice. Uphoff uses chaos theory to understand how detailed, long-term planning could not succeed in the inherently dynamic and uncertain situation at Gal Oya. He cites existentialism when he makes the observation that the project stakeholders need to "make choices and take responsibility for their actions in an uncertain world" (1992b, p. 409). Through phenomenology, Uphoff conceives of ways in which subjective factors such as "social energy" and "positive-sum" thinking transformed the objective situation of the project. Finally, he applies Einstein's theory of relativity to envision ways for divergent, clashing frames of reference among project stakeholders to be reconciled. This essay builds on Uphoff's work by continuing to explore the implications of multiple framing perspectives for participatory practice. This exploration of diversity is also an evaluation of the utility of a postmodern sensibility.

Postmodern writers such as Lyotard (1984) and Brown (1992) advocate abandoning metanarratives; that is, totalizing discourses that claim universal validity. As Brown (1992, p. 239) points out, "to the extent that various theoretical approaches articulate the vision and interests of different social groups, a single general theory is not desirable." The case for methodological pluralism is explored in this essay by mapping and evaluating alternative methodological streams in the literature on participation. The postmodern view is supported by the observation that each of the perspectives represented in the literature (positivist, pragmatic, hermeneutic and critical theory), continue to provide significant practical insights.

Another reason that a postmodern sensibility may be relevant to development managers is that it is in indigenous and other isolated communities that totalizing theories would be especially likely to make invisible what Liebman and Paulston (1994) refer to as "the mininarrative knowledge claims of cultural clusters" (p. 233). This concern for local diversity is more important if there is indeed significant diversity present among local community groups. If local communities all want pretty much the same thing for themselves, then there is less need for a sensibility promoting local diversity. But it may be that local conceptions of needs and desired ways of meeting these needs are locally constructed in small groups and hence may vary considerably across projects and regions.

This proposition is evaluated in seven different rural development communities in Honduras using a technique called participatory self-evaluation (developed by Uphoff, 1991, 1992a). The results of these evaluations have been summarized into social maps of the communities' project goals. The resulting maps show considerable variety in the values expressed by the community groups. This outcome suggests a uniquely postmodern justification of participatory development: that beneficiary participation makes projects more responsive to the local narratives of community groups.

Defining the field of inquiry examined in this chapter—participatory approaches to international rural development administration—is itself problematic. Oakley and Marsden (1984) acknowledge that working definitions of participation "are necessary if the process is to be understood at all," but they argue that "it is impossible to establish a universal definition of 'participation'" and, given its complex nature, "it is impossible to identify 'participation' as an actual 'social reality'" (pp. 18–19). Goulet (1989) defines participation in the context of the recent trend toward democracy in many developing countries: "Participation is newly conceptualized as a 'moral incentive' allowing the powerless poor to negotiate new 'material incentives' for themselves, and as a leverage point permitting successful micro actors to gain entry into macro arenas of decision making" (p. 165). An example of a more broadly accepted definition of participation would be Finsterbusch and Van Wicklin's (1989) version, in which "the term 'participation' is used to mean the contribution of beneficiaries to the decisions or work involved in the projects" (p. 575). These researchers further refine their definition by distinguishing five stages in which beneficiaries can potentially participate (project origin, design, redesign, implementation and maintenance). There is no need to settle on any particular definition, but it is interesting and surprising to find such a diversity of views on even the fundamental issues.

MAPPING TEXTS ON PARTICIPATION IN INTERNATIONAL RURAL DEVELOPMENT TEXTS

From a postmodern perspective, a variety of perspectives may be useful depending on the specific needs and circumstances of the situation. This orientation would seem to be especially important in the participation literature given the diversity of contexts and situations in the developing world. I examine postmodernism's glorification of heterogeneity through texts on participation in international rural development projects (hereafter referred to as participatory development or the participation literature). If texts from a variety of perspectives have something useful to contribute, this would argue against the modernist project of constructing one grand story of participatory development.

In evaluating the utility of these texts, I adopt a pragmatist perspective. Patton (1990) describes pragmatism as the approach used when the researcher has left the world of theory and is addressing "the quite concrete and practical questions of people working to make the world a better place (and wondering if what they're doing is working" (p. 89). Consistent with my pragmatist framing choice, I classify a work as useful if it can be effectively utilized by practitioners directly involved in social change efforts.

A central question in this effort is how to categorize the literature. I like the approach developed by Burrell and Morgan (1979) and later applied by Denhardt (1990), Paulston and Rippberger (1991), and Paulston (1992, 1993). Burrell and Morgan divided the field of sociology into four paradigms using two continua: ontological position (subjectivist versus objectivist) and change orientation (radical versus regulatory). The authors asserted that the knowledge space at intersection of each dimension could be classified as a

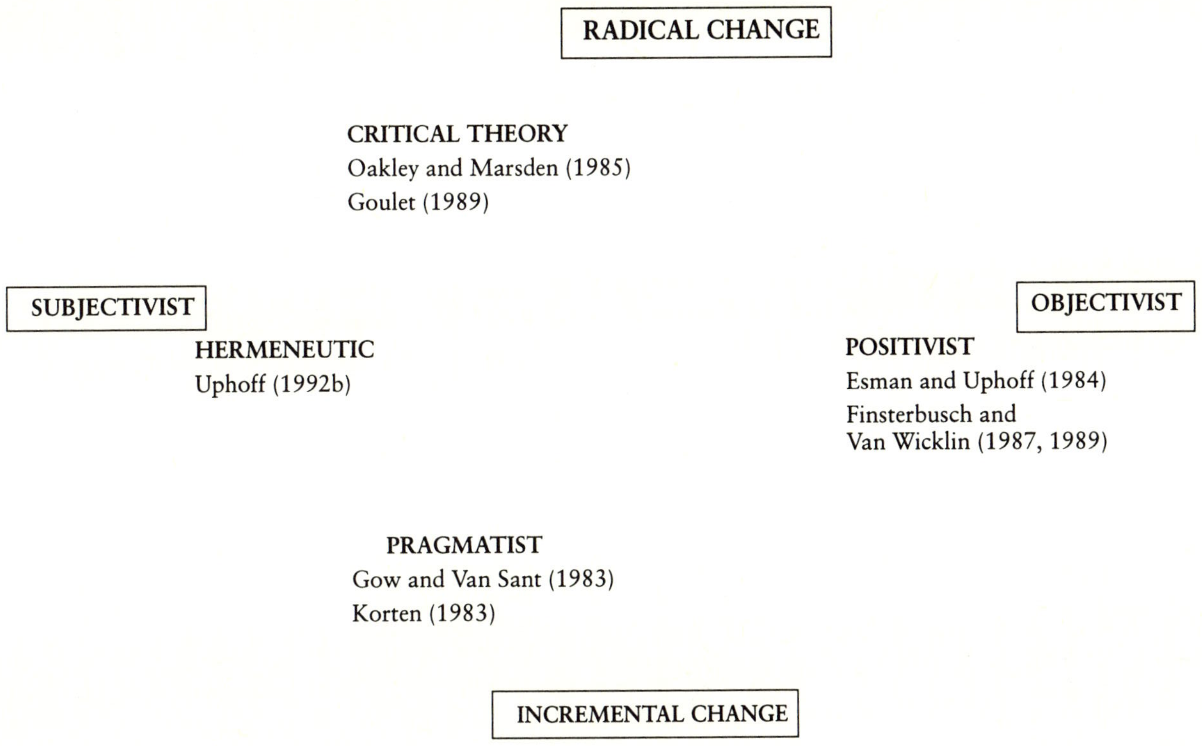

Figure 1. Ways of seeing participation in international rural development texts mapped as an intertextual field

separate paradigm (i.e., functionalist, interpretive, radical humanist and radical structuralist).

In this essay I retain the continua of ontology and change orientation, but I divide the literature according to "streams" or perspectives rather than paradigms (see Figure 1). The orientation of the text toward change is evaluated because change is a central dynamic and objective in participatory development. As such, there is a range of opinion in participation texts as to how fast, or radical, change efforts should be. The ontological continuum, where views of reality fall between subjective and objective poles, is utilized because the issue of whether or not the social world exists apart from individual cognition is a fundamental assumption guiding a variety of other epistemological and methodological choices in the research process. For example, I find that participatory development texts with an objectivist ontology (a modernist view of reality) tend also to be positivistic and nomothetic.[2] In utilizing the subjectivist-objectivist dichotomy, I acknowledge that such frames of reference are constructions of the mind and not absolute reality. Like Heshusius (1994) and the phenomenologists, I agree that understanding benefits from a narrowing of the gap between the inquiring subject and the object to be known. But at the same time, such constructs, especially when presented graphically in social maps, can help us to make sense of the multiplicity of thought in the postmodern world.

There are a number of other reasons why social mapping is particularly appropriate for the tasks of this essay. To evaluate the potential contribution of diversity within the participation literature, it is helpful to create a space where there is room for diversity and where no one discourse is privileged over any other. Mapping provides this space in a direct and literal manner.

Mapping is also useful for the present task because it contributes to the hermeneutic objective of deepening understanding. In exploring and comparing these texts, I am less interested in causality than in understanding and communicating the complex, multifaceted nature of participatory practice. A good review of the benefits of visual images over texts in the comprehension process is provided by Meyer (1991). He reports that the form or medium of the information input influences where in the brain it is encoded. The right hemisphere of the "brain codes sensory input in terms of images, the left hemisphere in terms of linguistic descriptions" (Levy, 1974; quoted by Meyer, 1991, p. 223). The two hemispheres of the brain seem to have different capabilities. The right hemisphere synthesizes over space rather than time, notes visual similarities to the exclusion of conceptual similarities and perceives form rather than detail. The left hemisphere has the opposite capabilities. The available research also suggests that utilizing both visual and written forms of communication facilitates heuristic search and improves recall. Finally, an advantage of graphic images is that they "allow for the simultaneous perception of parts as well as a grasp of interrelations between parts" (Meyer, 1991, p. 229). The implication of the available research is that interpreting and understanding information on a map is qualitatively different from understanding information in texts.

Mapping may also facilitate understanding by helping us to construct mental models. Johnson-Laird (1983) suggests that "human beings understand the world by constructing working models of it in their minds" (quoted by Huff, 1990, p. 13). Visual representation of the literature facilitates the construction of our mental models by providing a ready-made image of the literature and its interrelationships.

The map in Figure 1 should facilitate understanding of the variety of framing choices in the field and where each stands in relation to the others. The placement of each perspective reflects my opinion about the way ontology and change are addressed by texts within each perspective. The effort is provisional and ongoing. Many more texts from the participation literature could be added to this map. As additional texts are added, new streams will emerge and it will be possible to determine which perspectives are dominant and which are neglected. For my present purpose, I discuss the contributions of a few representative texts from each perspective in order to suggest what would be lost by use of only a single general theory, or metanarrative.

Positivist Texts

Some of the participation texts that utilize an objectivist ontology and an incrementalist change orientation can be classified as positivist. Historically, the Vienna circle of logical positivists were concerned with the problem of verifiability. They argued that "something is meaningful if and only if it is either verified empirically . . . or is a tautology of logic or mathematics" (Lacey, 1976, pp. 165–66). As a contemporary approach to social science, positivism is characterized by the use of deductive logic, hypothesis testing and research conducted from the standpoint of the observer rather than the participant. Positivism is a fundamental strand of modernist thought, providing the methodology for generating systematic knowledge to advance the human condition. A primary utility of the evaluated research within this stream is the production of cause-effect knowledge about the circumstances under which participatory methods can improve project success.

A classic work within the positivist tradition is Esman and Uphoff's comprehensive study of rural development, *Local Organizations: Interme-*

diaries in Rural Development (1984). The study is based on statistical analysis of 150 development projects from around the world, coded by the Rural Development Committee at Cornell. Esman and Uphoff find that the correlations between participatory orientation and local organization performance are all relatively high. The authors also address a question of central importance to development planners: What is the possibility for participation in hostile, authoritarian environments? Esman and Uphoff make the rather inspiring finding that local organizations operating under indifferent or adverse conditions have less success than average unless they are highly participatory and egalitarian. This finding provides a basis for arguing against the necessity of authoritarian vanguard approaches in hostile environments.

In their research, Esman and Uphoff (1984) utilize what they refer to as a "structural-reformist" perspective. This approach differs from structural-functionalism in that the structural-reformist mode utilizes the structural metaphor to describe a rural change strategy, whereas structural-functionalism is concerned with the way recurrent activities contribute to the maintenance of social equilibrium. Esman and Uphoff (1984) explain that the structural-reformist approach

> emphasizes the search for institutional and organizational changes that can cumulatively shift the balance of socioeconomic and political power. . . . Local organizations that strengthen rural people in large numbers through their own efforts are essential to the kind of incrementalism that this approach implies, building local power that can both limit and influence the actions of the state and of the private sector. (p. 56)

This impressive study succumbs to some of the totalizing, "one best way" discourse of empirical work of that era. Esman and Uphoff (1984), in arguing the merits of their incrementalist approach, assert that "measures to alter the structure of assets and exchange through the gradual buildup of local institutions offer the most solid prospects for developmental progress" (p. 56).

A problem for such universalizing approaches, as pointed out by Nicholson (1992), is that they must accomplish two mutually incompatible things: (1) incorporate all human activity, and (2) do so in a nontrivial way. In this case, the strategy of building up local organizations is so vague as to be nearly meaningless. Who would argue against such a strategy? This strategy describes the gradual mobilization of the North Vietnamese peasants by the Vietcong (Popkin, 1979; Migdal, 1974), as well as the participatory development efforts of the Interamerican Foundation (Carroll, 1992) and the World Bank (Bhatnagar and Williams, 1992; Cernia, 1988; Paul, 1987). The issue of the appropriate level of beneficiary political mobilization in a given social context is not so easily solved in one general, albeit insightful, account.

Like Esman and Uphoff (1984), Finsterbusch and Van Wicklin (1987, 1989) approach participation from a modernist, positivistic perspective. Finsterbusch and Van Wicklin (1987) critique the participation literature for having had an "advocacy nature" and for not being "very empirical" (p. 5). The problem with the participation literature, they argue, is that the dearth of reliable data across a broad range of projects has inhibited systematic analysis based on explicit hypotheses and appropriate measures to investigate these hypotheses. As part of their effort to bring some coherence to the diversity of the literature, Finsterbusch and Van Wicklin "elicit a general model of participatory development projects, deduce the central implicit hypotheses from this literature. . . , and statistically test these hypotheses from the empirical evidence provided by AID's series of 52 Impact Evaluation Reports" (Finsterbusch and Van Wicklin, 1987, p. 1). For their general model of participatory development, the authors use a systems framework to organize the large number of variables that could potentially influence project effectiveness and participation. As explained by Burrell and Morgan (1979) the systems metaphor is consistent with a positivist approach as it suggests a social reality with some underlying pattern of order.

A central question addressed by both Esman and Uphoff (1984) and Finsterbusch and Van Wicklin (1987, 1989) is the influence of participatory management approaches on project effectiveness. On this issue, the data analyzed by Finsterbusch and Van Wicklin represent a most difficult case. The fifty-two United States Agency for International Development (AID) projects evaluated in their sample are mostly large infrastructure projects and over half were begun before the U.S. Congress's New Directions legislation, which directed AID to increase beneficiary participation. In large infrastructure projects, it would be expected that factors other than beneficiary participation, such as the adequacy of funding or the appropriateness of the project technology, would be the significant predictors of project success. Finsterbusch and Van Wicklin test this hypothesis and find that beneficiary participation is more strongly correlated with project success in the smaller projects in their sample. But they report that overall, beneficiary participation is still correlated with project success, even in the older, large-scale infrastructure projects.

The utility of positivistic approaches for providing information about significant debates within the field is also illustrated by other issues addressed in Finsterbusch and Van Wicklin's research. Oakley and Marsden (1985) repeatedly emphasize the importance of projects not imposing an organization structure on beneficiaries, but rather let the organization structure emerge out of the group's own authentic process. This assertion is supported by Finsterbusch and Van Wicklin's research. In their sample of AID projects, they find that (1) the degree to which beneficiaries are organized, (2) the democracy of the beneficiary organization and (3) the indigenous origin of the beneficiary organization are all significantly correlated with the levels of beneficiary participation and the extent to which the project increased beneficiary capabilities. Contrary to Habermas's (1972) assertion that positivistic research is necessarily status quo-oriented, the research findings of Esman and Uphoff (1984) and Finsterbusch and Van Wicklin (1987, 1989) tend to support administrative approaches that shift power to the rural poor.

Pragmatist Texts

While the statistical correlations of Esman and Uphoff (1984) and Finsterbusch and Van Wicklin (1987, 1989) suggest that it is possible to utilize participatory approaches in factionalistic environments and in large infrastructure projects, the pragmatist texts by Gow and Van Sant (1983) and Korten (1983) provide a number of guidelines on how to actually accomplish this. The reviewed pragmatist texts summarize existing research and first-hand experience into practical guidelines on how to manage rural development with beneficiary participation.

Gow and Van Sant show their preference for a pragmatic perspective in the title of their 1983 article in *World Development*, "Beyond the Rhetoric of Rural Development Participation: How Can It Be Done?" In providing their recommendations, Gow and Van Sant construct the political environment of development projects as having an objective reality that functions with law-like consistency:

> leaders of most important factions—elite or otherwise must be included. . . . Some may be motivated to help their fellow villagers because of enlightened self-interest; others because they need a certain local constituency to support and implement their views. Whatever their motivation, local leaders are going to be represented —if not over represented—in any process of decision making. . . . The key is for leaders to be made accountable to a broad constituency regardless of their group of origin. (p. 435)

These recommendations sound reasonable, but to what extent are they really generalizable across the diversity of third world circumstances? Korten (1983) is more contingency-oriented when she proposes that if factions are strong, it may be necessary to develop separate groups. She recounts that the Bangladesh Rural Advancement Committee found "that the interests of the landed and the landless were sufficiently irreconcilable that the only solution was to develop separate groups" (p. 190). Depending on the circumstances, she also advocates the utilization of incentives and community organizers to unify people that must cooperate.

It seems that there is a division of labor among the various perspectives considered thus far. The quantitative studies with large sample sizes offer information about relevant causal relationships, while the "how-to" approaches provide clear guidelines for project management. Yet these incrementalist, objectivist approaches do not fill all of the intellectual space of participatory practice.

A Hermeneutic Text

Norman Uphoff describes his methodology in *Learning from Gal Oya: Possibilities for Participatory Development and Post-Newtonian Social Science* (1992b), as "barefoot hermeneutics." During field visits to the Gal Oya project in Sri Lanka between 1980 and 1987, Uphoff accumulated over 800 pages of field notes written in the first-person narrative form. At about the time he was getting ready to begin transforming these notes into a book, Uphoff discovered Bernstein's (1983) *Beyond Objectivism and Relativism*. As a consequence, he became involved in the philosophical literature on alternatives to positivist methodology and wrote *Learning from Gal Oya* with a hermeneutic methodology in which his field notes became a text to be interpreted.

In this work, Uphoff embraces both subjectivist and objectivist ontologies. Like Popper, Uphoff is concerned with explaining how subjective factors such as "purposes, deliberations, plans, decisions, theories, intentions and values" can influence the "physical world" (Popper, 1972, pp. 228–29; quoted by Uphoff, 1992b). One example of this approach is the distinction he makes between understanding the average tendency and attempting to obtain desirable outcomes significantly above the mean. This emphasis stands in contrast to the way modernity's grand theories "authorize us to describe and normalize" so as to find underlying causes and improve prediction (Hopenhayn, 1993, p. 94). Instead, Uphoff (1992b) argues that "the challenge of development work is how to make possible outcomes that are deemed desirable somehow more probable" (p. 299). A tactic that Uphoff advocates for achieving this objective is the mobilization of social energy. Uphoff asserts that the characteristics of social energy are better explained by post-Newtonian thinking than by traditional mechanistic frames of reference. Instead of gradual entropy,

> the physical and social system in Gal Oya . . . was now acquiring greater structure and productivity. . . . The Institutional Organizer program, with less than 5% of the total project investment, was transforming the situation in Gal Oya. We were witnessing a process of energization that itself needed explanation. (Uphoff, 1992b, p. 358)

Uphoff's efforts to explain this phenomenon result in impressive applications of chaos theory, phenomenology, existentialism and Einstein's theory of relativity.

Critical Theory Texts

The two critical theory texts are shown in the upper left of Figure 1 because they combine a subjectivist ontology with a radical change orientation. In these texts, the emphasis is not simply on understanding, as in Uphoff's (1992) hermeneutic text. These texts also present the ways in which ideologies, hopelessness and false consciousness enslave the poor. Burrell and Morgan (1979, p. 284) explain that proponents of a critical theory perspective "seek to reveal society for what it is, to unmask its essence and mode of operation and to lay the foundations for human emancipation through deep-seated social change." In following this political agenda, these works provide two contributions neglected by works in the other perspectives: (1) developing explicitly political change processes; and (2) bringing to the forefront the potential role of consciousness-raising in social change efforts.

In Oakley and Marsden's (1984) insightful *Approaches to Participation in Rural Development,* the knowledge framing style is decentering and critical. Their discussion of political power is presented in the context of a comparison between modernization and dependency theory, which, as was more prevalent in the mid-1980s, are presented as occupying all of the possible intellectual space in development theory. Modernization theory is critiqued for its naive, "idealistic" and "ahistorical" assumptions regarding the unity of interests within communities and nations, while "the search for more appropriate styles of development is fundamentally likened to what has been termed 'dependency theory'" (p. 7). Within this ' alternative" development path, the authors argue that it is difficult to "disassociate participation from its relation to power. . . . 'Participation' must be seen as an exercise of giving the rural poor the means to have a direct involvement in development projects. In other words they must be given the strength to be able to seek this direct involvement" (p. 64).

Like Oakley and Marsden, Goulet, in a 1989 article in *World Development*, makes the relation between political power and participation the central focus of his analysis. For Goulet, the problem is how to move from successful microlevel participation to "entry into the macro arenas of decision-making" (p. 165). More than any of the other reviewed works, Goulet analyzes the potential impact of community participation at the societal—rather than the project—level.

> The best indication of whether participation is purely ornamental or a vital element of strategy is the relative weight assigned to it in the overall development practice of a given society. Experience shows that numerous successful micro operations never expand beyond their initial small scale. Many others, although they may grow to achieve "critical mass," do not successfully resist being repressed, co-opted, or marginalized. The supremely difficult transition is, precisely, that which takes a movement from the micro arena to the macro without dilution or destruction. (pp. 168–169)

Oakley and Marsden (1985) share this view that "authentic" development involves participation of the poor at the macrolevel and they critique the Nepalese Small Farmer Development Program for failing to achieve any "effective participation the wider context of Nepalese rural society" (p. 42).

The question then becomes how best to bring about these macrolevel changes given the structural realities faced by the poor in developing countries. Oakley and Marsden (1985) critique the "official" (i.e., UN agency) literature for recommending such participation prerequisites as the "decentralization of government decision making by strengthening supporting delivery systems at the lowest level" (p. 65). Such strategies, according to Oakley and Marsden, are unrealistic, since

> the preconditions to participation as expressed . . . are not going to occur in the foreseeable future and the existing socio-political frameworks are not going to facilitate meaningful grass-roots participation. We must, therefore, consider a strategy that does not depend, for example, on bureaucratic decentralization or legislation to encourage local organizations, but which attempts to achieve participation in the context of existing administrative frameworks. (p. 65)

Oakley and Marsden see Freirian-style nonformal education and consciousness-raising as a means to "bring about effective participation without waiting for the structural changes generally indicated as indispensable"[3] (p. 65). They describe a number of characteristics of this pedagogy:

> The process of empowering for participation is essentially a non-formal educational activity. . . . A number of terms have been used to describe this educational process, the more common of which are "education for liberation" and "conscientisation" . . . It is a radical departure from the classical, formal educational approach, and it seeks to liberate individuals from the environment which constrains them. (p. 70)

Like Oakley and Marsden (1985), Goulet also develops a model of social change based on the power of consciousness-raising. In the Sarvodaya project in Sri Lanka, Goulet (1989) explains that consciousness-raising takes a form specific to the classical doctrines of Theravada Buddhism, in which "the goal of true development is that all individuals progress toward full enlightenment" (p. 169). Within this spiritually oriented value system, the concept of consciousness-raising for macrolevel change is retained. The aim of this project is a situation in which "there is no ruling class and the people are all powerful" (p. 170). In texts by both Oakley and Marsden (1985) and Goulet (1989), consciousness-raising can bring about societal change because both utilize a subjectivist ontology, in which "reality is socially constructed and sustained so that change in consciousness can constitute real change" (Gottlieb, 1989, p. 137).

The critical theory perspective has been found to be applicable to a wide range of third world contexts. Consciousness-raising has been used as a tool for empowering the poor to construct new collective identities through which they negotiated significant political changes in the Bhooma Sena movement in India (Oakley and Marsden, 1985), the Sarvodaya project in Sri Lanka (Goulet, 1989), and the Catholic Church's radio literacy program in Southern Honduras (Brockett, 1988).

In this section I have explored postmodernism's call for methodological pluralism by evaluating selections from a variety of perspectives within the participation literature. I have found a surprisingly broad range of perspectives within the field, each offering useful ideas not found in the others. The utility of this diversity suggests that a postmodern sensibility, which values the opportunity for each perspective to be heard, has merit within this field. In addition to the individual contributions of each perspective, the sum total of the many perspectives is also a benefit. This diversity of thought aids development practitioners by providing them with a range of cognitive conceptions for coping with the diversity of phenomena that they find in the field. The nature and extent of the diversity encountered in development projects is a matter taken up in the following section.

Developing a Postmodern Justification for Participation

Each of the perspectives reviewed so far conceives of the development problem in different ways and, to a large extent, offers different justifications for using participatory project management methods. In positivist and pragmatic texts, the macrolevel political situation is taken as given and participation is justified primarily according to individual-level capacity building and various project-level measures of success. Using a hermeneutic approach, Uphoff (1992b) is also focused on project-level success, but the benefits of participation are framed in terms of mobilizing social energy in order to make unlikely events (e.g., project success) more probable. Finally, from a critical theory standpoint, the current development path is viewed as fundamentally flawed. Oakley and Marsden (1985) and Goulet (1989) value participation as a means to raise the consciousness of the poor so that they may change national politics. In sum, there are a variety of justifications for participatory practice, depending on the assumptions of the particular perspective. In this section, I would like to add to this variety by developing and testing a postmodern justification for participation.

One reason that postmodern writers are antagonistic to the modernist project is their concern that modernity's totalizing metanarratives will smother the discourses of local and marginalized groups (Liebman and Paulston, 1994). For example, Hopenhayn (1993) critiques "the understanding of development as a progressive process of homogenization . . . and a pretension of cultural cohesion that proves anachronistic in light of the 'proliferation of variety' of the new times" (p. 97).

In view of this concern, a postmodern justification for beneficiary participation would be that such approaches respect local cultural diversity. If beneficiaries make the final decisions regarding project design, implementation, and evaluation, then it is highly likely that the project will be responsive to local culture. This justification is rooted in a subjectivist ontology in which ideas about desired project goals are locally constructed in small groups and hence vary considerably across projects and regions.[4] The alternative explanation would be that people are relatively homogeneous in their responses to their objective problems and hence there is broad agreement among beneficiaries across projects as to what constitutes a successful project. These alternative conceptions, which can be categorized as postmodernist and modernist, are evaluated among peasant farmers in rural Honduras.

The Peasants, the Peasant Movement, and Agrarian Reform

Seven different peasant groups agreed to participate in the participatory self-evaluation. Some of these groups are affiliated with organizations that are part of the organized peasant movement (*Central Nacional de Trabajadores del Campo* [CNTC] and *Organizacion Campesina Hondureno* [OCH]). The remaining groups are affiliated with the independent development project Prodai.

The Honduran peasant movement has developed primarily in response to the scarcity of land for peasant farmers. During the 1950s, the banana worker unions in the north and the educational activities of the Catholic Church in the south provided the organizational structure through which the peasants channeled their demands for land.

By the late 1960s, as the membership of the major peasant organizations grew to approximately 90,000 families, the peasant movement began to play an active role in Honduran politics. In the 1970s and 1980s, the peasant organizations, along with the unions, played key roles in largely unsuccessful efforts pressuring the political parties and several presidents to adopt democratic practices and end clientelism and corruption (Posas, 1992).[5] Although unsuccessful in their immediate aims, these democratic reform efforts suggest that the political activism of the popular sectors can be oriented toward the creation of a democratic political order, as promised by popular education movement intellectuals (e.g., Freire, 1970; Osorio, 1990).

The peasant movement's primary focus has been pressuring the government into implementing the nation's agrarian reform legislation. Between 1962 and 1980, about 36,000 rural families received land under the agrarian reform. This amount represented about 22% of the landless or land-poor families in the mid-1970s (Brockett, 1988). Since the rural population has been growing by 10,500 families per year, rural population growth has far outstripped the quantity of land redistributed through the agrarian reform program. As of 1984, there were roughly 125,000 to 150,000 landless rural families and an additional 90,000 land-poor families subsisting on less than two hectares (Ruhl, 1985).

Rural poverty exists side by side with low educational attainment levels. Roughly 70% of the rural population is illiterate and only 31% of the entering students finish primary school (Sijbrandij and Ooijens, 1988).

The high illiteracy levels create a scarcity of leadership and management resources within the organized peasant movement. This scarcity has been an impetus for the expansion of popular education programs at the community level. Freirian methods of dialogue and reflection are common within these popular education programs (Sijbrandij and Ooijens, 1988; Mausolff, 1994).

Those surveyed groups that are not part of the organized peasant movement are part of Prodai, a unique and independent development project initiated by peasants. Approximately twenty years ago, the Catholic Church provided literacy training to a small group of peasants so that they might teach Catholicism to other peasants in the surrounding communities. Over time, the peasants' goals for their work grew from being purely spiritual to also incorporating the physical well-being of their neighbors. Gradually they developed an impressive development project providing health, nutrition and organic agriculture training to roughly thirty communities in their region. The Prodai project, like the peasant movement projects, has only peasant promoters working directly with community groups. There are no university-trained extensionists or educators on their staff. The project differs from those of the peasant movement organizations in that it is not involved in a societal-level change effort. The project is committed to the use of dialogue and participatory decision making, but the explicitly Freirian methodologies of the peasant movement are not in evidence. The inclusion of Prodai in this study provides a form of comparison with the more political work within the organized peasant movement.

Research Methodology

My knowledge-framing perspective in evaluating the diversity of local values is a provisional positivism. I am not interested in promoting positivism as the only useful method for improving development practice. Some might find it paradoxical to utilize positivism to evaluate the applicability of postmodern values, but I see a postmodern frame as neither privileging nor excluding any discourse, including positivism. This study attempts to be positivistic in the use of (1) hypothesis testing with empirical data, and (2) a most similar systems research design that attempts to minimize alternative explanations for variations in local culture. In other ways, this research design does not fall within the boundaries of positivism. I cannot state a priori what result would constitute sufficient variability to reject the modernist assumption of cultural homogeneity. Also, the small sample size increases the likelihood that the outcome is just a chance result.

Most Similar Systems Design

I have selected peasant groups for involvement in this study according to the logic of a "most similar systems" research design (Przeworski and Teune, 1970). In this research design, the intent is to select cases that are all similar except with regard to the hypothesized causal variable. Under these conditions, any variations in outcome can be attributed to the hypothesized causal variable since other possible causal factors have been controlled for by the similarity of cases. In this study, the sample of rural Honduran peasant groups all appear to live in similar objective environments with regard to lack of land, food, health care and educational opportunity. The hypothesized variable that would explain any variation in group conceptions of means and ends is the cultural value system that has developed during each group's unique history.

Participatory Self-Evaluation

I gathered data on the variability of project goals by teaching each peasant group how to conduct their own evaluations, a process called participatory self-evaluation. Uphoff (1991) developed this method for the People's Participation Programme of the Food and Agriculture Organization of the United Nations (FAO) to enhance the management and group skills of community groups. The following steps were followed in completing the participatory self-evaluations for this study:[6]

1.   The group facilitator asked the group members for their goals and recorded their responses on a blackboard. The areas of evaluation included both group goals (e.g., punctuality, cooperation, broad participation) and project-level goals (e.g., agricultural production levels). Note that in some groups, the author facilitated the process, while in others, the group's usual extensionist or educator was utilized.

2.   The group facilitator developed statements that aided the measurement of these goals. Each goal was expressed in a positive form to facilitate consistency in scoring. For example, a women's group associated with the Honduran peasant movement was concerned about how little land it had. The positive statement for evaluation that they agreed to was "The group has made a sufficient effort to obtain more land."

3.   The facilitator presented a scoring system for the group to use in evaluating its performance in achieving each goal.

4.   The group discussed each goal statement until it reached a consensus on the score that most accurately reflected the current situation of the group or its projects. In some groups, a few group members were shy and

hesitant to participate. One facilitator broadened involvement by asking each member directly how they would score each dimension.

5.   The evaluation statements and the corresponding scores were recorded by a member of the group so that the group had ready access to this information for their next evaluation.

In this study, participatory self-evaluation is not just a self-evaluation tool for farmer groups, it is also a social science research tool. As an evaluation tool, participatory self-evaluation provides capacity building by directly involving project beneficiaries in the evaluation process. As a social science research tool, it can be categorized as a type of focus group research. Its application is limited primarily to gathering data around beneficiary conceptions of project goals. As a method for studying other cultures, one benefit is that, to a large extent, it allows "the expression of culture in its own terms" (Welch, 1993, p. 20; quoted by Liebman and Paulston, 1994, p. 235).

Study Results

During the participatory self-evaluations, the peasant groups chose a wide variety of group and project goals. The variance across the groups can be clearly seen in the maps in Figures 2 and 3. These maps show peasant responses according to common themes that emerge from the goal statements. The number of times one of these organizing themes is mentioned determines a group's location on the map. (Group names are duplicated when there are all-female or all-male branches of one movement.) The map in Figure 2 illustrates how the variation among groups within the same project is greater than the interproject variation. This map also shows that groups involved in the organized peasant movement are more alike than groups involved in the Prodai project. This pattern suggests that there may be some socialization taking place through involvement in the organized peasant movement. However, even among peasant movement organizations, there is considerable variety. The groupings shown on the map in Figure 3 indicate that gender is not a predictor of group values. Overall, the results support the hypothesis that a significant amount of the construction of cultural values takes place at the group level. This finding suggests that postmodernism's promotion of diversity and cultural relativism may have some relevance to the practice of development administration.

Conclusion

In evaluating the usefulness of postmodernism as a framing perspective for

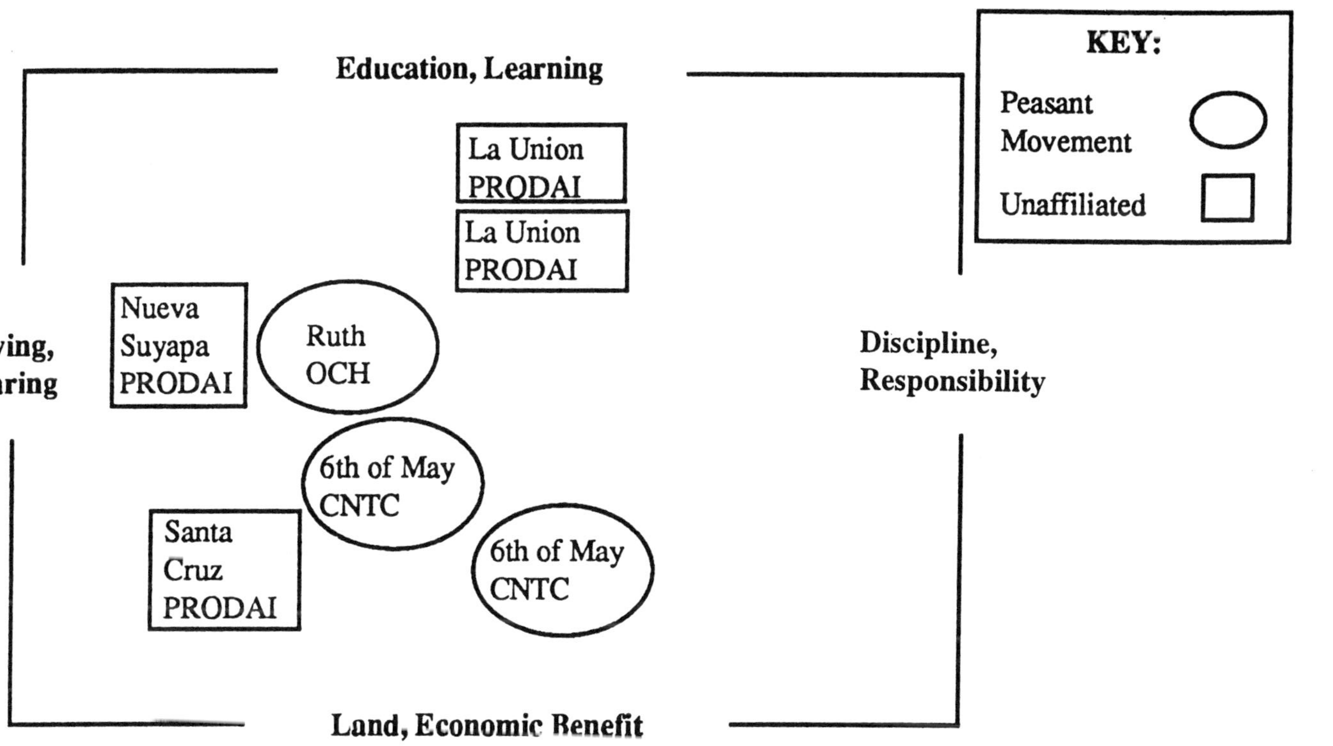

Figure 2. The social geography of seven rural Honduran community groups: peasant organization involvement

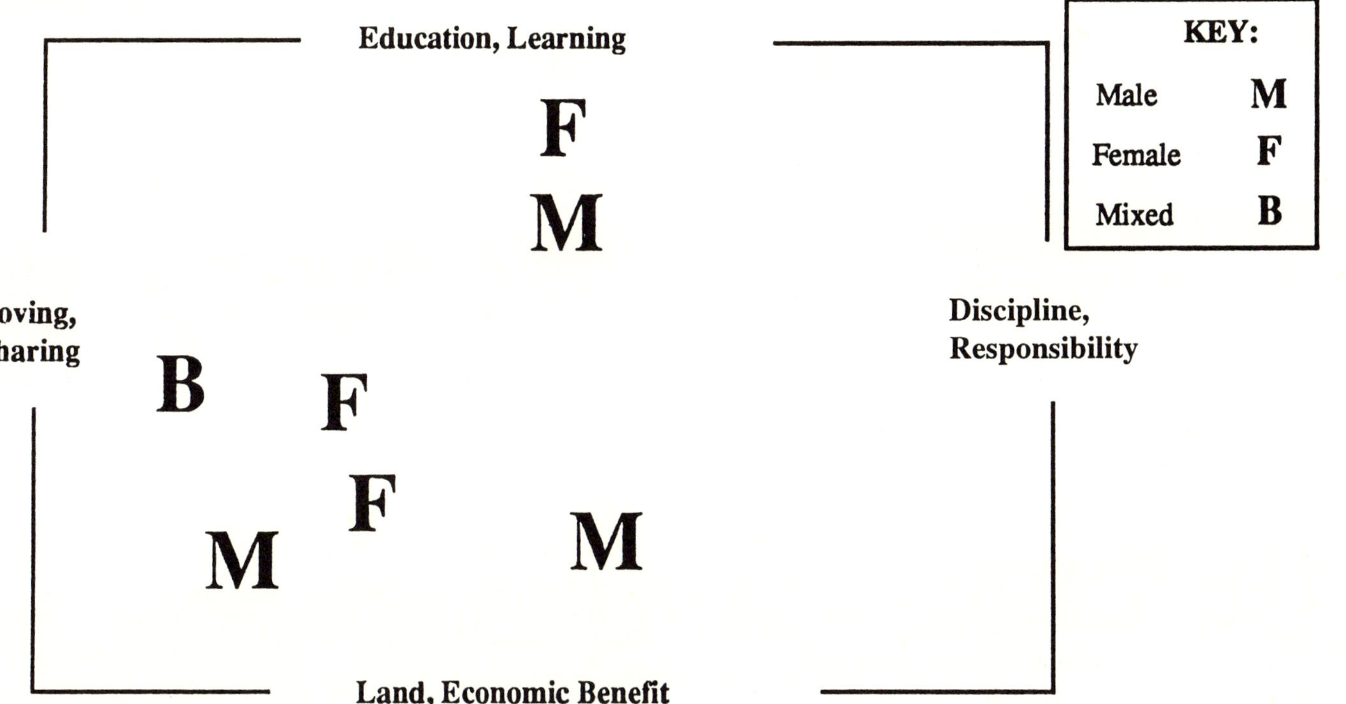

Figure 3. The social geography of seven rural Honduran community groups: gender

participatory rural development, I have not attempted to write from a postmodern perspective. Instead I have evaluated postmodernism from pragmatic and positivist perspectives. In the first section, I tested the case for methodological pluralism that follows from that critique. A textual analysis of recent work in the field of participatory development reveals that works from each of the represented perspectives provides useful knowledge and insights not generated in the others. Positivist works provide information about significant cause-effect relationships, while the pragmatist works offer useful applied suggestions on what to do in politicized environments. Uphoff's (1992b) work using a hermeneutic approach wrestles with the problem of how subjective factors such as social energy can be used to obtain unusually good project outcomes. With the two critical theory texts, the emphasis shifts from a project to a societal level. The primary contribution is, again, insight about how the subjective can influence the objective, but this time the subjective is Freirian-style consciousness-raising and the objective is societal-level change. The practical utility of the discourses of each of the paradigms supports the postmodern position of making room for all of the discourses, rather than trying to develop a single general theory of participatory development administration.

In the second part of this essay, I explored another aspect of postmodernism's "exaltation of diversity." In the participatory self-evaluations conducted in similar Honduran rural development projects, beneficiaries identified a wide range of project means and ends. This multiplicity suggests the need for a sensibility that values local cultural diversity. A justification for participatory development administration informed by postmodern concerns is that participatory approaches, like the postmodern perspective, seek to open spaces congenial to multiplicity and differences. The practical utility of this aspect of postmodernism suggests that practitioners can selectively appropriate from postmodern thought without having to buy into its more questionable elements.[7]

NOTES

1. I would like to thank the Center for Latin American Studies and the Graduate School of Public and International Affairs of the University of Pittsburgh for funding my field work in Honduras.

2. This observation is consistent with Burrell and Morgan's (1979) experience with foundational texts in sociology.

3. Paulston (1992) welcomes Freire's valuable contribution in bringing critical theory to education, but critiques Freire for "privileging the teacher-intellectual's views over those of the learners" (p. 198) in his practice. This top-down ideological orientation is, fortunately, not apparent in the Freirian-inspired change orientations presented by either Oakley and Marsden (1985) or Goulet (1989). Oakley and Marsden (1985) argue for an approach to participation in which there is an "absence

of any pre-determined models and the emphasis upon the emergence spontaneously of a relevant approach from below" (p. 67). Goulet (1989) is in agreement when he writes that "authenticity means locating true decisional power in non-elite people, and freeing them from manipulation and co-optation" (p. 168).

4. As noted by Edgar Schein in *Organizational Culture and Leadership* (1985), culture can be formed at a variety of levels, from national culture to subcultures within organizations. Schein explains that the problem is to determine "within a broader host culture—the unique features of a particular social unit in which we are interested." Similarly, in my study, I do not deny the importance of a national culture, or a specific culture within the organized peasant movement in Honduras. My focus, however, is on culture as it develops within small groups.

5. Honduras has had civilian rule off and on during this century. The most recent period of civilian government has been in effect since 1981. A constitution enacted in 1982 provides for the legal separation and independence of the branches of government (Rosenberg, 1989).

6. The method described here differs from that given by Uphoff (1991, 1992a) in that Uphoff has the project staff, rather than the beneficiaries, develop the list of goals.

7. See for example Baudrillard's argument (1984) that: "Playing with the pieces—that is postmodernism" (p. 20).

REFERENCES

Baudrillard, J. (1984). Games with vestiges. *On the Beach, 5,* 19–25.

Bhatnagar, B., and A. Williams. (1992). *Participatory development and the World Bank.* World Bank Discussion Papers. Washington, DC: World Bank.

Bernstein, R. (1983). *Beyond objectivism and relativism: Science, hermeneutics, and praxis.* Philadelphia: University of Pennsylvania Press.

Brockett, C. (1988). *Land, power, and poverty: Agrarian transformation and political conflict in Central America.* Boston: Unwin Hyman.

Brown, H. (1992). Social science and society as discourse: Toward a sociology for civic competence. In S. Seidman and D. Wagner (eds.), *Postmodernism and social theory: The debate over general theory* (223–243). Cambridge, MA: Blackwell.

Burell, G. and G. Morgan. (1979) *Sociological paradigms and organizational analysis: Elements of the sociology of corporate life.* London: Heinemann.

Carroll, T. (1992). *Intermediary NGOs: The supporting link in grassroots development.* West Hartford, CT: Kumarian Press.

Cernia, M. (1988). *Nongovernment organizations and local development.* World Bank discussion papers. Washington, DC: World Bank.

Denhardt, R. (1990). Public administration theory: The state of the discipline. In N. Lynn and A. Wildavsky (eds.), *Public administration: The state of the discipline* (43–72). Chatham, NJ: Chatham House.

Esman, M., and N. Uphoff. (1984). *Local organizations: Intermediaries in rural development.* Ithaca, NY: Cornell University Press.

Finsterbusch, K., and W. Van Wicklin. (1987). The contribution of beneficiary participation to development project effectiveness. *Public Administration and Development, 7,* 1–23.

Finsterbusch, K., and W. Van Wicklin. (1989). Beneficiary participation in development projects: Empirical tests of popular theories. *Economic Development and Cultural Change, 37,* 573–593.

Freire, P. (1970). *Pedagogy of the oppressed.* New York: Seabury.

Gerson, P. (1993). *Popular participation in economic theory and practice.* Baltimore: Johns Hopkins University Press.

Gottlieb, E. (1989). The discursive construction of knowledge: The case of radical education discourse. *Qualitative Studies in Education, 2*(2), 131–144.

Goulet, D. (1989). Participation in development: New avenues. *World Development,* 17(2), 165–178.

Gow, D., and J. Van Sant. (1983). Beyond the rhetoric of rural development participation: How can it be done? *World Development,* 11(5), 427–446.

Habermas, J. (1972). *Knowledge and human interests.* Boston: Beacon Press.

Heshusius, L. (1994). Freeing ourselves from objectivity: Managing subjectivity or turning toward a participatory mode of consciousness? *Educational Researcher,* 23(3), 15–22.

Hopenhayn, M. (1993). Postmodernism and neoliberalism in Latin America: Boundary 2. *The Postmodernism Debate in Latin America,* 20(3, special issue), 93–109.

Huff, A. (1990). Mapping strategic thought. In A. Huff (ed.), *Mapping strategic thought* (11–49). New York: John Wiley and Sons.

Johnson-Laird, P. (1983). *Mental models.* Cambridge, MA: Harvard University Press.

Korten, F. (1983). Community participation: A management perspective on obstacles and options. In D. Korten and F. Alfonso (eds.), *Bureaucracy and the poor: Closing the gap* (181–200). West Hartford, CT: Kumarian Press.

Lacey, A. (1976). *A dictionary of philosophy.* New York: Charles Scribner's Sons.

Levy, J. (1974). Psychobiological implications of bilateral asymmetry. In S. Dimond and J. Beaumont (eds.), *Hemisphere function in the human brain.* New York: John Wiley and Sons.

Liebman, M., and R. Paulston. (1994). Social cartography: A new methodology for comparative studies. *Compare,* 24(3), 233–245.

Lyotard, J.F. (1984). *The postmodern condition: A report on knowledge.* Translated by G. Bennington and B. Massumi. Minneapolis: University of Minnesota Press.

Mausolff, C. (1994). *The popular education programs of the Honduran peasant movement.* Unpublished essay. Pittsburgh, PA: Graduate School of Public and International Affairs, University of Pittsburgh.

Meyer, A. (1991). Visual data in organizational research. *Organization Science,* 2(2), 218–236.

Migdal, J. (1974). *Peasants, politics, and revolution: Pressures toward political and social change in the third world.* Princeton, NJ: Princeton University Press.

Nicholson, L. (1992). On the postmodern barricades: Feminism, politics, and theory. In S. Seidman and D. Wagner (eds.), *Postmodernism and social theory* (82–100). Cambridge, MA: Blackwell.

Oakley, P., and D. Marsden. (1984). *Approaches to participation in rural development.* Geneva: International Labour Office.

Osorio, J. (1990). Controversies and assertions of popular education in Latin America. Santiago, Chile: Secretaria General del Consejo de Educacion de Adultos de América Latina (CEAAL).

Patton, M. (1990). *Qualitative evaluation and research methods.* 2nd ed. Newbury Park, CA: Sage.

Paul, S. (1987). *Community participation in development projects: The World Bank experience.* World Bank Discussion Papers. Washington, DC: World Bank.

Paulston, R. (1992). Ways of seeing education and social change in Latin America: A phenomenographic perspective. *Latin American Research Review.* 27(3), 177–202.

Paulston, R. (1993). Mapping discourse in comparative education texts. *Compare,* 23(2), 101–114.

Paulston, R., and S. Rippberger. (1991). Ideological pluralism in Nicaraguan university reform. In M. Ginsburg (ed.), *Understanding educational reform in global context* (179–210). New York: Garland.

Popkin, S. (1979). *The rational peasant: The political economy of rural society in Vietnam.* Berkeley: University of California Press.

Popper, K. (1972). *Objective knowledge: An evolutionary approach*. London: Oxford University Press.

Posas, M. (1992). Los sindicatos y la construccion de la democracia. In *Puntos de Vista* (95–110). Tegucigalpa, Honduras: Centro de Documentacion de Honduras (CEDOH).

Przeworski, A., and H. Teune. (1970). *The logic of comparative social inquiry*. New York: John Wiley and Sons.

Rosenberg, M. (1989). Can democracy survive the democrats? From transition to consolidation in Honduras. In J. A. Booth and Mitchell A. Seligson (eds.), *Elections and democracy in Central America*. Chapel Hill: University of North Carolina Press.

Ruhl, M. (1985). The Honduran agrarian reform under Suazo Cordova, 1982–85; an assessment. *Inter-American Economic Affairs* 29 (2), 63–80.

Schein, E. (1985). *Organizational culture and leadership*. San Francisco: Jossey-Bass.

Sijbrandij, P., and J. Ooijens. (1988). La alfabetizacion popular y organizacion campesina; el proyecto con la ANACH. In A. van Dam, J. Ooijens, and G. Peter (eds.), *Educacion popular en America Latina*. La Haya, Holanda: Centro para el Estudio de la Educacion en Paises en via de Desarrollo (CESO).

Uphoff, N. (1991). A field methodology for participatory self-evaluation. *Community Development Journal, 26(4)*, 271–285.

Uphoff, N. (1992a). Monitoring and evaluating popular participation in World Bank assisted projects. In B. Bhatnagar and A. Williams (eds.), *Participatory development and the World Bank*. Washington, DC: World Bank.

Uphoff, N. (1992b). *Learning from Gal Oya: Possibilities for participatory development and post-Newtonian social science*. Ithaca, NY: Cornell University Press.

Welch, A. (1993). Class, culture, and the state in comparative education: problems, perspectives, and prospects. *Comparative Education, 29 (1)*, 7–27.

Listening to the Other

Mapping Intercultural Communication in Postcolonial Educational Consultancies

Christine Fox

Introduction: Dilemmas of Choice

Members of indigenous societies responsible for schooling and education in their own countries face some fundamental choices between what is often proffered as knowledge in industrialized systems, and what is understood as knowledge in its local cultural context. The choice is often seen as a contrast between linear, rational, fragmented forms of knowledge and a more spatial approach to knowledge. Yet a continuum of approaches, falling somewhere between these extremes, is very much evident in current educational theory internationally. This is no less so in so-called third world societies, where national education systems have relied heavily on international input to educational planning.

The question of whose knowledge, whose way of seeing educational reform is particularly apparent in South Pacific island countries that have not yet freed themselves entirely of colonialism, or at least neocolonialism (Finnegan, 1995). In other words, as Edward Berman points out (Berman, 1992), it is still quite common for the donor agent to transfer development schemes and specific projects without consultation with individuals in the recipient country or at best after only token consultation.

In spite of their enormous differences in cultures, contexts, locations and history, small Pacific states still tend to be recipients of curriculum-building projects that are bureaucratic, technological, scientific enterprises. Current trends in providing development assistance in education mirror other corporate developments in capital works programs. Tenders are put out to consultant firms, university departments and multinational private agencies who are vying for space and time in places they formerly knew or cared little about. There is a sense of contextlessness and timelessness about such educational ventures. The actual processes necessary to bring about change tend to be ignored. James Brown considers, as Stephen Klees did a few years earlier (Klees, 1986), that while many development projects fail at least in part due to the unsuitability of the project (Dyer, 1996), perhaps more importantly, they fail by ignoring the importance of managing change, and through the inappropriateness of the consultants' approaches (Brown, 1992). The managerial, positivist style of educational rationalism sits uneasily with the more fluid and in many respects more complex interpersonal relationships in most Pacific island cultural settings (Crossley, 1992).

Lasting change usually occurs as a result of a change in interrelationships. It may be the result of changing fundamental structures, changing values or changing ways of doing things. People's relationships with each other also change in order to bring about that modification or transformation (see Fullan, 1993; Goodlad, Soder and Sirotnik, 1990). Thus the role of a consultant is usually expected to be that of a change agent rather than a program deliverer. To consult is to ask advice from, to consider, to refer to something or someone for information. A consultant therefore should be someone who seeks and gives advice by communicating with her or his colleagues. Yet in many educational reform projects carried out across cultures, effective communication of ideas is often blocked, even among those who are anxious to break down barriers of perceived difference.[1] Authentic communication—that is, communication that claims to be free of coercion and open to all ways of seeing in context—appears difficult, if not impossible, to achieve.

It is fruitful to explore this continuum of ways of communicating through a social mapping method. In this chapter, some preliminary ideas are proposed, giving first a picture of historically situated approaches to intercultural communication, and, secondly, mapping the spatial relationships between interlocutors in specific modes of attempted authentic communication.

BLOCKS IN INTERCULTURAL COMMUNICATION

Whether consultants can communicate effectively with indigenous groups, and vice versa, rests not on whether ethnic cultures and ways of thinking are actually compatible. After all, intercultural interests, such as feminist values or working class values, may be stronger identifiers than ethnicity. Similarly, one and the same person may be versed in both so-called Western knowledge as well as indigenous knowledge. The question is whether power interferes with the expression of cultural values and cultural knowledge to such an extent that authentic communication is rendered impossible. For surely, if the two cultural groups were perceived as equally powerful, and their values and beliefs able to be equally expressed without coercion, there would be scant reason why an "authentic" discussion about those differences could not take place—that is, with sincerity, truthfulness within one's own position and a willingness to accept and engage in different expressive discourse practices.

The historical record shows that the web of mutual ignorance and apparent incompatibility can, over time, become untangled and eventually understood. The illusion is not that intercultural communication exists, but that the construction of an embodied identity creates a barrier. Attitudes change. New language is created. New metaphors develop to translate new meanings more adequately. There are indeed ways of discovering how people can filter meaning through their own cultural world views in order to "get into," as Corson says, another world view (Corson, 1995, p. 185).

A social map may seem an appropriate way to identify world views to try to see more graphically where the connections are made or broken (Carrier and Carrier, 1995). On the other hand, it would be dangerous to make assumptions of cultural or ideological differences or to fix one set of cultural "ways of knowing" as if it were somehow located in a different time and space, and therefore incompatible with other ways of knowing that have not been contextualized, gendered or placed in a social or "class" perspective. The usefulness of mapping in these intercultural situations must therefore be treated with extreme caution.

The idea of incompatibility tends to come mainly from an ideological construction of unequal relationships of power between two cultures. As Homi Bhabha says:

> An important feature of colonial discourse is its dependence on the concept of "fixity" in the ideological construction of otherness. Fixity, as the sign of cultural/historical/racial difference in the discourse of colonialism, is a paradoxical mode of representation: it connotes rigidity and an unchanging order as well as disorder, degeneracy and daemonic repetition. . . . The stereotype . . . is its major discursive strategy. (1994, p. 66)

John Thompson has put forward a comprehensive view of how the ideological construction of Otherness operates. He classifies these ways of operating as including the legitimation of the current relations of domination, the dissimulation or concealment of this domination, a unification process that embraces powerful individuals in a collective identity, the fragmentation of the various nondominant groups of "other" or the reification of the unequal state of affairs as if it were permanent, natural and outside of time (1990, pp. 60–65).

The illusion of incompatibility is therefore related more to politics and power than to a genuine attempt to look at the dynamics of intercultural communication. The expression of ethical concepts may be culturally constructed, but they are neither universally fixed, nor are they beyond the realms of engagement and debate interculturally.

Those who have worked in educational development projects will be aware of the dangers of generalization, and will know some consultants who have achieved excellent rapport with their counterpart educators. They will also know about other consultants who have portrayed a colonial-style arrogance in their dealings with people from cultures different from their own (Epstein, 1995). The problematic nature of the role of a consultant cannot be reduced to the anecdotal. It concerns ideological orientations, and may involve a severe misuse of power and loss of trust. The issue for some might be a matter of rival justices, those of human rights on the one hand, and those of utility on the other (MacIntyre, 1988). However, in the case of the dominant "orthodox" model of development planning, the claim of utility is tenuous, and more likely illustrates a case of economic opportunism.

How, then, can the social cartographer avoid the dangers of duplicating the negative effects on international relations of those global geographic mapping projections used as instruments of imperial control over oppressed populations? In Tally's chapter in this volume, the Foucauldian argument that mapping had an enormous impact on the Western colonization of Asia and Africa is a case in point. Tally rightly points out the usefulness of Foucault's analysis of power through his spatial depiction of relationships captured so effectively in the Panoptic diagram of *Discipline and Punish* (Foucault, 1977). Similarly, it should be possible to analyze the power relations, and the ideological relationships depicted in different approaches to intercultural interaction.

How easily the visual can represent relationships of unequal power (Mitchell, 1994). Take, for example, typical paintings of early "culture contact" in the Pacific: The arriving ship with its tall sails rests in a distant harbor, the sun's rays lighting up the center of the picture where the stately ship lies, its upright and lordly British or other uniformed conquerors on board gazing down at the darkened shores; "hordes" of "natives," usually depicted in a bowed position, or in disarray, sometimes brandishing futile weapons, always out of the sunlight, appear lower in the picture or in a corner, and are always relationally minuscule and indistinguishable compared with the large figures, whose features clearly are painted and who are supposedly in command of the situation. This is a classic example of Charles Hampden-Turner's argument that value judgments are implicit in cataloguing a perceived status quo from the dominant perspective (Hampden-Turner, 1981). It is symbolic of maps drawn with Europe or the United States in the center of the map, without the advantage of a Peters projection, so that the countries of the "south" appear as either smaller, or marginal, in relation to the central power.

Mapping world views, as Paulston has done, creates its own sense of "state of the art," where movements coming from a number of orientations are seen as occupying a spatial surface, in terms of criteria for locating each theory. In this volume, Paulston and Liebman call such a map a Derridean momentary crystallization of the space claimed by social and ideological ways of seeing. Does, perhaps, this art of mapping offer a useful tool for the study of intercultural communication?

HISTORICAL TRENDS

Several general trends with regard to dominant cultural beliefs about other cultures are discernible in the literature. I have classified these as the 4 Ds, for they concern who has been seen to be culturally *deficient, disadvantaged, different* or *dominated*. The fifth D, that of *dialogue*, creates the conditions for developing authentic communicative action interculturally.

The negative features of the first three stages can be seen clearly as expressions of racism (deficiency paradigm), inequity (disadvantage paradigm), or intolerance (difference paradigm). The fourth category has been challenged as being somewhat static, and not addressing the multiple contexts of power and ideology (cultural reproduction paradigm). The last category is transformative, emancipatory and, as Freire would say, concerned with conscientization (Freire, 1972).

The trends can be categorized in terms of typical interactional manifestations by subordinate and dominant cultures, as I have outlined in Figure 1.

Paradigms of intercultural interaction	"Typical" interactional manifestations by subordinate culture	"Typical" interaction manifestations by dominant culture
Deficiency	isolation, silence	racism, ignorance
Disadvantage	failure, "acting white"	inequality, injustice
Difference	tokenism, "acting black"	intolerance, curiosity
Domination	resistance	cultural reproduction
Dialogue	transformation, emancipation	transformation, emancipation

Figure 1. Trends in approaches to intercultural education

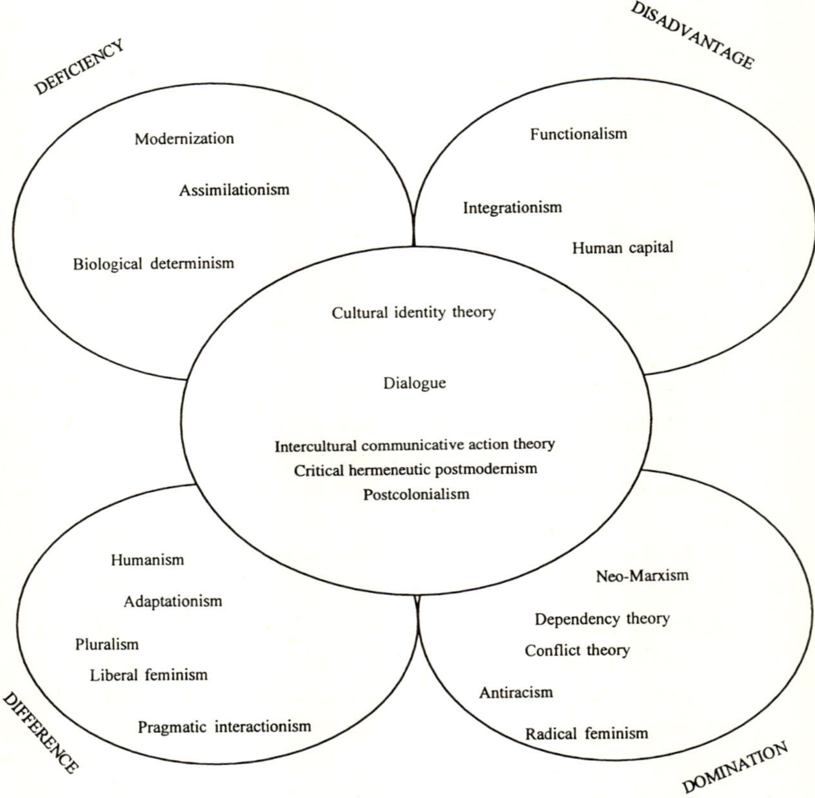

Figure 2. Mapping theories of intercultural interaction

There are problems with this type of modernist dichotomy, since, while each paradigm has historical antecedents, its articulation is also discernible at any one moment within any individual at any level of inner and outer consciousness. If any of the first four paradigms is adopted while at the same time unacknowledged and unchallenged, authentic communication cannot take place, since each approach contains an assumption of unequal power, be it political, linguistic, social, economic or psychological. It is therefore more useful to try to approach the process by mapping these concepts in their historical and ideological contexts, as is illustrated in Figure 2.

Concepts of deficiency were prevalent in most early writings about culture contact. Groups of people in "the colonies" were considered to be childlike, less intelligent or less able to reason. The literature is full of accounts of so-called primitive people, and the notion of primitive carries on today, in less obvious ways, such as educators trying to develop programs for students with a supposed cultural deficit. English literature syllabuses for students in many African schools were until very recently exclusively concerned with UK authors; African authors were not included (Lewin, 1985).

The idea of disadvantage has changed its meaning over the years, away from the deficit paradigm towards a socioeconomic definition, but a confusion generally remains between the oppressive nature of poverty, and the transformatory nature of learning critically. The term "disadvantage" tends to link economic poverty with intellectual and educational poverty and even poverty of cultural spirit: the same deficient concept of not as "good" as the dominant culture.

Cultural difference was a way of defining intercultural contexts. Differences were acknowledged, but the question remains, different from what? The answer, is invariably, different from the norm, the dominant culture, the culture with power. Typical classroom programs illustrating this paradigm are those social studies units where another culture is studied superficially, out of context, as a curiosity or, perhaps even worse, as a "problem" culture with only wars, famines, disease and poverty to describe it.

The fourth paradigm, which gained widespread credence in the late 1970s, was the increasing attention paid to the ideological concept of dominance, including structural discrimination, racism, sexism and the institutionalization of power favoring the dominant culture. This was an important and critical stage, incorporating the concept of cultural reproduction, and has been followed at least in some circles by a determination to act, to empower the so-called powerless.

The fifth paradigm is that of a process of dialogue—the critical process of making meaning, of sharing meanings, and of building bridges across those multiple realities and multiple truths. It involves what Jürgen Habermas calls "ethical discourse" (Habermas, 1990). It has been referred to by Cameron McCarthy (1990), who says "We need to emphasize the symbolic, signifying, languaging dimensions of social interaction and their integral relationship to both systems of control and strategies for curricular reform" (p. 10). From this perspective, then, of the symbolic and signifying dimensions of social interaction, the tool of social cartography is necessary.

Filters in Intercultural Communication

Blocks to communication can be overcome provided there is an intention to act communicatively rather than strategically (Habermas, 1984, 1987). In intercultural communication, a number of filtering processes take place, either consciously or unconsciously (Fox, 1992). These processes are "peppered" with metaphors, a metaphorically visual way of constructing meaning. Paulston and Liebman (in this volume) talk of developing a "visual dialogue"—a way of constructing metaphors to locate the discussion. A social map is itself a metaphor of the language of space.

Metaphor creates meaning through the use of pictures projected into the minds of the interlocutors, who are seeking to describe the relationships between ideas and objects. Given different cultural backgrounds, it is not surprising that these relationships are very revealing of how people are perceiving their reality. With practice at listening carefully, the hearer can anticipate different interpretations and possibly evidence of bias through cultural filtering. For example, a consultant may use a number of military metaphors in relation to her or his task, indicating the strategic nature of their vision of "delivery," "going in" or "setting targets." On the other hand, the consultant may use metaphors connoting care or collegiality, using more reflective words.

Filtering processes can be mapped, as illustrated in Figure 3. The processes overlap and intersect; they are all-encompassing, yet they are discrete. They should be depicted as four-dimensional processes, because they relate to time as well as space, language, body, aesthetics, ethics and hermeneutics.

Different cognitive classifications of the world are ways of seeing that can be mapped, shared, disputed, remapped and even agreed upon. As Val Rust emphasizes elsewhere in this volume, it is important not to fall into the trap of looking for metanarratives of different ways of understanding, expressing concepts and defining time and space or relationships. What is interesting, however, is to conduct some text analysis of particular moments of an intercultural dialogue, to discern intentions and assumed identities, and to locate moments of miscommunication. In the context of educational

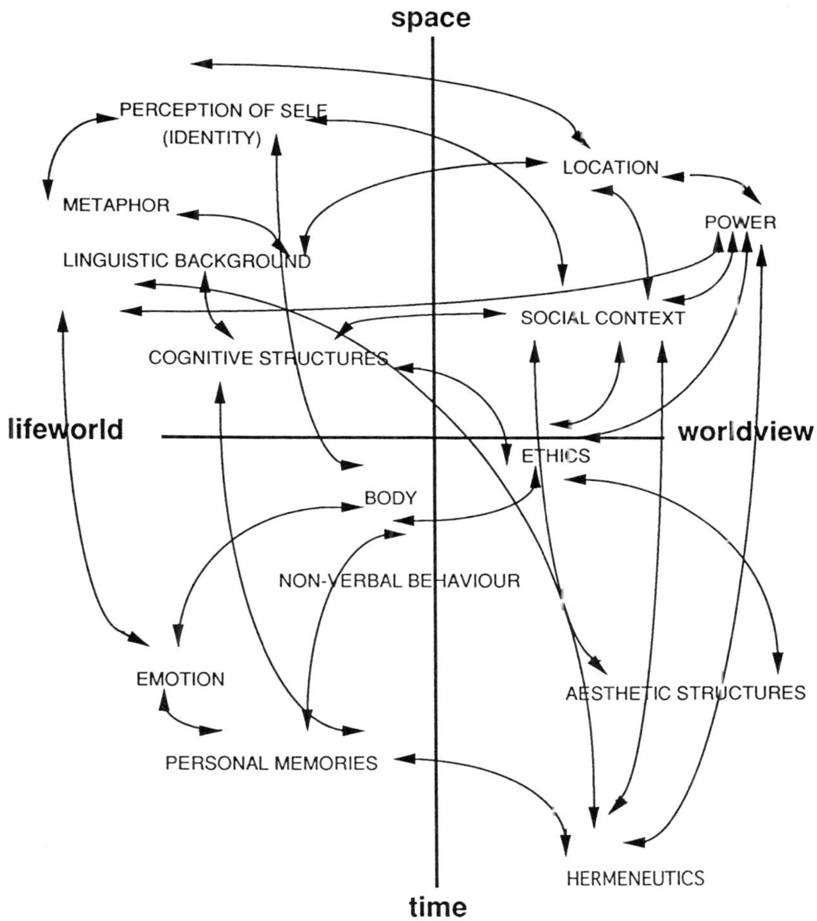

Figure 3. Mapping the filters of intercultural communication

consultancies, the kinds of difference may refer, for example, to the classification of the concepts "meeting," "workshop" or "assessment." A meeting taking place in a corporate organization usually has clear intentions, structures and outcomes. A meeting may also describe a meeting of minds for discussion and illumination only, based on an entirely different set of assumed cultural behavioral and operational norms. Normative behavior is indeed based on certain cognitive formulations.

In a study carried out by the author in the Pacific, an interesting miscommunication occurred between consultant and indigenous educator that was recorded and is partly reproduced below (Fox, 1992). The text is part of an exchage during a workshop held to discuss a new textbook being written by the indigenous educators with the assistance of a consultant. The task was to write and illustrate a chapter on jobs for graduating students. The consultant had classified the jobs into "indoor" jobs and "outdoor" jobs. The indigenous educators evidently had other classifications in mind, such as urban and rural, subsistence and commercial, village and town.

For the outdoor section, some new items were being written, which included taking part in a fisheries project, commercial farming, building construction, logging, copra cutting, harvesting cocoa, milling timber, boat building, curing and drying pandanus leaves and looking after a poultry farm. The following exchange concerns a suggested picture of a man in a garden, with a tractor in the picture. The Pacific Island participant wanted to take the tractor out of the picture.

> *Consultant (C):* But we are only interested in finding out about their interest in outdoor activities.
>
> *First Pacific Island teacher educator (P) :* We want the picture to look like subsistence farming, not commercial farming.
>
> C: All we want to know [is] if they are interested in working in a garden. It doesn't matter if it's commercial or subsistence.
>
> *First P:* But if we put in a tractor we are encouraging them to go into commercial farming. Villagers do not use tractors.
>
> *Second P:* Those are outdoor jobs—two different types of operations going on there. The student can choose which one he likes best.
>
> C: What we're looking at is the skills involved. Planting, etc. It doesn't matter if it's commercial or village. Just the skills.
>
> *First P:* We need a good variety. If we put both, they are close together. We've also got nursery work. I suggest we leave one out or combine them into one.

Second P: Could we change that commercial farming to ploughing a farm?

C: We have to fit in only a few extra pictures. We have to cut down. *The final agreement was to use two pictures: one with, one without the tractor.* (p. 374)

The consultant was looking at skills, at general classifications of "indoors" and "outdoors." But these were not relevant to the Pacific educators, given the far more significant cultural and economic differences denoted by jobs for commercial or subsistence farming . Yet the consultant failed to come to grips with this classification. Had the consultant reflected on the situation, she would have had an opportunity to discover the ethical, linguistic, categorizational, metaphorical dimensions of the interaction, and possibly found other ways to resolve the dilemma.

The mapping process enables the researcher to locate different interpretations of everyday events, both past and present, along dimensions of time and space that are neither linear nor fixed. Intercultural communication relies very much on context, both cultural and situational. "The collective background and context of speakers and hearers determines interpretations of their explicit utterances to an extraordinarily high degree" (Habermas, 1984, p. 335). Consultants are caught up in linear, fixed time, with project deadlines, limited moments in another culture and little likelihood of continuity in the change process. Members of indigenous societies are immersed in more continuous (but not necessarily linear) time frames, deeply committed to the evolving educational context, and expect a long, even cyclical, process of revisiting changes.

In the study mentioned above, another consultant made the following observation about time. Again, this consultant failed to appreciate how very culturally contextualized he was being. This text was recorded during an interview held with the consultant about his time on the Pacific project after he had returned home to Australia (Fox, 1992).

They are not a people who face open disagreement squarely. They will go all directions of sideways to avoid direct disagreement in a meeting, and so the easy agreement, apparent agreement. . . .

Because it's rush rush hurry hurry we've only got an hour, those disagreements are pushed aside and those are the reasons why things don't get done. . . . You go away and nothing happens because fundamentally there has been disagreement that you haven't had time to talk about. . . .

It's very difficult for a [Pacific Island person] to say "I don't think we should do that." . . . . And so I probe a bit and shut up a bit and I'm still aware that the watch is ticking and we haven't got all day. . . . (p. 364)

The problem here is not only about different perceptions of time. It is about culturally appropriate behavior. A further filtering process concerns the recognition of those normative behaviors and actions that are different across cultures. They relate to perceptions of body, of aesthetics, of anticipated emotional responses to certain behaviors, which can be mapped as in Figure 3. Sensitivity to these behaviors is an essential communicative task, whether it be forms of politeness, greeting, personal space, order of talking, eating behaviors, the use of silence, or ways of using body language (Fox, 1992).

This notion of exchanging understandings varies significantly from the notion of adaptation, which is often discussed in many works on intercultural communication (Kim, 1988; Anderson, 1994). Generally, theorists have been more concerned with individual adaptation and individual communicative competence rather than with interrelationships in an intercultural encounter. Again, an adaptation process tends to be unidirectional, not reflecting the interrelationships plotted in Figure 3. Yet the text above indicates how necessary it is to translate the implications rather than expect one group to adapt (see Taylor, 1994).

The ability to listen to the language of these cultural filters is frequently commented on by indigenous educators. A survey of Pacific Island attitudes toward consultants from outside (Fox, 1992) revealed the following list of attributes of a "good" consultant as being:

a good listener . . .
an active listener . . .
someone who understands us . . .
someone who contributes ideas we can really use . . .
someone who knows what we need . . .
a sympathetic person . . .
an honest person . . .
someone who doesn't always say "I", "I", "I" . . .
someone who sits down with us and is not condescending . . .
someone I can feel free to discuss things with . . .
someone who will observe and assist before they exchange their expertise . . . (p. 355)

It is not difficult to see this text as a list of prerequisites for effective communicative action. The Pacific Island educators volunteered numerous observations that directly related to communicative action. They noted that miscommunication tended to occur when consultants were not aware of cultural values, attitudes and practices, as well as ways and hierarchies of speaking. They did not see this as an impossible situation, since they felt that anyone who wished to would be able to discover these differences; or, in other words, they would have the ability to filter these cultural differences. What they called impediments to communication included ignorance of other cultures and lack of involvement or concern with the outcomes for their particular situation. There seems to be an unspoken understanding that the process of "filtering" is an active one.

ETHICAL AND POLITICAL FILTERS

One complex filtering process concerns ethical and political interpretations of action and interaction. The power carried by a consultant is combined with trust that the work is for the good of the recipients and their situation. Many studies indicate that such trust is often missing (Berman, 1992), particularly as much of the "good work" is thinly disguised paternalism locating the recipients as "Other" and subaltern. John Beverley, in this volume, emphasizes the dangers of mapping the Other as if the cartographer is desirous of fixing the Other in, at best, a different, but more likely a subaltern category. The categorization out of context of Otherness as subalternity assumes that same "fixity" that Bhabha (1994) decries; it is the very labeling that Spivak (1990) so vociferously condemns as evidence of norming domination and elite status. As Beverley implies, "Otherness" is a reflection of the perspective of the person who speaks from an elite status. The academic elite are in danger of creating their own communicative dichotomy of Other/Us, which reflects the structural dualism of the economic and political power of the dominant over the subordinated. They deny the flexibility and the movement of interrelationships where such distinctions are not valid, are consciously set aside, or are not relevant to the particular context. The question is an ethical one.

CONCLUSIONS

Beverley may have a case against mapping in the sense that to draw a map is indeed to locate and fix one position in relation to another. A map by its very nature must try to identify a place for everything. The problem, however, may be more about locating the cartographer than of constructing the maps. Beverley explores the possibility of making the shift from objectivity to solidarity, but is skeptical of doing so in situations where power over, and exploitation of, the "Other" radically differentiates the participants. He laments the situation of the "absence or difficulty or impossibility of representation of the subaltern" and looks forward to new forms of social empowerment and transformation.

The utility of social mapping as used in this chapter to explain intercultural relationships may go some way to depicting the very social empowerment and transformation that Beverley seeks. Just as Foucault was able to use the visual description of the Panoptic diagram to explain power relations, so it is important to use metaphor and various forms of linguistic classification to depict communicative and ethical relationships.

Charles Taylor sees the self as having a "plurality of visions" (1985, p. 26). Thus the question of rightness is not determined wholly by one culture or another, nor are people locked into particular positions or ideologies by being "Other." There is a danger of being determined by cultural relativity. Gayatri Spivak aptly summarizes this point when she objects to being positioned as "a third world" person, or "a woman," or a "postcolonial": "we who are from the other side of the globe very much fight against the labeling of all of us under that one rubric [third world], which follows from the logic of neo-colonialism" (1990, p. 114).

Edward Said (1978, 1983) has described one kind of projected vision as "Orientalism," which he sees as "a kind of Western projection onto and will to govern over the Orient" (1978, p. 95). And Seyla Benhabib notes that "the logic of binary oppositions is also a logic of subordination and domination" (1992, p. 15).

Nevertheless, it would be a great pity if a poststructuralist view of multiple voices, multiple interpretations, and no "one way" ethical solution were taken to mean that there were no need to pursue the notion of reaching a mutual understanding. By recognizing the fluidity of making meaning across cultures, by not allowing one set of meanings to swamp another and by being able to map ideas and locate ideological stances, the paths are open in a number of directions for discovering how people filter meaning through their own cultural world views and engage in authentic intercultural communication.

Note

1. J. Mayo (1993) cites a view from the village on this dilemma: "I the villager have never seen you, have never known you, have never talked with you. I hear my needs spill out of your mouth. Forgive me, but you must stop talking for me" (p. 60).

References

Anderson, L. (1994). A new look at an old construct: Cross-cultural adaptation. *In-*

ternational Journal of Intercultural Relations, 18(3), 293–328.

Benhabib, S. (1992). *Situating the self.* Cambridge, MA: Folity.

Berman, E. (1992). Donor agencies and third world educational development, 1945–1985. In R. Arnove, P. Altbach, and G. Kelly (eds.), *Emergent issues in education: Comparative perspectives* (57–74). Albany, NY: SUNY Press.

Bhabha, H. (1994). *The location of culture.* London: Routledge.

Brown, J. (1992). Educational innovation in developing countries: Some considerations for the international consultant. *Canadian and International Education, 21*(1), 44–54.

Carrier, J. and A. Carrier. (1995). Every pictures tells a story: Visual alternatives to oral tradition in Ponam society. In R. Finnegan and M. Orbell (eds.), *South Pacific oral traditions,* 195–214. Bloomington, IN: Indiana University Press.

Corson, D. (1995). World view, cultural values and discourse norms: The cycle of cultural reproduction. *International Journal of Intercultural Relations, 19* (2), 183–196.

Crossley, M. (1992). Collaborative research, ethnography and comparative and international education in the South Pacific. In R.J. Burns and A.R. Welch (eds.), *Contemporary perspectives in comparative education* (171–192). New York: Garland.

Dyer, C. (1996). Primary teachers and policy innovation in India: Some neglected issues. *International Journal of Educational Development, 16* (1), 27–40.

Epstein, I. (1995). Comparative education in North America: The search for others through the escape from self? *Compare, 25* (1), 5–16.

Finnegan, R. (1995). Why the comparativist should take account of the South Pacific. In R. Finnegan and M. Orbell (eds.), *South Pacific oral traditions* (1–29). Bloomington, IN: Indiana University Press.

Foucault, M. (1977). *Discipline and punish: The birth of the prison.* New York: Pantheon.

Fox, C. (1992). *A critical analysis of intercultural communication: Towards a new theory.* Ph.D. dissertation, Sydney University, Australia.

Freire, P. (1972). *Pedagogy of the oppressed.* Harmondsworth, England: Penguin.

Fullan, M. (1993). *Change forces.* London: Falmer.

Goodlad, J., R. Soder, and K. Sirotnik, eds. (1990). *The moral dimensions of teaching.* San Francisco: Jossey-Bass.

Habermas, J. (1984). *The theory of communicative action.* Vol. 1. Translated by T. McCarthy. Boston: Beacon.

Habermas, J. (1987). *The theory of communicative action.* Vol. 2. Translated by T. McCarthy. Cambridge, MA: Polity.

Habermas, J. (1990). Discourse ethics: Notes on a program of philosophical justification. In S. Benhabib and F. Dallmayr (eds.), *The communicative ethics controversy* (60–110). Cambridge, MA: MIT Press.

Hampden-Turner, C. (1981). *Maps of the mind.* New York: Collier.

Kim, Y. (1988). *Communication and cross-cultural adaptation: An integrative theory.* Clevedon, England: Multilingual Matters.

Klees, S. (1986). Planning and policy analysis in education: What can economics tell us? *Comparative Education Review, 30,* 574–607.

Lewin, K. (1985). Quality in question: A new agenda for curriculum reform in developing countries. *Comparative Education, 21*(2), 117–134.

MacIntyre, A. (1988). *Whose justice? Which rationality?* London: Duckworth.

Mayo, J. (1993). *The third channel.* New York: UNICEF.

McCarthy, C. (1990). *Race and curriculum: Social inequality and the theories and politics of difference.* London: Falmer.

Mitchell, W.J.T., ed. (1994). *Landscape and power.* Chicago: The University of Chicago Press.

Niranjana, T. (1992). *Sitting translation: History, post-structuralism, and the colonial context*. Berkeley: University of California Press.

Paulston, R. (1993). Comparative education as an intellectual field: Mapping the theoretical landscape. *Compare, 23*(2), 101–114.

Said, E. (1978). *Orientalism: Western conceptions of the Orient*. Harmondsworth, England: Penguin.

Said, E. (1983). *The world, the text and the critic*. Cambridge, MA: Harvard University Press.

Spivak, G. (1990). *The post-colonial critic: Interviews, strategies, dialogues*. New York: Routledge.

Taylor, C. (1985). *Human agency and language: Philosophical papers*. Vol. 1. Cambridge, MA: Cambridge University Press.

Taylor, C. (1991). *The ethics of authenticity*. Cambridge, MA: Harvard University

A Ludic Approach to Mapping Environmental Education Discourse

Jo Victoria Nicholson-Goodman

> *crimson flames tied through my ears*
> *growing high and mighty traps*
> *countless fires on flaming roads*
> *using ideas as my maps*
> *we'll meet on edges soon, said i*
> *proud 'neath heated brow*
> *ah, but i was so much older then—*
> *i'm younger than that now*

Bob Dylan, My Back Pages

Genesis

Questions of what environmental education (EE) should aspire to teach have engendered a flurry of dialogue in recent decades. Today, an ever expanding EE discourse explores new ways of seeing the relationship between humans and the rest of the natural world, humans and science-and-technology communities, humans as social beings in enclaves variously competing with each other for resources or attempting to work together to protect an increasingly ravaged planet. The implications of the questions raised spread themselves over a broad range of disciplinary fields, even leading to the creation of new fields of inquiry, particularly in the natural and social sciences. The result is a wildfire of ideas and applications for educators to consider as they prepare to educate youth to cope with the "realities" of the future. Friction between varying truth-and-value choices forms the combustible core, with political, philosophical and sociocultural knowledge claims feeding the flame. Within this scenario, educators are expected to choose programs and applications that can assist their students to make "wise" choices for the future.

In this chapter I argue that the methodology of social mapping, as it makes visible sometimes bewildering relationships between old and new ways of seeing within the field of EE discourse, may serve: both as an aid to clarifying truth-and-value choices implicit in the discourse, and as a means of orienting advocates and practitioners alike to the range of perspectives that currently compose its vague and rapidly expanding boundaries. This cartography of ideas serves, as well, as a ludic approach to truth-and-value conflicts—that is, a "playful" form of resistance to any totalizing discourse that would seek to silence all others (Paulston, 1995). This chapter portrays the process of social cartography with reference to a range of pronouncements that variously revolve around or direct attention to the intertextual field of EE discourse.

In the field of EE discourse, I argue that a central orb of EE concerns is dichotomized perceptually into two aspects, *risk* and *relationship*, which serve as overarching themes, each with a variety of issue orientations informing their attendant truth-and-value choices. Risk attends to hazards created by human interaction with the rest of the natural world; relationship attends to the effects of meaning on these interactions.

In my map, each aspect is portrayed as one half of the orb. I use the notion of dichotomy in its astronomical sense: that is, the phase of the moon (or an inferior planet) when half of its disk is visible. The wholeness of the orb itself is not in question, although it is perhaps beyond precise human articulation—it is the wholeness of our human perception that is problematized here.

Intersecting each aspect, I see two dimensions that inform the direction discourse takes. The first is an epistemological dimension involving truth choices, polarized at one end by a vision of the world as material (materialist) and at the other by a vision of the world as immanent (immanence). The second is an axiological dimension involving value choices, polarized at one end by an ecocentric outlook (expanding "community" to embrace nonhuman beings as equals worthy of consideration) and at the other by an anthropocentric outlook (operating from a perspective that favors human needs and interests above all others).

What results from this intersection are four areas that inform current EE concerns: ecology, deep ecology, scientific humanism and theology. Seeing EE discourse in this way may help unravel the conundrum of the current multiplicity of EE perspectives while allowing space for further expansion of insight. My intent is to present a descriptive narrative and conceptual mapping of various perspectives within the orb.

Important elements of this work include the reader, whose interaction with this text will create its new text life; my own persona as mapper;

the intertextual fields that inform both the discourse that I select to represent EE discourse and social cartography as a postmodern-sensitive epistemological practice; and, finally, a style of inquiry that is based on a continual intersubjective interviewing of experiences with people, programs and literary communications as "texts" which may be "decoded." Within this style of inquiry, I serve as a translator, acknowledging that my translation is itself a text that may be decoded (Paulston, 1994). The point of this inquiry is to make both visible and comprehensible truth-and-value bases of alternative perspectives of EE, producing a "visual dialogue" (Paulston, 1993) of the debates.

The mapping process serves as well, for me, as a *koan*, a device from Eastern traditions that may produce an "attitude of open attention and contemplation," a key feature of ecological consciousness (and therefore, I think, of consciousness about ecological consciousness). It is a feature that requires the "direct action of turning and turning, seeing from different perspectives and from different depths" (Devall and Sessions, 1985, p. 10). As an aid to seeing my place relative to the mapping in this chapter, I will share some pertinent highlights of my persona as mapper.

I spent a considerable portion of my adult life in lay ministry, first as a "fundamentalist," then as a "charismatic" Christian, serving as teacher, counselor and preacher. While teaching full-time in a fundamentalist Christian school, I reentered the world of "secular" education, which resulted in my becoming a social studies teacher. At this time, questions of social consciousness, which had been my focus prior to my Christian experience, took on new life and meaning. Additionally, concerns related to human cognition of and relationship with nature posed special problems and insights for me, leading me to pursue a master's degree in science education as well.

The period of wrestling with varying constructions of reality that ensued inaugurated a series of cognitive and spiritual transitions, and moved me to seek answers by dialoguing with teachers from various religious faiths, along with leaders and members of ecospiritual communities and with Native American elders, all of whom enhanced my understanding of the breadth of the questions involved.

I met these people while serving as a speaker and facilitator for an interfaith group. I joined the group to share both my experiences as a Christian and the questions raised by transition, and found there a movement that sought to create space for unity within difference and to celebrate difference within unity. I learned, again, to value my own spiritual "song," but also to see with different sight what I thought had been explored to exhaustion. The process of walking the circle of human experience, and stopping to see with new vision at varying points along the way, has shaped not only the form, but also the essence of my thoughts and actions. For me, *this process is the map*.

Having spoken from a position of personal self-disclosure, I now turn from my celebration of my right to difference and shift voices, allowing academic discourse to shape the spatial interplay of the elements of the text before us. I begin with an examination of selected major EE perspectives, and then briefly introduce four points of postmodern sensibility as groundwork for the exploration of the usefulness of social cartography. Finally, I map what I see as the shifting fields of EE discourse.

THE INTERTEXTUAL FIELD OF EE CONCERNS: MAJOR PERSPECTIVES

This section identifies several major perspectives in the EE literature, which I see as constituting four areas of the intertextual field: ecology, deep ecology, scientific humanism and theology. Ecology focuses on the whole community of beings, living and nonliving, and their interactions as equals, while deep ecology favors higher gestalts that look at ecological issues from a deeper questioning of meaning. Scientific humanism focuses on the community of humans and their interaction with each other and the rest of the natural world, while theology looks at the deeper questioning of the meanings of human existence and of the character of the natural world as a whole.

Each perspective is identified by reflecting on its position in relation to truth choices, which amount to ways of knowing environment, and value choices, which amount to ways of seeing environmental issues. For each, I provide a short descriptive narrative, some of its representative "talk," and its defining characteristics as represented above.

Faulconer (1993), for example, provides a *historical perspective*, which looks at EE discourse from the point of view of an epistemological evolution that proceeds from conservation education in decades past to the more current focus on ecology. Analyzing shifts both in language usage and reliance upon sources, she enunciates changes from "wise-use" talk in the 1950s to ecological talk, which "expanded the environmental education conversation" to make space in the discourse for "the ethical and philosophical nuance of deep ecology," along with a "striking shift" from reliance upon governmental sources of information to an attendance upon social literature for this kind of information (p. 13).

Referring to "the litany of environmental problems" as "a wild debate," she connects "the language used to define ecological issues as crucial" (p. 1), concluding that

Education has the choice to open the door to a new model of environmental education founded upon an ethic of ecocentricity that accepts the moral imperative to help students construct a value system that works for the natural world, or to unwittingly perpetuate a set of beliefs that actually works against the creation of a truly sustainable society. (p. 14)

I locate Faulconer's text near the meeting place between risk and relationship (ignoring issues of immanence, but seeking moral and philosophical value construction), giving greater value to the ecocentric view, within the field of ecology (see Figure 1).

Views and pronouncements from within the field of science education provide us with additional perspectives. While I acknowledge that there are many noteworthy science education reforms currently transforming the field (e.g., Project 2061, Scope and Sequence, and "hands-on" science), I have selected the call for "science literacy" and the Science-Technology-Society (STS) reforms because they interplay most directly with EE discourse debates.

Shortland (1988) promotes a *science literacy perspective* to bridge the gap between specialized practitioners and the lay public. For Shortland, the "heart of the matter" is "the way in which ordinary people . . . relate to the world of science as they experience, meet and confront it" (p. 315). Sketching arguments from the 1950s, he adds individual and political arguments of his own for science literacy, listing the essentials of relations between science as a field, scientists as a community, society as a whole and the lay public at the core of each argument.

Each argument makes a materialist claim about these social relations, a few of which follow: technologically based market economies depend upon scientific excellence for economic prowess and technology compliance to foster and sustain consumer demand; military prowess depends upon a supply of trained scientists and engineers for war technologies; the spread of American ideology depends upon the ability to export science technologies to developing countries; and a lack of science literacy produces a cultural response of fear mixed with adulation, unhealthy both for the culture and for the science community.

For Shortland, there is no question that materialist science—science as we have known it—is at the root of the success and well-being of our culture, economy, world status, sense of what it is to be human and knowledge relations. His only discontent can be found in the fact that not enough of us know enough about science. Science is seen here as evolving knowledge, and the scientist as humanist wanting to impart this knowledge.

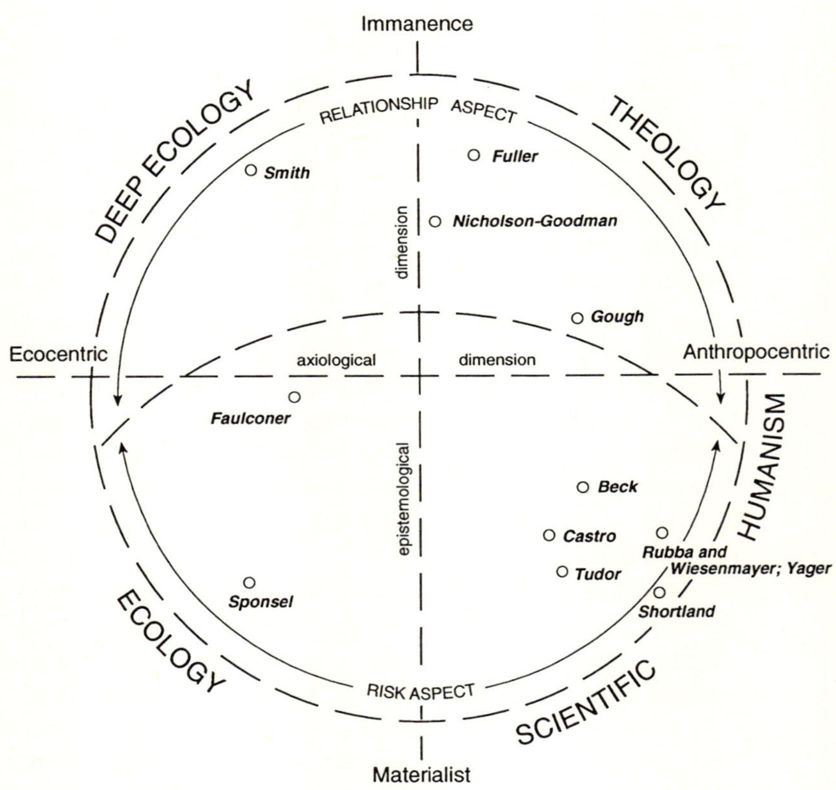

Figure 1. A conceptual mapping of major perspectives within EE discourse

This position has encountered critique from postmodern views of science, and his conclusion is telling:

> Yet in recent years, ethical arguments for the promotion of the public understanding of science have become rather unfashionable. Partly . . . this may be because they have attracted a great deal of critical attention from moral and social philosophers, but partly, also, it is because the scientific humanism upon which they commonly rest has gradually withered in the face of growing public concern about what have come to be seen as the totally unwelcome social implications of much scientific research. (p. 311)

I locate Shortland's text at the very base of the materialist dimension of ways of knowing environment, and at the extreme end of the anthropocentric dimension of ways of seeing the issues, even though he hasn't spoken to environmental issues. His stance on scientific knowledge and relations places him on my map within the field of scientific humanism because of his silence on these issues.

While Shortland contends, implicitly, that the truth-and-value choices science makes are suffering from a lack of public awareness or support, and Faulconer contends that EE discourse is currently speaking the language of ecology more than the language of science, I question what this means for science education as it is impacted by EE discourse.

Texts from the *Science-Technology-Society* perspective make this connection explicit, while echoing many of the values and concerns of scientific humanism. Rubba and Wiesenmayer (1988) offer the notion that

> the goal of STS education is to help students develop the knowledge, skills and affective qualities to take responsible action on the myriad of STS issues facing humankind. That goal is congruent with the superordinate goal of EE . . . recent EE research on responsible environmental behavior was used to generate a goal structure for STS education. . . . (p. 38)

Hitching its wagon to the goal structure of EE (which involves developing in citizens "the ability and desire to take responsible environmental actions" (Rubba and Wiesenmayer, p. 41), the STS movement takes a different avenue of approach than traditional science:

> Science is becoming something to experience; it results in student ac-

tions; it is becoming central to the school program; it is visible in the community. Science for S/T/S teachers and students is not learning the material found in textbooks and further elaborated by teachers; it is no longer a matter of information acquisition; it is no longer information alien to living. (Yager, 1990, p. 198)

Called a "paradigm shift for the field of science education" (Hart and Robottom, 1990, cited in Yager, 1993, p. 145), STS means "dealing with students in their own environments and with their own frames of references. It means moving into the world of application, the world of technology, the world where the student makes his or her own connections to living and to the traditional disciplines" (Yager, 1993, p. 148).

The STS perspective favors materialist science, but would attempt to democratize it and take it off its pedestal, making it relevant to the public. Science is seen here as useful knowledge, and public ownership is encouraged, once "knowledge, skills and affective qualities" (Rubba and Wiesenmayer, p. 38) have been imparted by the experts.

Knowledge claims of the STS perspective may be more apt to reflect personal truth-and-value choices made by students, but the question of ownership of knowledge lies at the core of this perspective of EE, and ownership still belongs to the science community.

I locate Rubba and Wiesenmayer's and Yager's texts within the risk aspect, favoring a material, anthropocentric outlook, in the field of scientific humanism, but closer to relationship than Shortland (see Figure 1).

Beck (1992) questions this ownership of knowledge, presenting us with a *radicalized science perspective*. He argues that one of the pressing consequences of scientific and industrial development is the creation of risks and hazards previously unknown to humankind. The locus of the problem, Beck argues, is in the "constitutive role assigned to science and knowledge" (p. 2).

Beck asserts that a radicalization of rationalization, a reflexivity operationalized through the critique of science, can help us deal with what he sees as the "*effets pervers*" of the "risk society" and possesses a moral claim equal to that of modern science. He notes that "the post-modern critique has exposed how modernity itself imposes constraints of a traditional kind—culturally imposed . . . around the quasi-religious icon of science" (pp. 2–3). The cultural form of this iconization of science is scientism: "The culture of scientism has in effect imposed identity upon social actors by demanding their identification with particular social institutions and their ideologies, notably in constructions of risk . . ." (p. 3).

I locate Beck's text at the center of the risk aspect, favoring a materialist vision, and at the center of the anthropocentric outlook, in the field of scientific humanism (see Figure 1).

Social constructions of risk, and the way these constructions affect knowledge claims of the scientific community, form a core issue in EE discourse. The breakdown of our trust in scientific risk constructions leads us in many different directions. But even within this scenario, we struggle for consensus and for collective action, both as educators and as individuals.

A prime example of this struggle emerges through a *sustainable development perspective*, which espouses balancing economic imperatives with environmental protection. At the 1990 International Environmental Education Conference, participants were told that a new world order was in the making, requiring new social, economic and political structures, and that environmental educators "should work toward visualizing new forms of an international 'federation,' whereby resources are managed through a global authority" (Tudor, 1991, p. 11).

Tudor's tone celebrates global consensus while acknowledging individual difference as she reports on Sibly's comments, which addressed these issues:

> As environmentalists, educators or otherwise, we all come from different political, spiritual and philosophic backgrounds. Where we've come from affects where we are going. We can't all hold the high moral ground. There are many pathways to reaching the ultimate goal. Let's be tolerant of alternative approaches. (p. 12)

Tudor claims consensus that EE is "at a new crossroads in its evolution," and that we are witnessing an "ideal time to reexamine all aspects of the environmental movement in terms of sustainable development" (p. 11). She qualifies her position by acknowledging that "a major leap of insight was required to consider environmental education in terms of global management goals," but goes on to state that in order to "move environmental education in a positive direction, some common ground must be found on a global level" (p. 14).

I locate Tudor's text within the risk aspect, close to the materialist vision, and at the center of the anthropocentric outlook, in the field of scientific humanism (see Figure 1).

Sponsel (1987) advocates a *cultural ecology perspective* that resonates with the same global consensual tone. She indicates the usefulness of this field in addressing environmental problems and education, "providing a broader cross-cultural and diachronic perspective on human-environment interactions and environmental problems" (p. 31). Categorizing the world's cultures into equilibrium societies, which keep population and consumption within the bounds of the local ecosystem, and disequilibrium societies, which have the opposite tendency, she sees that

> The main implication . . . is that a disequilibrium condition is developing for planet Earth as a whole as equilibrium societies become Westernized, modernized, developed. . . . From this perspective, cultural ecology provides a critique of the environmental ideas, actions, and consequences of disequilibrium societies . . . and a corresponding defense of equilibrium societies. (p. 36)

Seeing industrialization and even civilization as "experiments in cultural adaptation that have yet to stand the test of time," she suggests that we may be seeing their failure. Her stance appears to espouse a return to equilibrium as progress.

I locate Sponsel's text within the risk aspect close to the materialist pole, at the center of the ecocentric outlook, in the field of ecology (see Figure 1).

Castro's (1993) speech to participants in the Earth Summit in 1992, revealing a *differential development perspective*, placed the blame for risks now facing humankind squarely on the shoulders of consumer societies, arguing that an unjust economic order created this risk for all, and that we must now recognize "the ecological debt of the industrialized countries" (pp. 19–20). The primary factor in the deterioration of the global environment from this perspective is the

> model of economic behavior created by the most developed societies and spread by them to the rest of the world—using their own power and the influence of their mechanisms for shaping public opinion. A lifestyle based on the irrational zeal to consume and an absurd pilfering of resources is the main enemy of the environment in these times. (p. 20)

Imploring delegates to see to it that "indiscriminate exploitation of the environment is not accentuated . . . by indifference to the right of three-fourths of humanity to develop," Castro pleads for "a recognition of different degrees of responsibility and the establishment of fair and preferential treatment that would give the underdeveloped countries access to the resources and technology necessary to achieve such a goal" (p. 10).

I locate Castro's text within the risk aspect, close to the materialist pole and close to the anthropocentric pole, in the space of scientific humanism (see Figure 1).

A *fictive narrative perspective* emerges in Gough (1993), who regards both science education and EE as "storytelling practices." Arguing that "the characteristic discourse of much contemporary science and EE rarely encompasses the narrative complexities . . . that are needed to . . . make problems of human interrelationships with environments intelligible . . . and to conceptualise postmodern scientific understandings of 'nature' and 'reality'" (p. 9), he proposes (as does Liebman in his first chapter in this volume) that the problems and concepts involved are more apt to be modeled appropriately by literary fiction, especially the genre of science fiction.

> The narratives of both science and EE typically include representations of the phenomenal world ("reality") which may include descriptions and interpretations of the global conditions and circumstances which appear to make EE necessary and desirable. (p. 10)

Seeing the fact versus fiction myth of modern science as "part of a story that has been fashioned to rationalize the strategies used by scientists to produce facts," Gough asserts that "a signal feature of the narratives of modern science" can be seen in its "failure to accept that the creation of meaning in the world is a human and communal responsibility" (p. 11). Further,

> The majority of people in modern Western societies have abrogated their responsibility for "singing the world into existence." Instead, they accept uncritically the world that Bacon, Descartes, Newton and others "sang" into existence—a world that presents itself as a disenchanted machine of structures and systems. The reenchantment of the world requires our participation in the creative reconstruction of a language which foregrounds our kinship with nature. (p. 13)

Gough raises here an intriguing knowledge claim, one that heralds the singing into being of all knowledge. He suggests that the human, communal "song," however it is fashioned, is a fictive narrative, and that all knowledge, then, consists of storytelling.

I locate Gough's text within the relationship aspect at the low end of the immanence pole and at the center of the anthropocentric outlook, in the space of theology (see Figure 1).

Fuller (1988) presents an *aesthetic perspective*, connecting art with those disciplines that concern our knowledge of the natural world. Contending that the thinking of early aesthetic philosophers "was evinced by natural forms and phenomena as much as by the objects created by artists," he argues that modernism "drove a wedge between the pursuit of art and the study of nature" (p. 11).

Surveying the work of Ruskin and Bateson, among others, he details the ramifications of this wedge for aesthetic ways of seeing nature and ourselves. Interestingly, the language of ecology as saving grace from science emerges here:

> Ruskin was dogged by a sense of the *failure of nature* . . . the feeling that nature had been reduced to the grey lifeless monochrome that had so repelled him as a young man . . . soon after he admitted that there was no God, Ruskin began to detect what he was to describe as "The Storm Cloud of the 19th Century," and its accompanying "Plague Wind" which led him finally to his bleak vision that this "failure of nature" (which he felt had somehow been brought about by the blasphemous actions of men) was leading to ultimate annihilation, to Ruskin's vision of blanched sun-blighted grass and blinded man. (Fuller, p. 24; Fuller's italics)

Fuller's vision reveals the angst that underlies much of environmentalism:

> I think that we live in a world where both nature and art are widely seen as having lost their meaning. . . . We live in an anti-aesthetic, or anaesthetic, environment whose paradigm is, in a sense, the grey monochrome. . . . Ruskin's perception that the actions of men might be leading to a *real* failure in nature, and to a potential annihilation of human life, may turn out to be his most prophetic insight in this era of acid rain, ecological devastation, and potential nuclear winter. (p. 24)

Fuller sees the deep issue as the question of immanence:

> Gregory Bateson was no more a believer than I am. But he once pointed out how the erosion of the concept of divine immanence was leading men and women to see the world around them as mindless, and hence as not worthy of ethical or aesthetic consideration. This led them to see themselves as apart from nature. . . . Bateson went on to argue that, if we wanted to survive, we needed an "ecology of

mind," or a recognition of the fact that if nature is not itself the product of mind, then "mind is imminent in the total evolutionary structure." (pp. 24–25)

How does Fuller resolve the human aesthetic dilemma created by the intrusion of the "Modernist wedge"? He calls for a return to *theoria*, "the response to beauty of one's whole moral being" (p. 15):

> We have left the Garden of Eden. . . . We live, all of us . . . on the periphery of a potential desert. . . . We have become peculiarly ill at ease in the nature that nurtures us, constantly worried that through our own actions we will cause it to fail, certain that no God resides within the rocks and trees to save and console, sure that not much is for the best in this, our only possible world . . . insistence upon realising an aesthetic response . . . [may bear] witness to that irrepressible "impulse in the human breast" to affirm beauty in, and unity with, the natural world, *regardless*. (pp. 28–29; Fuller's italics)

I locate Fuller's text in the aspect of relationship, close to the immanence pole, right of center on the ecocentric-anthropocentric dimension, in the field of theology.

Smith (1993), who presents a *resacralized nature perspective*, admits that it was his intellectual and spiritual journey that caused him to search out and critique what he saw as fundamental principles underlying modern industrial civilization and its transmission in the schools. Urging us to see these principles for what they are, he also exhorts us to exchange them for others that are more supportive of the planet and its human communities. One of these principles, our human sense of meaning, Smith contends, is

> closely related to even more fundamental mythologies around which we construct our sense of place in the universe. Given the environmental catastrophes that have come to dominate the news, it is time to replace the myth of material progress with a myth that more realistically acknowledges the fragility and interdependence of the systems that sustain life on this planet. In a very real sense, our continued existence as a species may depend on our ability to reinfuse the physical world with meanings and values that transcend human understanding and needs. In other words, we must reenchant or resacralize the world. (p. 56)

While he acknowledges that "such an epistemology need not require the abandonment of scientific perspectives" (p. 56), his reconstruction of sense of place in the universe speaks of an ecologically oriented cosmology that can inform EE pedagogy.

I locate Smith's text in the relationship aspect, close to the immanence pole and to the ecocentric pole, in the space of deep ecology (see Figure 1).

I have outlined here ten perspectives selected to represent the fuller spectrum of EE discourse. I have attempted to allow space for each voice, and at the same time to pinpoint their relative truth-and-value choices for the purpose of informing those called upon to choose from the menu of EE options. This is, I believe, concordant with a postmodern sensibility. At this point, it may be helpful to see this sensibility as it influences our ways of seeing social meaning in relation to education.

POSTMODERN SENSIBILITY

Postmodernism as a sensibility is useful as a mediator in understanding the process of social cartography, and this mapping exercise particularly, in four ways, as it (1) makes space for diversity; (2) levels the playing field of perceptions; (3) highlights differences in ways of seeing knowledge claims; and (4) allows for ludic play in identity formation. It may be helpful to examine these four aspects of postmodern relevance to this mapping exercise in order to appreciate my placement of the ten perspectives outlined above. *Postmodernism and Education* (Usher and Edwards, 1994) provides us with a useful summary of this relevance.

First, modernist notions of EE tend to favor preservation of traditional choice systems, such as those predominantly embraced by the hegemony of advanced industrial cultures, while postmodernist notions tend to "deprivilege" these, making space for a wide-ranging alternative menu of choice systems that may open us to other ways of knowing.

Second, I see postmodernism as a perceptual space that makes possible a level playing field for new knowledge claims. New ways of knowing and diverse visions from those who were silenced in the modernist context are given equal voice. Knowledge claims emanating from those who were formerly marginalized, or made invisible by the dominant modernist notion of history as progress, emerge to expand the field of discourse. Understanding basic provisos of postmodern thought may enhance appreciation of this need.

Third, the postmodern "now" may be seen as a landscape which features arguments over historicity, aesthetic style and the nature of knowledge. It finds a centering influence in paradox, which makes it appear undefinable and therefore inarticulable (Usher and Edwards, 1994, p. 7). This perceptual space, then, highlights differences in ways of seeing knowledge as each new way of knowing sheds light on alternatives, expanding the possibilities. It is knowledge—or knowing—and its related value choices that are my focus here. My rationale for such a definitive focus is based on the goal of EE discourse itself, which attends largely to answering this question: What should EE aspire to teach?

Finally, we will need to consider what this current deconstruction of traditional and foundational knowledge claims means for individual actors:

> Postmodernity . . . "is marked by a view of the human world as irreducibly and irrevocably pluralistic, split into a multitude of sovereign units and sites of authority. . . ." The parallel and related search for a "true" or authentic self gives way to an aestheticization of everyday life in a "playfulness" where identity is formed by a constantly unfolding desire expressed through choices of lifestyle. Thus in postmodernity, the decentering of knowledge is parallelled by the decentering of the subject. (Usher and Edwards, 1994, p. 12)

This playfulness—that is, the ludic play of identity formation occasioned by postmodern sensibility, mirrors and is mirrored by the ludic play of environmentalism itself as "utopic praxis" (Marin, 1994) and is therefore an integral feature of a mapping exercise based on postmodern sensibility.

Postmodern sensibility, however, cannot be useful as an island floating on some inarticulable plane. It serves us better as it mediates our ways of seeing a particular task or process, and I have introduced it here, albeit briefly, as a means to making clear the usefulness of social cartography in professional context. I now move to a consideration of this mapping methodology within the field of comparative education.

SOCIAL CARTOGRAPHY

Noting that "knowledge constructs in comparative education . . . have become increasingly diverse and fragmented," Paulston (1993) examines differences between old and new ways of seeing and knowing in the field, concluding that "Today, no one world view or way of knowing can claim to fill all the space of vision or knowledge." He further prophesies for this field "an extended period of learning to work together as a diverse yet interactive global community of scholars," calling for "goodwill, translation and cognitive maps to help us see a shifting theoretical landscape" (p. 101).

Providing a history of changing representations of knowledge,

Paulston sees the present state of practice as an "emergent heterogeneity . . . a new and confused terrain of disputatious yet complementary communities" in which "the use of knowledge becomes more eclectic and reoriented by new ideas and new knowledge methods . . . for example, interpretations, simulations, translations, probes, and conceptual mapping" (1993, p. 105). We are thus presented with a scenario in which collaboration and goodwill are not only warranted, but may well prosper.

Theory mapping is depicted as a "kind of cognitive art or 'play of figuration' to help orient comparative educators as they face challenging new intellectual and representational tasks." A constructionist stance is taken, where "social and intellectual worlds may be uttered and constructed in different ways according to different principles of vision and division." Theory mapping serves as an aid to self-knowledge and we are, in fact, exhorted to realize that "failing to construct the space of positions leaves you no chance of seeing the point from which you see what you see" (Paulston, 1993, p. 105). This is a reflexive practice involving an interviewing of texts and their value and power relations. More importantly, it creates space for the interviewer to ironically "open out" her own value and power orientations as they play into the map's construction.

Liebman and Paulston (1994) have worked to develop a methodology, drawing from several discourse fields, that assists in bringing into view cognitive variations of the spatial interplay of ideas and their value bases within an intertextual field (Paulston, 1993, 1994, 1995; Liebman and Paulston, 1994) to produce, as noted above, a "visual dialogue." I introduce this methodology, for the purposes of this chapter, in terms of two facets that strike me as both essential and illustrative: rationale and process.

For Paulston (1995), rationale is situated within a postmodernist conceptualization of "now," in which we, as educators, struggle to find ways through new terrain:

> Today, long dominant goals and assumptions underlying modern theories of education and society are undergoing a ravaging subversion. Post-structuralist, post-modernist, post-patriarchal, post-Marxist . . . theories push forth new ways of seeing grounded in, paradoxically, anti-essentialist and anti-foundationalist ideas. Social relations and basic notions of reality and knowledge production undergo fragmentation, and many find themselves confused and disoriented in a shifting intellectual landscape with new knowledge communities speaking seemingly incomprehensible research languages. Surprisingly swift and unexpected, this rupture is also im-

ploding the study of educational change. Now no metanarrative, or
grand theory . . . can credibly claim to fill all the space of truth. . . .
Given this opening up of ontological and epistemological pluralism,
how are we as educators and scholars to move past our present unsettling aporia into a post-modern space of heterogeneous knowledge relations with their promise of renewed intellectual energy for
our time? (p. 137)

Paulston's storytelling about the "now" that confronts us constitutes
a statement of belief, a prosaic and descriptive narrative in its own right. It
serves to help us "locate" his methodological text within a postmodern sensibility, and to alert us to value and power relations that result from that
location. It forms a self-referential reflexive narrative that attempts not to
mask but to reveal its orientation.

This is a methodology whose process is created expressly, then, to confront and adapt to a postmodern mode of being in the world without collapsing into postmodernity as a totalizing discourse. This process engages
in ludic play in a distinctive approach to seeing and knowing from various
points of view.

One crucial element of social cartography consists of Bourdieu's notion of "*habitus*," which views intellectual fields as systems composed of
"'durable, transposable dispositions' produced by a dialectical interaction
with objective structures and actors' views of the world" (Paulston, 1995,
p. 138). Intellectual fields, then, in this view, are not removed from other
struggles for social meaning, but mirror them and in fact participate in these
struggles. To display or "reveal" these dispositions, another element,
Barthes's notion of text as an arrangement of "social space" such that it
"leaves no language safe or untouched" (Paulston, 1995, p. 138) is useful.
In short, text is that which is problematized (or problematizes itself) concordant with provisos of postmodern sensibility. Social cartography, then,
is a postmodern-sensitive methodology that constitutes an interpretive approach to intertextual comparison of education discourse.

This methodology has given me the opportunity here to draw together
into a "unified" whole some of the many voices and visions embodied in
EE discourse. It has, in short, allowed a recognition of both unity and difference, and has provided a process for articulation of truth-and-value dimensions of various ways of seeing the choices before us.

Mapping of EE Texts

The conceptual map (Figure 1) of the EE texts discussed above covers the ten perspectives outlined. It is a preliminary work in preparation for a more comprehensive and fully researched phenomenographic map to be undertaken later. Liebman and Paulston (1994) explain:

> The conceptual map develops perceived relationships within or between categories. Unlike the research based phenomenographic map, the conceptual map is more open to the mapper's ideas and world view, particularly where the mapper observes from inside as a participant in a particular world view, taking in all views that comprise the environment of which the mapper's world view is one [Doll, 1989, p. 247]. The intensity of research and references is not as vital to the conceptual map. (p. 14)

The central orb of EE concerns is dichotomized into two aspects, risk and relationship, which result from the intersection of two dimensions. The first dimension is epistemological, polarized by a vision of the world as materialist at one end, immanent at the other. The second dimension is axiological, polarized by a valuing which is ecocentric at one end, anthropocentric at the other. The result is a four-directional schema that yields four subfields informing EE discourse: ecology, deep ecology, scientific humanism and theology. I have placed each of the ten perspectives according to their relative truth-and-value choices within one of the directions of discourse.

What I see as I "walk the circle" of the map is that the complexity and multiplicity of truth-and-value choices found in EE discourse are somewhat simplified without being reduced. The notion that these choice-systems could be dichotomized into ecocentrism versus anthropocentrism alone is reductionist in nature. More is involved, and has emerged here through the mapping process itself, which is a cognitive interplay of ideas. I hope that further mapping of this discourse will open out new spaces and new dimensions, leading to greater insight within the field. This map simply seeks to derive a sense of intended meanings which can provide a view of interrelated ways of seeing EE discourse.

CONCLUSION

Social mapping provides an intriguing and challenging reflexive process for constructing meaning in the face of complexity. Postmodern sensibility encourages us to see diversity and to open ourselves to alternative ways of seeing. Self-knowledge and self-referencing play crucial roles in this process as well. The mapping process is not, however, without its problems.

Harley (1988) explains:

Maps will be regarded as part of the broader family of value-laden images. Maps cease to be understood primarily as inert records of morphological landscapes or passive reflections of the world of objects, but are regarded as refracted images contributing to dialogue in a socially constructed world. . . . Maps are never value-free images. . . . Both in selectivity of their content and in their signs and styles of representation maps are a way of conceiving, articulating, and structuring the human world which is biased towards, promoted by, and exerts influence upon particular sets of social relations. (p. 278)

While this mapping exercise has been conducted with the intent of walking the circle of ways of seeing, it is of some importance that I locate myself also on the map in Figure 1, acknowledging that my mapping is one of many ways of seeing the interrelations of this discourse. Although I embrace aspects of *all* of the subfields formed by the discourse selected here, believing that all paths lead to *some* truth, I locate myself at the center of the relationship aspect, slightly to the right of the meeting place between ecocentric and anthropocentric outlooks, in the area of theology.

REFERENCES

Barthes, R. (1979). From work to text. In Harris, J. (ed.), *Textual strategies: Perspectives in poststructural criticism* (48–63). Ithaca, NY: Cornell University Press.

Beck, U. (1992). *Risk society: Toward a new modernity*. Translated by M. Ritter. London: Sage.

Bourdieu, P. (1989). Social and symbolic power. *Sociological Theory*, 7, 14–25.

Castro, F. (1993). *Tomorrow is too late*. Victoria, Australia: Ocean.

Devall, B., and G. Sessions. (1985). *Deep ecology*. Salt Lake City: Peregrine Smith.

Faulconer, T. (1993). *Situating beliefs and trends in environmental education within the ecological debate*. Paper presented at the meeting of the American Educational Research Association, Atlanta, GA, April, 1993.

Fuller, P. (1988). The geography of Mother Nature. In D. Cosgrove and S. Daniels (eds.), *The iconography of landscape* (11–31). Cambridge, England: Press Syndicate of the University of Cambridge.

Gough, N. (1993). Neuromancing the stones: Experience, intertextuality, and cyber punk science fiction. *Journal of Experiential Education*, 16 (3), Dec., 9–17.

Harley, J.B. (1988). Maps, knowledge, and power. In D. Cosgrove and S. Daniels (eds.), *The iconography of landscape* (277–312). Cambridge, England: Press Syndicate of the University of Cambridge.

Liebman, M., and R. Paulston. (1994). Social cartography: A new methodology for comparative studies. *Compare*, 24(3), 233–245.

Marin, L. (1994). *Utopics: The semiological play of textual spaces*. Translated by R.A. Vollrath. Highlands, NJ: Humanities Press International.

Paulston, R. (1993). Mapping discourse in comparative education texts. *Compare*, 23(2), 101–114.

Paulston, R. (1994). *Ways of seeing educational change practice*. Handout for Council of Graduate Students in Education talk, University of Pittsburgh, 9 February.

Paulston, R. (1995). Mapping knowledge perspectives in studies of educational change.

In P.W. Cookson and B. Schneider (eds.), *Transforming schools* (137–180). New York: Garland.

Paulston, R., and M. Liebman. (1993). *The promise of a critical postmodern cartography*. Research Report no. 2, Occasional Paper Series. Pittsburgh: University of Pittsburgh, Department of Administrative and Policy Studies.

Rubba, P.A., and R.L. Wiesenmayer. (1988). Goals and competencies for precollege STS education: Recommendations based upon recent literature in environmental education. *Journal of Environmental Education, 19*(4), 38–44.

Shortland, M. (1988). Advocating science: Literacy and public understanding. *UNESCO, 38*(4), 305–316.

Smith, G.A. (1993). Schooling in an era of limits. *Holistic Education Review, 6 (2),* 45–59.

Sponsel, L.E. (1987). Cultural ecology and environmental education. *Journal of Environmental Education, 19*(1), 31–42.

Tudor, M.T. (1991). The 1990 International Environmental Education Conference: A participant's perspective. *Journal of Environmental Education, 22*(2), 11–14.

Usher, R., and R. Edwards. (1994). *Postmodernism and education.* London: Routledge.

Yager, R.E. (1990). The Science-Technology-Society Movement in the United States: Its Origins, Evolution, and Rationale. *Social Education, 54*(4), April–May 1990, 198–201.

Yager, R.E. (1993). Science-Technology-Society as reform. *School Science and Mathematics, 93*(3), 145–151.

SOCIAL MAPPING

THE ART OF REPRESENTING INTELLECTUAL PERCEPTION

Martin Liebman

Social mapping[1] offers comparative education and the social sciences a facility similar to what some mapped geographies convey: personal insight, comfort and possibly relief when an identifiable landmark, icon or name on the map permits an individual to affirm "I am here," whether the map reveals a network of roads or the floor plan of a building (see Wilford, 1981, pp. 7–13). Social mapping, clearly a member of the larger, long-recognized cognitive mapping family, extends cognitive mapping beyond its earlier concerns—"knowing" geographic location and arranging information "in a [merely] hierarchial order of importance" (Gold, 1984, p. 277).

The maps presented in this chapter example the possibilities social mapping extends to comparative education and other research domains where interests focus on exploring and understanding human interactions, social movements and ideas. Ranging from ingenious to romantic to analytic, their origins and purposes varying significantly, these examples of social mapping practice present the mappers' perceptions or understandings of knowledge; each map presents some portion of its creator's view of the world, although not necessarily the mapper's worldview (Thrift, 1996).

RICHARD EATON'S STRATEGIC MILITARY MAPS

Richard Eaton's (Brigadier General, US Army, Ret.) maps uniquely exhibit both individual geographic perceptions alongside understood or ingrained political-social associations or estrangements. Peter Gould and Rodney White (1986) explain that Eaton invited staff college officers representing several nations to evaluate other nations they considered strategically significant. The resulting maps revealed the staff officers' global military impressions: geographic locations, physical areas, military support or threat and strategic strengths (see the legend accompanying Figure 1).

Gould and White's analyses find that the US officer's map (Figure 1),

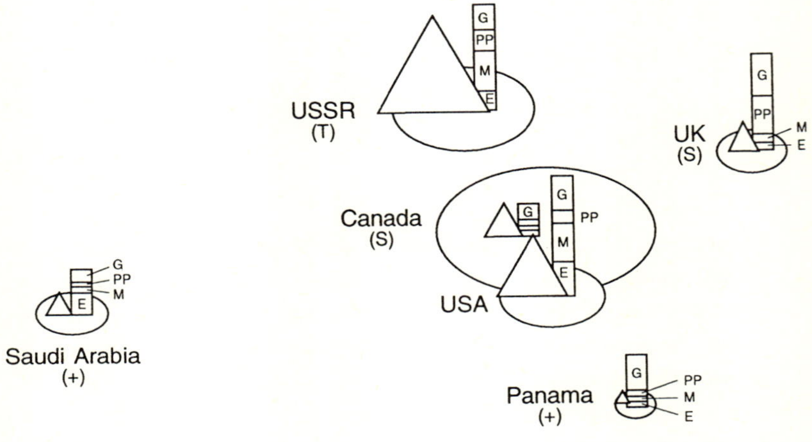

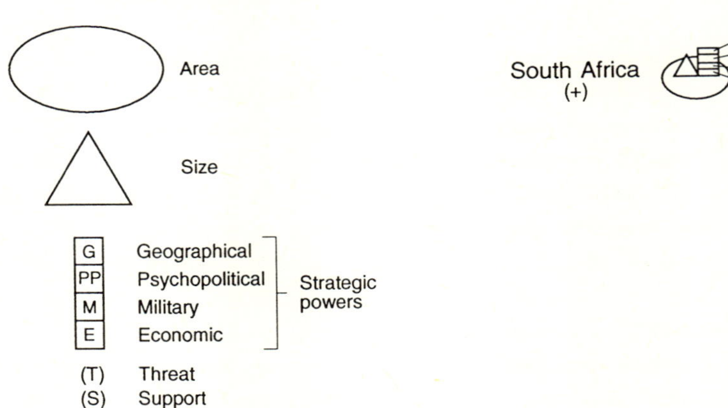

Figure 1. US officer's global military map

in particular, accurately defines the US and the USSR's geographic relation to each other (via the polar route) and to the United Kingdom.

The US officer, however, misrepresents the Saudi Arabian and South African land masses and populations, and the Canadian and UK populations. Gould and White note the US officer's map omits China's potential threat, a role the former USSR holds exclusively; this officer shows the other mapped nations are supportive militarily. Generally, the US officer's world view is one of relative national security.

Quite different is the map depicting the Venezuelan staff officer's world view (Figure 2). This participant in Eaton's study greatly compresses the distance between countries, distorting the nations' geographic positions. Indicating a large number of hostile threats, this officer sees the world political arena consisting essentially of nations threatening Venezuela's security.

Gold and White's analyses compare their knowledge with these officers' perceptions; they weigh real geographic distance and position against impressions of the world political arena, strategic threats and military preparedness. While considering real distance as statute or nautical mile measurements, as do Gould and White, we may also measure these miles as obstacles constraining movement, as may these staff officers. When the US staff officer includes the former USSR as a threat but not China, his logic might consider miles and threats as Soviet ICBMs and Chinese boots, with the former seeming more immediate than the latter. The point is that Gould and White's analyses neglect human considerations that comparative education or sociology might consider. Specific among these are cultural experiences influencing fear and bias. This might better explain the perceptions the Venezuelan staff officer maps.

Richard Eaton created this map series using ethnographic research-generated information. These maps' significance is their recommendation that we can map individual awareness. While not conforming to cartographic precision, these maps illustrate that only imaginative limits bound the model's usefulness as a comparative instrument. The question remains: How may we characterize such maps and situate them in comparative studies if they are not traditional cartography?

Several years ago, environmental planner Joseph Seppi, who contributes another chapter in this book, dispelled the notion that social maps exhibit any similarity to cartographic map making. He coined the word *cognographers* to describe social mappers—*Cogno* from cognizance, the perception or the scope of knowledge, and *graphers*, the mappers. Eaton's maps are cognography; they are the mapper's perceptions of comparative global spatial relationships. We can also consider Eaton's maps as geopolitical

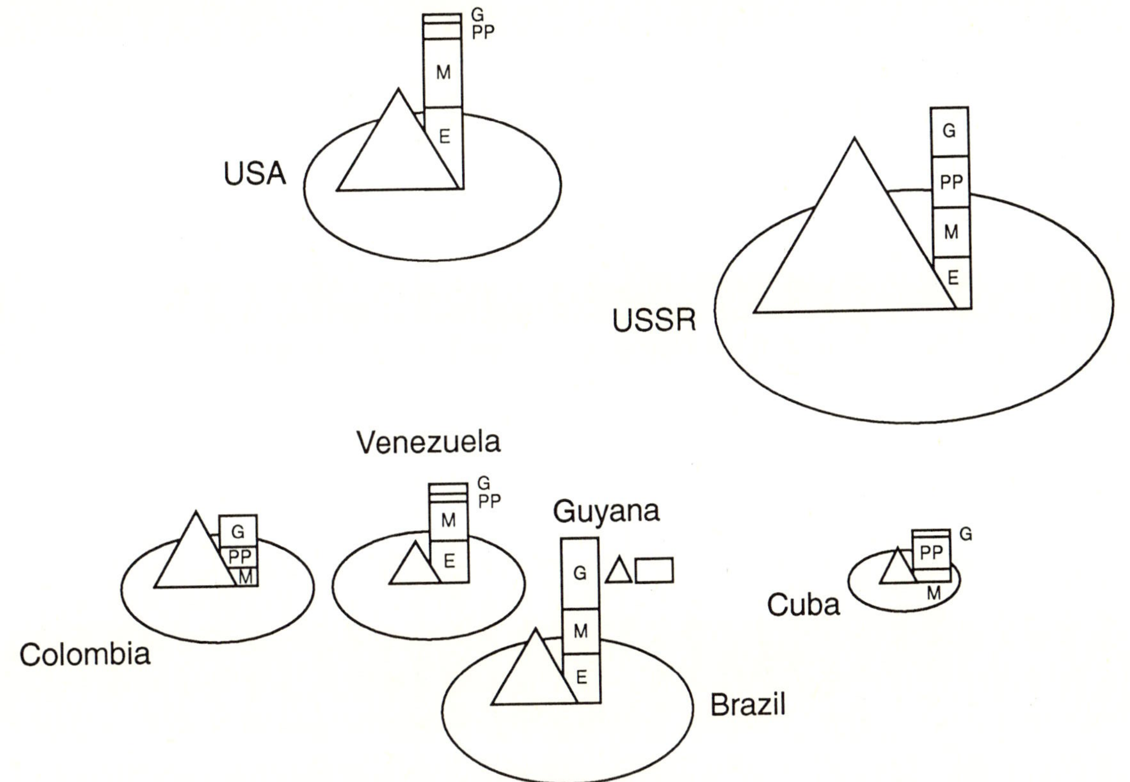

Figure 2. *Venezuelan officer's global military map*

cognographs, ally-adversary cognographs or other associative labels, deriving these labels from the information the maps contain, from Eaton's creative intentions or from the readers' reactions to the maps. As with any text, an open field is extended to analysis and interpretation.

Quite different from Eaton's mapping project, the terms *cultural-ography* and *exegetic-ography*, respectively, better describe the next two maps presented in this chapter. These maps offer singularly individual perspectives of the conflict and debate absorbing intellectual communities in the fields of comparative education and the social sciences today.

Julie Graham's Cultural-ographic Island

A whimsical freehand representation of relationships, Julie Graham's (1992) map portrays an isolated island containing intellectual and cultural regions (Figure 3). Her map demonstrates Jean Baudrillard's (1990) suggestion that we create space as a commodity and then alter it "so as to turn it, as opposed to that other space without limits, into a space with limits, with a rule of play, an arbitrariness. Basically, the scene is about the arbitrary, which makes no sense in terms of normal space" (p. 29). Although an anti-essentialist Marxist, Graham maps an essentialist world. This map is a caprice and as such opens itself to a capricious analysis.

In that temper I offer that a map true to her anti-essentialist/overdeterminist philosophy reveals the universe imploding and exploding simultaneously. At every infinitesimal instant, each imploding/exploding subatomic particle interacts with every other similar particle; each particle simultaneously forms every conceivable gaseous, liquid, geologic and biologic construct the universe contains. This universal swirl better maps Graham's anti-essentialist perspective: "Every aspect of reality participates in constituting the world and, more specifically, in constituting every other aspect" (Graham, 1992, p. 142). This analysis considers that Graham's map results not from her anti-essentialist perspective but from the open conflict consuming the time and talents of persons opposing her philosophy; rather than mapping her ideal world, Graham maps the less-than-desirable world others create, a world that includes interminable and determined landscapes, geocultural domains complete with concertina wire, impassible borders and imprisoned peoples.

> The landscape of this region [Graham's anti-essentialist nirvana] is very different from that of classical Marxism where Dick Peet dwells. The institutions and rules are different as well. Over the years, the differences between the two regions have contributed to a variety of

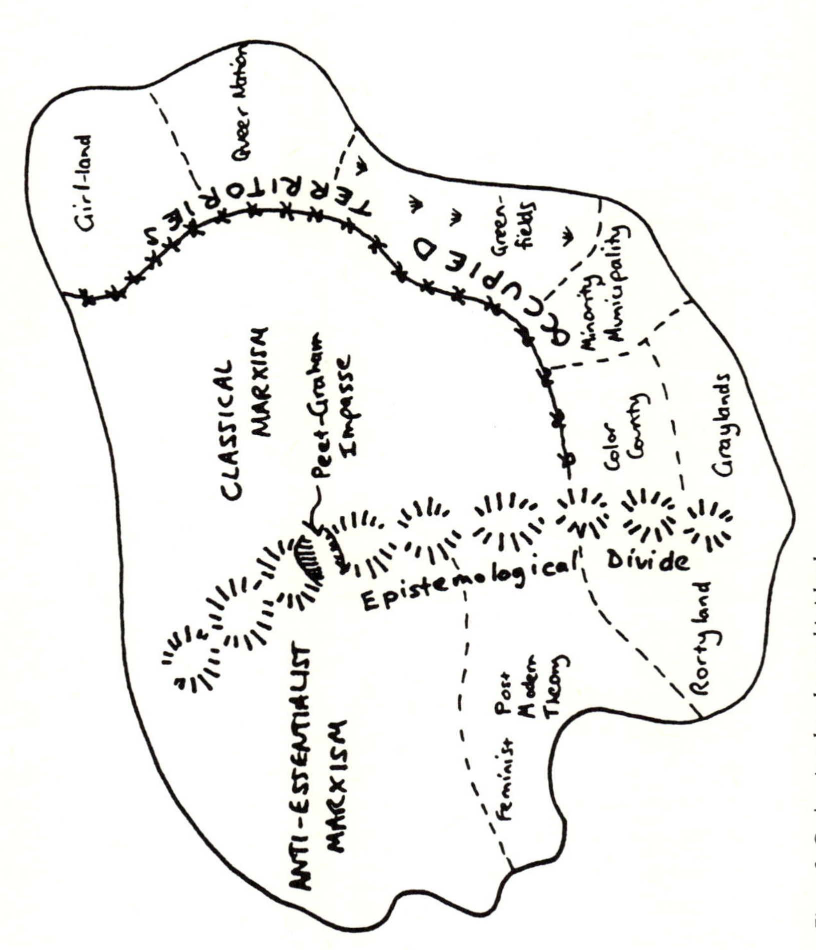

Figure 3. Graham's cultural-ographic island

misunderstandings and even outright battles between their inhabitants. None of these battles have been decisive, leading to the suppression or annexation of the foe. Nevertheless, it is [Graham's] impression that Peet's region is losing population [denizens daring the Peet-Graham Impasse, seeking deliverance in the better life] while [Graham's] is experiencing a small but significant gain. (Graham, 1992, p. 144)

Barring a revelation of theory tendering confused island inhabitants delivery to Graham's philosophical conviction, the island's physical considerations force a number of conclusions, among them that nature (rather than belief) may erode the Epistemological Divide and the island's beaches. The citizens of Greenfields could then assume the island's administration with a "we told you so" mien following the ecological calamity.

Graham's article characterizes Dick Peet's essentialism "as relatively stable over time and similar over space, permitting the theorist to make generalizations and to apply the knowledge of one spatial or temporal location to the interpretation of another" (1992, p. 150), a social theory sustaining her map's geofixity but not her philosophy. It is into this essentialist domain that Graham's map and the preceding playful analysis romp while heedful of her own anti-essentialist penchant.

Graham's map illustrates H.V. Wiseman's (1966) point of view that the most important behavioral component of the spatial dimension pertinent to sociopolitical organization is human or group territoriality. Edward Soja (1971) defines territoriality as "a behavioral phenomenon associated with the organization of space into spheres of influence or clearly demarcated territories which are made distinctive and considered at least partially exclusive by their *occupants* or *defiers*" (p. 19, emphasis added). Graham organizes space and bestows or denies territory to social theories, offering exclusive citizenship to her island domain while concurrently functioning as occupant and defier in the context of her island's political organization.

Important to this chapter is Graham's emergence in the mapping arena, although she arrived unaware of her map's significance at the time social mapping experienced its generation. The agent of that awakening next engages this chapter.

Rolland Paulston's Exegetic-ography

Discussing his impressions of spatial relations, the finite and the infinite, Albert Einstein (1982) said: "We may without difficulty impart more depth and vigor to . . . ideas by carrying out special imaginary constructions."

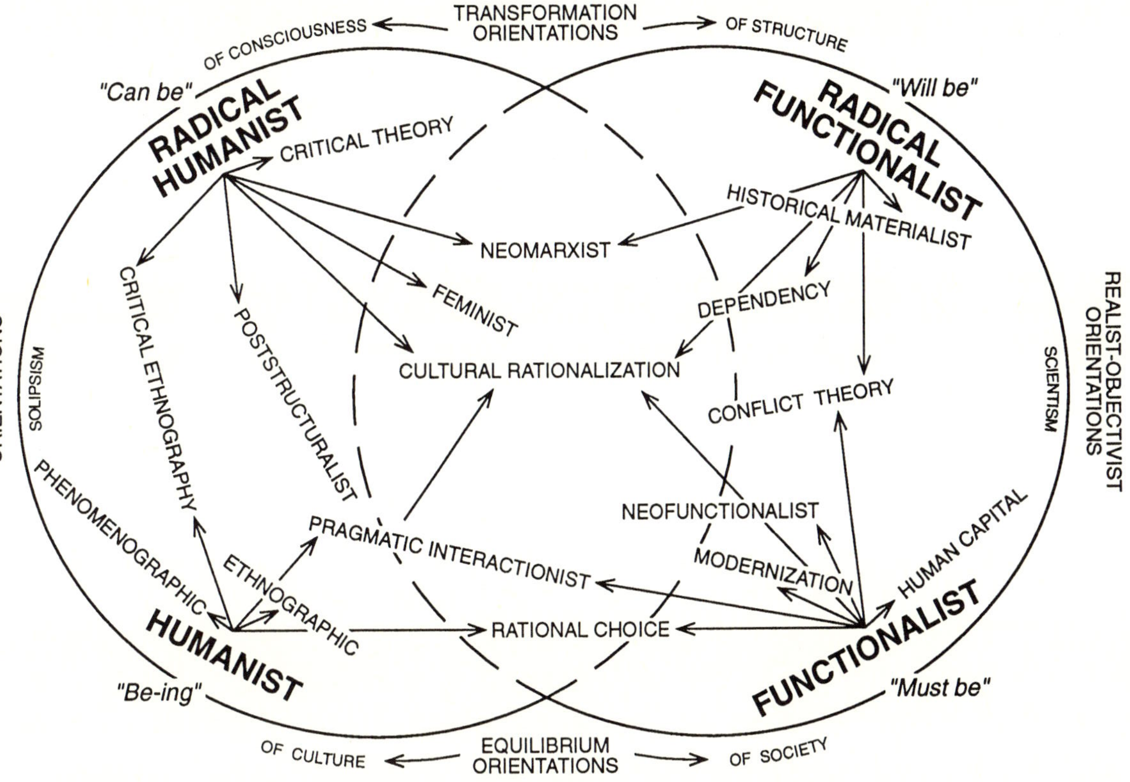

Figure 4. Paulston's macro-mapping of paradigms and theories in comparative and international education texts

Rolland Paulston (1994) forwarded this insight to comparative education studies specifically and the social sciences generally when he presented "a macro mapping of paradigms and theories in comparative and international education texts seen as an intellectual field." (Figure 4). Deriving the four paradigmatic nodes (functionalist, humanist, radical functionalist, radical humanist) from analysis of comparative and international education texts, Paulston arranges a juxtaposition of sixteen theories (human capital, rational choice, poststructuralist, etc.) within the field, denoting the interparadigmatic and intertheoretic borrowing and interaction of ideas with arrows. Several theory models offer insights applicable to Paulston's imaginative efforts. The following sections of this chapter survey several theory models regarding Paulston's map as well as Eaton's and Graham's maps where applicable.

SET THEORY AND FUZZY LOGIC

A recent entrant to the theory-system realm is fuzzy logic. Degrees of reality, ambiguous or nebulous scales play in arenas previously consisting of static, determined, calculated sets. Fuzzy logicians hold that nothing endures, that measurements are not absolute. The observation of any system is relative to measurement units and methods, an idea presented in the analysis of Eaton's maps.

Daniel McNeill and Paul Freiberger (1993) discuss four ideas attributed to set theory that apply to fuzzy systems, ideas with which Paulston obviously dealt, consciously or unconsciously, when determining what information he should map. *Complement* is the set of groups that belong. *Containment* is the discovery of "what groups belong to what other groups." *Intersection* is the area(s) where some members belong to more than one group. *Union* is the area where membership is open, where every group belongs (pp. 24–26).

The complement of Paulston's map includes cited comparative and international education texts. The four paradigmatic nodes define containment. Intersection occurs where arrows suggest a particular theory common to two paradigmatic nodes (these theories are critical ethnography, pragmatic interactionist, rational choice, conflict theory and neomarxist). Union is that area where the four paradigmatic nodes join: cultural rationalization.

On the surface, Paulston's textual domain appears neither fuzzy, nor vague, nor gray-scaled. Containing more than sixty labels and arrows graphically represented within dual hemispheres, the map appears clear, specific and high-contrast. Fuzzy logicians, however, would question the map's crispness. To what degree is each theory viable among university faculties in terms of classes taught from each perspective or in terms of texts, chapters, pages and words authors devote to each theory, favorably and unfavorably? Mapping with these questions in mind conceivably might alter the picture significantly, contrasting the map's meanings and symbols with any number of modifiers.

McNeill and Freiberger note that fuzzy logic "rests on the idea that all things admit of degrees . . . all come on a sliding scale" (p. 12). What differentiates little and big? When does a hill become a mountain or a pond become a lake? The social peripheries and geographic ends of Graham's island are not quite as crisp as her map reveals; determined, tenacious boundaries and landscape soften with the fuzzing of terrain and domain. "Fuzzy logic," McNeill and Freiberger write, "reflects how people think. It partly models our sense of words [and meanings], our decision making [and consequent actions], or recognition of sights and sounds. It unveils a corner of intuition" (p. 12). Not ironically, the basic ambiguity of Richard Eaton's strategic cognitive maps demonstrate fuzzy logic. While likely not governing the map making process, fuzzy logic certainly guides our understanding when, as map readers, we apply modifiers to the mapped concepts.

THEORIES OF ENVIRONMENTS AND DOMAINS OF ACTION

Social mapping employs a graphic method to question and answer how people organize within a bounded environment. Social mapping may also attempt to organize people. Adele E. Clarke (1991) suggests this point/counterpoint potential obtains in organizational theory studies; the former, how people organize themselves, prompts the sociology of social movements, and the latter, how people organize others, motivates the sociology of management (p. 119). In either case, "through grappling with identifying and understanding the structural dimensions of social organization, one can plot sociological maps . . . and ultimately study the actions and interactions of the many kinds of collective actors in given situations" (pp. 119–20). Clarke diagrams six organization theory models. Eaton, Graham and Paulston each work within at least one of the several models Clarke profiles.

Richard Eaton's strategic ally/adversary military maps, generated following ethnographic investigations involving military officers attached to their respective nations' staff colleges, clearly correspond to Clarke's *network model* (Figure 5). This model analyzes the "density or quantity, linkages (e.g., shared participants), and sometimes the content of the communications and exchanges between entities" (Clarke, 1991, p. 127). The "positional attributes of participants" concern the connective natures of the mapped entities and their movement within the model's boundaries. Network model

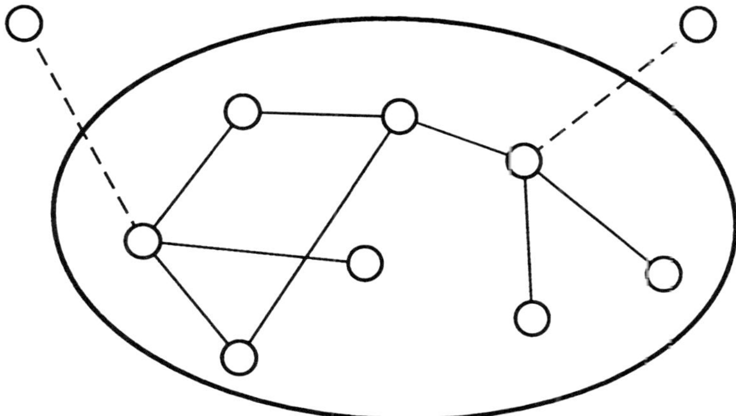

Figure 5. Clarke's network model

boundaries result from complementary selective evaluation, a fluid proposition permitting change at a variety of levels and swiftness simultaneously. The fluid nature of network model boundaries seems critical to military readiness, the mission motivating the actions of the officers Eaton surveyed.

Julie Graham's map embodies Clarke's *population ecology model* (Figure 6). According to Clarke, this model "capture(s) aspects of the broader social system of which a population of organizations are a part during the process of 'natural' selection" (1991, p. 124). Clarke observes that population ecology models contain significantly analogous organizations. Inhabited with many cultural and philosophical organizations populating the contentious campus microcosm, Graham draws attention to the "natural" selection of will, the human ability to select and choose allegiances and the freedom to create or choose cultural and philosophical assumptions, similar to Clarke's finding that population ecology models focus on the emergence, evolvement and dissipation of organizations and their competition for limited resources.

Clarke's population ecology model and Graham's map exhibit similar boundaries, a boundary Graham draws as the natural coastline of an island. Similarly apparent and consistent with Graham's map is Clarke's description of the population ecology model's strengths and weaknesses. One apparent strength of Graham's map is what Clarke calls "its attempt to grasp the wider ecological domain" (1991, p. 124), with Graham's meeting the challenge of representing a diverse population. The weaknesses inherent in population ecology models similarly appear in Graham's map. Clarke calls these "conceptual limitations." First is the limited vision the map imposes, a constraint of environment. The second weakness is the conspicuous restriction on social com-

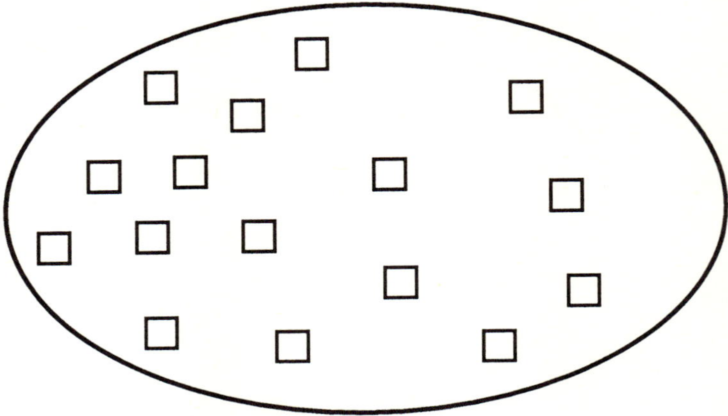

Figure 6. Clarke's population ecology model

petition—conflict, cooperation, negotiation, and exchange. Graham exaggerates this limitation by focusing almost exclusively on conflict.

Paulston's map declines categorizing within any single model Clarke considers, sharing instead characteristics with descriptions of the areal field, functional field, organizational field and social worlds/arenas theory models. Although a geographic-bound field working within naturally occurring communities, *areal field models* examine social structures, emphasizing inhabitant's collaborative natures as survival techniques. Areal field theorists view collaboration as negotiating the natural environment, thus expanding the perception that all work within the field is across a horizontal plane void of hierarchy; Paulston represents paradigms and theories void of a constructed hierarchy.

Functional field models eschew geographic boundaries, their focus instead relating ideological spheres or other determined domains. Functional fields additionally contain smaller "sectoral boundaries," placing related spheres within the more comprehensive field the full boundary delineates.

Organizational field models include the "totality" of relevant actors, a comprehensive overview such as Paulston's "macro mapping of paradigms and theories in comparative and international education." The organization field model's competition and communication components are also evident in Paulston's map. Competition and communication are also basic criteria in the population ecology model (Graham's map) and the network model (Eaton's map). The organizational field model requires an empirically defined structure (such as texts produced within given intellectual domains), and must exhibit four main facets found in Paulston's map: "increased interaction among the organizations," represented by the arrows; "interor-ganizational patterns of domination and coalition," represented by the grouping of theories near the four paradigm nodes; "increased information load," represented by the production of texts by advocates and adversaries writing within the sixteen theories; "and the development of a mutual awareness of involvement in a common enterprise," generally, cultural acknowledgment and intellectual interaction (Clarke, 1991, p. 126). Organization field models commonly involve the study of competitive domains. Paulston's map informs the competitive shaping of minds and policy.

Social worlds/arenas theory, an outgrowth of the areal field model, considers the work of actors committed to a particular arena (Clarke specifically notes social movements and ideologies). The authors Paulston studied commit their texts to the arena of ideas, the intellectual fields composing international and comparative education studies. This bounding enables empirical conclusions of whom to place within the model or on the map, actors whose works represent contemporary thinking on matters of "conflict, competition, cooperation, exchange, and negotiation" (Clarke, 1991, p. 128).

Conclusions

The maps presented in this chapter, particularly that of Rolland Paulston, clearly receive special status as innovative prototypes, vanguards of a dynamic method offering comparative education and the social sciences the capacity to view and understand the development, growth, and philosophic relationships driving cultural perspectives. Citing Jean Baudrillard, Adele E. Clarke, Daniel McNeil and Paul Freiberger, and Edward Soja, this study shows how Eaton, Graham and Paulston, knowingly or not, incorporated already well-developed and recognized sociologic models to their respective mapping methods.

This chapter illustrates how mapping social theories conveys a personal sense of social or cultural belonging, where logic may or may not play a role when the map reader confronts a mapper's situating of self to others. Future mapping efforts may incorporate some ideas introduced by the maps presented here. These maps should animate the imaginative potential of new maps rather than pattern the social map's limits.

Note

1. While Rolland Paulston prefers the phrase "social cartography," as he uses in this book's title, I prefer "social mapping," suggesting a softer, more flexible application consistent with the evaluation of the mapping examples illustrated in this chapter.

REFERENCES

Baudrillard, J. (1990). *Revenge of the crystal: Selected writings on the modern object and its destiny, 1963–1983*. Translated and edited by P. Foss and J. Pefanis. London: Pluto.

Clarke, A.E. (1991). Social worlds/arenas theory as organizational theory. In D. R. Maines (ed.), *Social organization and social process: Essays in honor of Anselm Strauss*. New York: Aldine de Gruyter.

Einstein, A. (1982). *Ideas and opinions*. New York: Crown.

Gold, P.C. (1984). Cognitive mapping. *Academic Therapy* 9(3), 277–284.

Gould, P., and R. White. (1986). *Mental maps*. Boston: Allen and Unwin.

Graham, J. (1992). Anti-essentialism and overdetermination—A response to Dick Peet. *Antipode,* 24(2), 141–146.

McNeill, D., and P. Freiberger. (1993). *Fuzzy logic*. New York: Simon and Schuster.

Paulston, R. (1994). Comparative and international education: Paradigms and theories. In T. Husén and N. Postlethwaite (eds.), *International Encyclopedia of Education*. Vol. 2 (923–933). Oxford: Pergamon.

Soja, E. (1971). *The political organization of space*. Washington, DC: Association of American Geographers.

Thrift, N. (1996). *Spatial formations*. London: Sage.

Wilford, J.N. (1981). *The mapmakers*. New York: Knopf.

Wiseman, H.V. (1966). *Political systems—some sociological approaches*. New York: Praeger.

IV Mapping Debates

"Koyemsi, or Mud-head kachinas, appear in almost every Hopi ceremony as clowns, interlocutors, announcers of dances, drummers and in many other roles. Koyemsi are usually the ones that play games with the audience." Illustration and description from a postcard sent by George E. Marcus to Rolland G. Paulston, 3 April 1995.

Introduction

John Beverley's essay, "Pedagogy and Subalternity: Mapping the Limits of Academic Knowledge," notes that readers of *I, Rigoberta Menchú* will recall that it begins with a strategic disavowal of both literature and the liberal concepts of the authority of private experience that literature can engender: "My name is Rigoberta Menchú. I am 23 years old. This is my testimony. I didn't learn it from a book, and I didn't learn it alone." It would, Beverley contends, be yet another version of the native informant of classical anthropology to grant testimonial narrators like Rigoberta Menchú only the possibility of being witnesses, not the power to create their own narrative authority and negotiate its conditions of truth and representativity. This would be a way of saying that the subaltern can of course speak, but only through the institutionally sanctioned authority—itself dependent on and implicated in the power relations that produce subalternity—of the journalist, ethnographer, or academic who alone has the power to decide what counts in the narrator's "raw material" and to turn it into literature.

What a text like *I, Rigoberta Menchú* forces readers to confront is the subaltern not only as a "represented" subject but also as agent of a transformative project that aspires to become itself hegemonic. Although we can enter into relations of solidarity with it, this project is not our own in any immediate sense and in fact involves structurally a contradiction with our position of relative privilege and authority. In other words, the very idea of "studying" the subaltern is catachrestic—or paradoxical—in a way that points to a new register of knowledge where the power of the university to understand and represent the world breaks down or reaches a limit. The idea of representing or studying the subaltern must itself confront the dilemma of subaltern resistance to and insurgency against elite conceptions.

Beverley argues that recognizing this situation involves learning how to work against the grain of our own interests and prejudices—a process Gayatri Spivak calls "unlearning privilege," which involves undoing the authority of the academy and knowledge centers at the same time that we continue to participate fully in them and to deploy their authority as teachers, researchers and theorists. This aim distinguishes a subalternist perspective from alternative projects of postmodernist "cognitive mapping" such as those of cultural studies or multiculturalism. Although animated by theoretical and political concerns coming from feminism, Marxism, deconstructionism, postcolonial thought and the practice of the new social movements, cultural studies tends to have a purely descriptive relation to the emerging "scapes"—to borrow Arjun Appadurai's term—of global culture it seeks to map. As such, it is from Beverley's point of view complicitous in producing discursively what could be called an "academic sublime"—an aesthetic-cognitive remapping that may become itself a functional element of late capitalist culture, just as the Romantic sublime was in the nineteenth century. The subalternist perspective, by contrast, does not claim to represent ("map," "let speak," and so on) the subaltern or alterity; it registers instead the way in which the knowledge we construct and impart as academics is structured by the absence or difficulty or impossibility of representation of the subaltern and alterity.

Patti Lather's chapter, "Postcolonial Feminism in an International Frame: From Mapping the Researched to Interrogating Mapping," traces the transnationalization of contemporary theoretical discourse through the event of her experiences as a Fulbright lecturer in New Zealand. Her title comes from Gayatri Spivak's 1981 essay on the circulation of French feminist theory across the world, in which Spivak deconstructs speaking for a generalized West and cautions that the appropriation of French theory by Anglo-American intellectuals often elides issues of the epistemic violence of contemporary forms of theoretical imperialism. Grounded in Spivak's deconstruction of this form of Eurocentrism, Lather's interest in a postcolonial feminism is delineated via her New Zealand experience where she both positioned herself and was positioned by others as something of a cheerleader for poststructuralism. The chapter proceeds by exploring the implications of postmodernism for emancipatory projects, sketches postcolonial tensions in her New Zealand experience and concludes with three mappings of modernity/postmodernity, one in the context of research methodology, with duly noted cautions regarding the limits of such binary modeling.

Crystal Bartolovich's essay, "Mapping the Spaces of Capital," argues that "mapping" (along the lines of Fredric Jameson's "cognitive mapping"), understood as an attempt to render visible relations among elements in a complex social whole, is an appropriate gesture as long as its limitations are scrupulously taken into account. To get at these limitations, one of her classes examined the "totality" debate in Marxist/post-Marxist theory, and considered its implications for any project of social mapping. This theoretical project was accompanied by attention to the discourse of "globalization" in contemporary media and business publications. By yoking a theoretical project to concrete analysis in this way, Bartolovich used the classroom to pursue with her students a version of Gramscian "analysis of situations" that provides a possible linkage between "mapping," "pedagogy" and "social change" invoked in the title of this collection of essays.

Robert T. Tally, Jr., in his chapter "Jameson's Project of Cognitive

Mapping: A Critical Engagement," examines Fredric Jameson's concept and practice of cognitive mapping and focuses on whether maps are or are not, in John Beverley's words, "inevitably bound up with domination." First, Tally traces the project of cognitive mapping through its various representations in Jameson's work on postmodernism. Then, by examining recent critical perspectives, he takes up the question of mapping in relation to the "subaltern." That is, given the force of cartographic representation exerted over so-called marginal or subaltern groups and individuals, is mapping a possible, or even desirable, counterpractice? Much of the apparently anti-mapping arguments, or at least those arguments that focus on the somewhat repressive effects of mapping, owe a great deal to Foucault's genealogical work. Thus, Tally looks at Foucault's spatial analysis of power, and at Deleuze's contributions to Foucault's critical project. He concludes that one cannot either dismiss or embrace mapping as a useful critical practice *tout court*, since the effects of mapping permeate the social, and even global, relations that characterize present political economic situations. Rather, Tally argues that a combination of Jameson's cognitive mapping and Foucault's spatial analytic of power—what he calls a "project of cartographics"—can provide a useful tool, or set of tools, for the educator and social critic.

In the concluding chapter, "Social Cartography, Comparative Education, and Critical Modernism: Afterthought," Carlos Alberto Torres contends that the implications of postmodernity as a sociocultural context may have fundamentally transformed the nature of social and cultural reproduction in education. Torres's position is advanced from what he sees as the critical-theory side of the opposition between postmodernist and critical social theory. This involves siding with the critical realists in assuming that despite poststructuralist critiques of representation, it is still possible to theorize about social reality, albeit in more self-reflexive and less totalizing ways. He also sides with the critical theories of postmodernity that take seriously both the need to revise—as good historicists—our theoretical constructs to take into account emergent and novel features of social and cultural life.

Taking Latin American education as an illustration, Torres argues that understanding the theoretical and political problems of Latin America requires the greater conceptual sophistication of a critical modernist perspective. His analysis maps out new and old concepts that may be employed in understanding the complexities of Latin American education, including the notions of unequal and combined development of Latin American educational systems, the notion of an organic crisis, the presence of new subjects of education, the role of the state in the context of neoliberalism, and the presence of fractures and "borders" in modern education.

Torres concludes that despite some critical considerations, mapping and social cartography offer interesting avenues for research. However, as a heuristical tool for comparative education, mapping should face the challenges of both positivism and constructivism to demonstrate its epistemological value, practical utility and ethical-political worth.

PEDAGOGY AND SUBALTERNITY

MAPPING THE LIMITS OF ACADEMIC KNOWLEDGE[1]

John Beverley

Are maps inevitably bound up with domination? The idea of "cognitive mapping" that animates this collection involves the possibility of building more egalitarian and respectful forms of understanding and relations between ourselves and the people and phenomena we posit as our objects of study. But cognitive mapping also responds to the new epistemologies of capital itself, the generation of new forms of public policy demanded by the need to administer and discipline an increasingly multicultural US population and a heterogenous transnational working class, and the central role of the university, as itself an eminently transnational institution, in producing the "knowledges" appropriate to these phenomena. As Robert McNamara's recent memoirs confirm, the failure of US strategy in the Vietnam War—at a time of tremendous expansion of higher education in this country—was one of the first indications of the problems caused for public policy by the incomprehension or misunderstanding of subaltern classes or social groups by dominant academic methodologies and disciplines.

In the succinct definition of Ranajit Guha, the founder of the collective of South Asian historians known as the Subaltern Studies Group, the word *subaltern* is "a name for the general attribute of subordination . . . whether this is expressed in terms of class, caste, age, gender and office or in any other way" (Guha, 1988, p. 35). "In any other way" might surely be understood to include the distinction between "educated" and "not (or partially) educated" that the indoctrination into the procedures and results of academic knowledge confers, both in metropolitan and in colonial and postcolonial contexts. How can it be said then that one can "know" the subaltern from the standpoint of academic knowledge, when that knowledge involves precisely the othering of a subaltern subject?

There is a passage in Richard Rodriguez's autobiographical book, *Hunger of Memory*, that captures, perhaps inadvertently, this sense of the way in which academic knowledge is itself implicated in the social construction of subalternity, and in which, vice versa, the emergence of the subaltern into hegemony necessarily disrupts that knowledge. *Hunger of Memory* tells the story of how the narrator's apprenticeship as a Chicano "scholarship boy" majoring in English, first at Stanford and then at Berkeley as a graduate student, gave him the chance to transcend his parochial (in his view), working-class, Spanish-speaking family background. Returning from college to his old neighborhood in Sacramento to take a summer job, Rodriguez observes of his fellow workers:

> The wages those Mexicans received for their labor were only a measure of their disadvantaged condition. Their silence is more telling. They lack a public identity. They remain profoundly alien. . . . Their silence stays with me. I have taken these many words to describe its impact. Something uncanny about it. Its compliance. Vulnerability. Pathos. As I heard their truck rumbling away, I shuddered, my face mirrored with sweat. I had finally come face to face with *los pobres* (1983, pp. 138–39).

What Rodriguez means by *los pobres* is, of course, what Guha and the Subaltern Studies Group mean by the subaltern. In fact, I know of no more exact description of the production of subaltern identity as the "necessary antithesis"—the phrase is Guha's—of a dominant subject than this brief passage, built on a binary of verbal fluency–power/mutism–subalternity. The passage itself enacts the dichotomy. Though it is not without marks of conflict and irremediable loss that its neoconservative admirers often tend to overlook, *Hunger of Memory* is ultimately a celebration of the power of the university, the traditional humanities curriculum in literature, and English writing skills in particular, to give a "'socially disadvantaged' child," as Rodriguez describes himself, a sense of self and agency.[2]

By contrast, readers of *I, Rigoberta Menchú*, which is also an autobiographical text about how one negotiates between subaltern and elite status (Menchú won the Nobel Peace Prize in 1992), will recall that it begins with a strategic disavowal of both literature and the liberal concept of the authority of personal experience that literature can engender: "My name is Rigoberta Menchú. I am 23 years old. This is my testimony. I didn't learn it from a book, and I didn't learn it alone" (Menchú and Burgos, 1984, p. 1).

What *Hunger of Memory* and *I, Rigoberta Menchú* share, beside the fact that they are autobiographies, is a coincidental connection to Stanford University. The decision to include *I, Rigoberta Menchú* in one of the tracks of the Stanford basic humanities program was a key issue in the great debate over multiculturalism and political correctness some years ago, with the much publicized interventions of Dinesh D'Souza, in his best seller *Illiberal Education*, and William Bennett. The scandal was not so much in the use of *I, Rigoberta Menchú* as a document or oral history from the world of the subaltern: Western culture in general and anthropology in particular thrive on reports of or from subaltern Others. It was, rather, in placing the text at the center of a canonic set of readings for undergraduates at a university whose primary function is to reproduce local, national and transnational elites.[3]

When Gayatri Spivak makes the apparently paradoxical claim that the subaltern cannot speak (Spivak, 1988b), she means that the subaltern cannot speak in a way that would carry any sort of authority or meaning for us without altering the relations of power and knowledge that constitute it as subaltern in the first place. *Richard Rodriguez* can speak (write), in other words, but not as a subaltern, not as Ricardo Rodríguez, and not (despite the fact that the United States is now the fifth largest nation of the Spanish-speaking world) in Spanish. It would be yet another version of the native informant of classical anthropology to grant narrators like Rodriguez or Menchú only the possibility of being witnesses, not the power to create their own narrative authority and negotiate its conditions of truth and representativity. This would be a way of saying that the subaltern can speak, but only through the institutionally sanctioned authority—itself dependent on and implicated in the power relations that produce subalternity—of the professional journalist or ethnographer, who alone has the power to decide what counts in the narrator's raw material and to turn it into a literary and/ or ethnographic narrative. By the same token, however—because not only *our* purposes count in relation to it—bringing a text like *I, Rigoberta Menchú* into the classroom or into the canon, as in the case of the Stanford Western Culture requirement, cannot be in itself the remedy that the "situation of emergency"—to borrow an idea of René Jara's—that motivated its production as a text in the first place requires (in Menchú's case, the genocidal counterinsurgency war waged against the highland Indian population by the Guatemalan army in the early 1980s), although it can certainly contribute to the international play of political and social forces that pertain one way or another to that remedy.

It is important to note that the "silence" of the subaltern, its aquiescence or "vulnerability," in Rodriguez's image, is only so from the perspective of the elite status he feels he has attained, his narrative authority, so to speak. It is what norms domination and elite status, just as, as

Spivak puts it, "subaltern practice norms official historiography."[4] *Los pobres* also have lives, selves, narratives, cognitive remappings. Their silence in the face of Rodriguez is strategic: They do not trust him, they sense perhaps that, despite his mestizo features, he is not one of them, that he is a *letrado*—a word that in Latin American Spanish suggests an agent of the state or the landowners. If their narrative were to be somehow produced as a text for us, it would be in effect *I, Rigoberta Menchú*.[5] And if such a narrative were in turn admitted into the hegemony—for example, used, like *I, Rigoberta Menchú*, as a core humanities curriculum reading at Stanford— it would give lie to Rodriguez's claim to difference and authority, a claim based precisely on his mastery of the codes of Western literature he has learned as a English major at Stanford.

Guha's deeply engaging study of peasant rebellions in nineteenth-century India, *Elementary Aspects of Peasant Insurgency in Colonial India*, makes it clear the subaltern marches under a banner inscribed, in effect, with the words of the Sermon on the Mount: "The last shall be first and the first shall be last."[6] To access the peasant rebel as a subject of history requires then a corresponding epistemological inversion, what Guha, almost as if he were responding directly to Rodriguez, calls a "writing in reverse." The problem is that the fact of these rebellions is captured precisely in the language (and the corresponding cultural assumptions) of the elites—both native and colonial—the rebellions are directed *against*. Thus, Guha argues,

the historical phenomenon of insurgency meets the eye for the first time as an image framed in the prose, hence the outlook, of counterinsurgency—an image caught in a distorting mirror. However, the distortion has a logic to it. That is the logic of opposition between the rebels and their enemies not only as parties engaged in active hostility on a particular occasion but as the mutually antagonistic elements of a semi-feudal society under colonial rule. The antagonism is rooted deeply enough in the material and spiritual conditions of their existence to reduce the difference between elite and subaltern perceptions of a radical peasant movement to a difference between the terms of a binary pair. A rural uprising turns thus into a site for two rival cognitions to meet and define each other negatively. It is precisely this contradiction which is the key to our understanding of peasant rebellion as a representation of the will of its subjects. For that will has been known to us only in its mirror image. Inscribed in elite discourse it had to be read as a writing in reverse. Since our access to rebel consciousness lay, so to say, through enemy country, we had to seize on the evidence of elite consciousness and force it to show us the way to its Other. (Guha, 1983, p. 333)

Guha means by the "prose of counter-insurgency" not only the records contained in the nineteenth-century colonial archive, but also the use (including the use in the present) of that archive to construct academic knowledges (historical, ethnographic, literary, and so on) that purport to represent these peasant insurgencies and place them in a teleological narrative of state formation. He is concerned with the way in which "the sense of history [is] converted into an element of administrative concern" in these narratives. Since, as the passage from *Hunger of Memory* suggests, the subaltern is conceptualized and experienced in the first place as something that lacks the power of (self-) representation, "by making the security of the state into the central problematic of peasant insurgency," Guha claims that these narratives necessarily deny the peasant rebel "recognition as a subject of history in his own right even for a project that was all his own" (Guha, 1983, p. 3). By contrast, what a text like *I, Rigoberta Menchú* forces us to confront is the subaltern not only as a represented subject but also as an agent of a transformative project that aspires to become hegemonic or dominant itself. Although we can enter into relations of solidarity with this project, it is not our own in any immediate sense and in fact involves structurally a contradiction with our position of relative privilege and authority in the academy.

In other words, the very idea of studying the subaltern is catachrestic or self-contradictory, in a way that points to a new register of knowledge where the power of the university to understand and represent or map the world breaks down or reaches a limit, and where the idea of representing the subaltern must itself confront the dilemma of subaltern resistance to and insurgency against elite conceptions. Recognizing the nature of this paradox means learning how to work against the grain of our own interests and prejudices—a process that Spivak calls "unlearning privilege" and that involves undoing the authority of the academy and knowledge centers at the same time that we continue to participate fully in them and to deploy their authority as teachers, researchers, planners and theorists [7] This aim distinguishes the subalternist perspective from alternative projects of postmodernist cognitive mapping, such as the idea of transnational cultural studies developed by Arjun Appadurai and his colleagues in the journal *Public Culture*, now centered at the University of Chicago. From its roots in the work of the British Communist Party historians like E.P. Thompson and Christopher Hill as well as the Birmingham Centre, cultural studies inherits a sense of popular, and then pop or mass, culture—that is, the kind of culture that traditionally did not count in academic discourse, or counted only as designating the essential alterity of the subaltern—as a form of subaltern agency. The distinction between high and low culture, in other words, and the decision on the part of cultural studies to transgress it, represents not only a functional differention of cultural spheres, but also the social antagonism between elite and subaltern groups and classes.[8] Likewise, cultural studies was animated by the same cluster of theoretical and political concerns as subaltern studies (feminism, Marxism, deconstructionism, poststructuralism, postcolonial discourse, and the like).

However, in its current process of academic institutionalization, cultural studies runs, in my opinion, the risk of becoming a primarily descriptive register for the emerging "scapes"—to borrow Appadurai's own term—of global and local cultures it seeks to map. As such, it may be complicit in producing discursively what I have called elsewhere (apropos the work of Stephen Greenblatt and the New Historicism) a "postmodernist tourist sublime" (Beverley, 1993, p. 44). The phrase is only partly ironic: I have in mind the capacity of cultural studies to produce an aesthetic-cognitive remapping of spheres of academic knowledge in a postnational register in ways that may become or are in fact becoming themselves a functional element in the hegemony of transnational capitalism, just as the Romantic sublime did in the nineteenth century. It is worth noting that various articulations of transnational cultural studies, including the one I have been connected with, the Inter-American Cultural Studies Network, have attracted the interest and support of the Rockefeller Foundation, whose liberal and "vanguard" identity as a foundation involves, at least in some measure, anticipating the future requirements of that hegemony.[9]

The project of subaltern studies, on the other hand, is necessarily a partisan and a negative one.[10] It renounces the cognitive reach (and the possibility of instrumentalization of its findings) of cultural studies in order to locate itself on the dividing lines where the relation between domination and subordination continues to be produced, lines that extend into the academy itself. As such, it is something like a secular version of the "preferential option for the poor" of liberation theology, and it shares with liberation theology the essential methodology of what theologian Gustavo Gutiérrez calls "listening to the poor."[11] Subaltern studies is not so much a question of finding new and more powerful techniques for information retrieval on the subaltern, nor of reinterpreting the historical past for its own sake,[12] as of dismantling the relationships that construct the elite/subaltern distinction in the first place.

It is by now a commonplace that this recognition entails moving from an epistemology of objectivity to a model of teaching and scholarship that would see these as forms of solidarity work. A persuasive statement of this idea—one, in any case, that many contributors to this collection seem to share—is made by Richard Rorty in his essay "Solidarity or Objectivity?", where he writes:

> There are two principal ways in which reflective human beings try, by placing their lives in a larger context, to give sense to those lives. The first is by telling the story of their contribution to a community. This community may be the actual historical one in which they live, or another actual one, distant in time or place, or a quite imaginary one, consisting perhaps of a dozen heroes and heroines selected from history or fiction or both. The second way is to describe themselves as standing in an immediate relation to a non-human reality. The relation is immediate in the sense that it does not derive from a relation between such a reality and their tribe, or their nation, or their imagined band of comrades. I shall say that stories of the former kind exemplify the desire for solidarity, and that stories of the latter kind exemplify the desire for objectivity. (Rorty, 1985, p. 3)

Spivak has made it clear, however, that in making the shift from objectivity to solidarity we cannot simply disavow representation—representation as both "speaking for" and "speaking about"—in favor of allowing the subaltern to "speak for itself."[13] And there is a sense in which the (necessarily?) liberal political articulation Rorty gives his antifoundationalism is also, as the 1960s slogan has it, "part of the problem," because it assumes that conversation, irony, mutual understanding, tolerance and the like are possible across power/exploitation divides that radically differentiate the participants. The Latin American philosopher Enrique Dussel notes that:

> when Rorty argues for the desirability of "conversation" in place of rational epistemology, he does not take seriously the asymmetrical situation of the other, the concrete empirical impossibility that the "excluded," "dominated," or "compelled" can intervene effectively in such a discussion. He takes as his starting point "we liberal Americans," not "we Aztecs in relation to Cortés," or "we Latin Americans in relation to a North American in 1992." In such cases, *not even conversation* is possible. (Dussel, 1995, p. 75 n. 15; Dussel's italics).

Gutiérrez insists, moreover, that the passage from objectivity to solidarity must begin with a relation of "concrete friendship" with the poor, that it cannot be simply a matter of taking thought or "dialogue," or for that matter of romanticizing or idealizing the poor (for, as the Gospels have it, the poor are also "poor in spirit," and, in any case, would prefer not to be poor). He concludes that the consequences for us of a preferential option for the poor are symbolized by the structure of an asymptotic curve: We can approach in our work, personal relations and politics closer and closer the world of the subaltern, but we can never actually merge with it, even if, in the fashion of the Russian *narodniks*, we were to "go to the people."

Those of us who are involved in the project of subaltern studies are often asked how we, who are (in the main) white or upper-caste and/or- class academics in top- or middle-level research universities, can claim to represent the subaltern. But we do not claim to represent ("map," "let speak," "speak for") the subaltern: what would be the point after all of representing the subaltern *as subaltern*? We seek to register instead the way in which the knowledge we construct and impart as academics is structured by the absence, difficulty or impossibility of representation of the subaltern. This is to recognize, however, the fundamental inadequacy of this knowledge and of the institutions that contain it, and therefore the need for general social change in the direction of a more radically democratic and nonhierarchical social order. What is at stake in an inchoate way in the challenge of subaltern studies is imagining the form of a new and potentially hegemonic project of social empowerment and transformation that (in a way I see modeled by the relation of feminist theory to the practical demands and tactics of the women's movement) necessarily passes through the "enemy country" of academic knowledge, rather than simply being represented in or by it: that is, the possibility of a new kind of politics.

NOTES

1. The ideas in this piece derive in part from a previous collection of my essays on the state of literary studies today, *Against Literature* (Beverley, 1993), and an essay, "Writing in Reverse: On the Project of the Latin American Subaltern Studies Group," in a special issue of the journal *Dispositio* (Beverley, 1994). I'm also indebted to the contributions of the participants in my seminar on the work of the Subaltern Studies Group at the University of Pittsburgh in winter 1995.

2. "Once upon a time I was a 'socially disadvantaged' child. An enchantedly happy child. Mine was a childhood of intense family closeness. And public alienation. Thirty years later I write this book as a middle-class American. Assimilated" (Rodriguez, 1983, p. 3).

3. On the Stanford debate, see Pratt, 1992.

4. "(T)he arena of the subaltern's persistent emergence into hegemony must always and by definition remain heterogeneous to the efforts of the disciplinary historian. The historian must persist in *his* efforts in this awareness, that the subaltern is

necessarily the absolute limit of the place where history is narrativized into logic. It is a hard lesson to learn, but not to learn it is merely to nominate elegant solutions to be correct theoretical practice. When has history ever contradicted that practice norms theory, as subaltern practice norms official historiography in this case?" (Spivak, 1988a, p. 16).

5. "testimonio [testimonial or ethnographic life stories] is a fundamentally democratic and egalitarian form of narrative in the sense that it implies that *any* life so narrated can have a kind of representativity. Each individual testimonio evokes an absent polyphony of other voices, other possible lives and experiences" (Beverley, 1993, p. 75).

6. Guha's epigraph for his book is a passage from Buddhist scripture, which he translates from the Sanskrit as follows (Guha, 1983, p. i):

(Buddha to Assalayana): What do you think about this, Assalayana? Have you heard that in Yona and Kamboja and other neighboring *janapadas* there are only two varnas, the master and the slave? And that having been a master one becomes a slave; having been a slave one becomes a master?

7. For a practical example of what this means in Spivak's own work, see her recent essay, "Responsibility" (Spivak, 1994), which concerns a development project by the World Bank and others to impose at great expense a new flood drain control system on the lowland peasants of Bangladesh.

8. See Aronowitz (1993) on this point. The idea of the "people" as a more heterogeneous agent of social struggle than the "class" was articulated in different ways in the period of Popular Front Communism. Guha, whose own roots as an activist and a historian are in both Gramsci and Mao, clarifies that he uses "the terms 'people' and 'subaltern classes' . . . as synonymous" (Guha, 1988, p. 44).

9. One should note here too the risk of a specifically Marxist quietism in Fredric Jameson's "cognitive mappings" of the cultural landscape of late capitalism, which seem to deflect the possibility of counterhegemonic agency to the fufillment of the logic of globalization itself. Mike Davis makes the point that Jameson's famous evocation of the Bonaventura Hotel in Los Angeles as a form of postmodernist "hyperreality" in his postmodernism essay effaces the surrounding activity of immigrant community groups and trade union struggles in the neighborhoods immediately adjacent to the hotel (Davis, 1988).

10. The category that defines subaltern agency in Guha's *Elementary Aspects* (1983) is Negation.

11. The phrase is from a series of lectures Gutiérrez delivered on *The New Evangelism* at the Pittsburgh Theological Seminary in May 1993.

12. This is the limitation of Florencia Mallon's otherwise quite perceptive critique of the work of the Latin American Subaltern Studies Group in a recent issue of the *American Historical Review* (Mallon, 1994). For Mallon, the question is how a subalternist perspective allows her to do a new kind of social history of peasants and their relation to the state in nineteenth- and early-twentieth-century Latin America, not how it puts into question the power relations she herself as an academic historian is implicated in in the present.

13. That is in fact the main theme of "Can the Subaltern Speak?" (Spivak, 1988b).

REFERENCES

Aronowitz, S. (1993). *Roll over Beethoven: The return of cultural strife.* Hanover, NH: Wesleyan University Press.
Beverley, J. (1993). *Against literature.* Minneapolis: University of Minnesota Press.
Beverley, J. (1994). Writing in reverse: On the project of the Latin American subal-

tern studies group. *Dispositio, xix*, 46.

Davis, M. (1988). Urban renaissance and the spirit of postmodernism. In E. Ann Kaplan (ed.), *Postmodernism and its discontents* (79–87). London: Verso.

Dussel, E. (1995). Eurocentrism and modernity. Introduction to the Frankfurt lectures. In J. Beverley, J. Oviedo, and M. Aronna (eds.), *The postmodernism debate in Latin America* (65–76). Durham, NC: Duke University Press.

Guha, R. (1983). *Elementary aspects of peasant insurgency in colonial India*. Delhi: Oxford University Press.

Guha, R. (1988). Preface. In R. Guha and G. Spivak (eds.), *Selected subaltern studies* (35–36). New York: Oxford University Press.

Gutiérrez, G. (1993). *The new evangelism*. Lecture series, Pittsburgh Theological Seminary.

Mallon, F. (1994). The promise and dilemma of subaltern studies: Perspectives from Latin American history. *American Historical Review, 99*(5), 1491–1515.

Menchú, R., and E. Burgos. (1984). *I, Rigoberta Menchú: An Indian woman in Guatemala*. London: Verso.

Pratt, M.L. (1992). Humanities for the future: Reflections on the Western culture debate at Stanford. In D. Gless and B. Herrnstein Smith (eds.), *The politics of liberal education*, 13–31. Durham, NC: Duke University Press.

Rodriguez, R. (1983). *Hunger of memory*. New York: Bantam.

Rorty, R. (1985). Solidarity or objectivity? In J. Rajchman and C. West (eds.), *Postanalytic philosophy* (3–19). New York: Columbia University Press.

Spivak, G. (1988a). Subaltern studies: Deconstructing historiography. In R. Guha and G. Spivak (eds.), *Selected subaltern studies* (3–34). New York: Oxford University Press.

Spivak, G. (1988b). Can the subaltern speak? In C. Nelson and L. Grossberg (eds.), *Marxism and the interpretation of culture* (271–313). Urbana: University of Illinois Press.

Spivak, G. (1994). Responsibility. *boundary 2, 21*(3), Fall 1994, 19–64.

Postcolonial Feminism in an International Frame

From Mapping the Researched to Interrogating Mapping[1]

Patti Lather

Theory is no longer naturally "at home" in the West . . . this privileged place is now increasingly contested, cut across, by other locations, claims, trajectories of knowledge articulating racial, gender, and cultural differences. But how is theory appropriated and resisted, located and displaced? How do theories travel among the unequal spaces of postcolonial confusion and contestation? What are their predicaments? How does theory travel and how do theorists travel? Complex, unresolved questions.

James Clifford, "Notes on Theory and Travel"

This chapter traces the transnationalization of contemporary theoretical discourse through the event of my experiences as a Fulbright lecturer in New Zealand in 1989. My title comes from Gayatri Spivak's 1981 essay on the circulation of French feminist theory across the world, in which Spivak deconstructs speaking for a generalized West and cautions that the appropriation of French theory by Anglo-American intellectuals often elides issues of the epistemic violence of contemporary forms of theoretical imperialism. Grounded in Spivak's deconstruction of this form of Eurocentrism, my interest in a postcolonial feminism is delineated via my New Zealand experience, where I both positioned myself and was positioned by others as something of a cheerleader for poststructuralism. The chapter proceeds by exploring the implications of postmodernism for emancipatory projects, sketches postcolonial tensions in my New Zealand experience and concludes with three mappings of modernity/postmodernity, one in the context of research methodology, with duly noted cautions regarding the limits of such binary modeling.

I wanted to go to New Zealand to experience academic work in a more "leftist" climate while I worked on a book about the implications of the various feminisms, neo-Marxisms and poststructuralisms for critical practices of research and pedagogy (Lather, 1991). While there I visited six campuses and consulted with groups that spanned nursing, sociology, education and women's studies. I engaged in much discussion, both formal and informal, about this thing I was then calling postmodernism/poststructuralism/deconstruction, terms I used somewhat interchangeably. Taking my manuscript to New Zealand provided me with the opportunity to present a fairly finished document to colleagues from another part of the world, to test its reception outside of the United States context where poststructuralism was beginning to move into the field of education as a theoretical adequation of contemporary times, of "New Times."[2]

Positioned, then, as an advocate for poststructuralism in an academic environment more steeped in critical perspectives than the United States, I encountered some memorable reactions. At the 1989 New Zealand women's studies conference, a keynote speaker termed poststructuralism a "virus" that threatened the coherence and effectivity of feminist work in the world. More than a few self-identified structuralist Marxists raised grave doubts about this latest theoreticism. "Attacks on the dialectic" were not well received and the relativisms presumed attendant upon poststructuralism were especially troublesome for the neo-Marxists. Many feminists branded it a "male conspiracy" which both represents in obtuse jargon what feminists had already formulated and serves to mark the panic of the decentered white male intellectual. Both feminists and neo-Marxists raised concerns along the lines of feminist political scientist Joan Cocks's (1989) caution regarding the collapse of ethics into aesthetics. She observes that there is a tendency toward depoliticization of interests on the part of thought preoccupied with shades of gray in the social world, a kind of mental masturbation that, while great fun in terms of language play, siphons off intellectual energy that could be better spent. Positioning postmodernism as "another set of outside ideas," many New Zealand academics across spectrums of race and sex expressed concern about its effect on the development of indigenous theory, theory grounded in the New Zealand struggle to unlearn neocolonialism and move from "little England" toward a "bicultural society" of Pakeha, Maori and Pacific Islanders.[3] Many Maori, for example, regarded postmodernism as "just more white academic theory" (e.g., Smith, 1992).

On the other hand, as I spoke throughout New Zealand, some others and others at some times viewed these theoretic movements as a way to get "unstuck," a way to think and act outside of the logic that limits leftist work in education in the face of both the right-wing discourses and the "crisis of difference" that have seized the imaginations and meaning-making of so much of the populations of both the United States and New Zealand. From this perspective, in the face of both the reassertion of the right and the "uprising of the marginalized," the available codes for conceptualizing radical political praxis seem inadequate: "One needs another language besides that of political liberation," Derrida answered an interviewer who asked, "Can the theoretical radicality of deconstruction be translated into a radical political praxis?" (Kearney, 1984, p. 122; see also Derrida, 1994; Spivak, 1994). Western logocentrism, with its dependence on oppositional relations with Otherness, its assumptions of self-presence and its pretensions toward mastery, totalization and certitude, has begun to implode. The emancipatory projects are, in varying degrees, inscribed in such logic, adrift and looking for bridges from existent political codes to codes more adequate in terms of "harbour[ing] a future of meaning" (Kearney, 1984, p. 110).

What does this mean for those of us who do our intellectual work in the name of social justice? After first exploring the implications of postmodernism for the emancipatory projects, I sketch some postcolonial tensions from my New Zealand experience in order to reflect on their implications for the development of research methodologies that interrupt present systems of dominance.

POSTMODERNISM AND THE EMANCIPATORY PROJECTS

Any effort at definition domesticates, analytically fixes and mobilizes pro and contra positions. Lattas (1989) suggests that a way to diffuse this is to get rid of the progressivist idea of history encoded in the "post" of postmodernism (p. 92). Not a position to be marked off, but a series of debates and interacting ideas, postmodernism is other than a "successor regime" (Harding, 1986, p. 142). Rather than a "progressive" development, it is movements of contradictory ideas and practices that "cross each other and give rise to something else, some *other* site" (Derrida, quoted in Kearney, 1984, p. 122). Rather than some historical/epistemological "break," some radically new conceptual framework, I see postmodernism as a reconfiguration and intensification of modernism that results in a conjunction that shifts our sense of who we are and what is possible. Borne out of the uprising of the marginalized, the revolution in communication technology, the fissures of a global multinational hypercapitalism and our consequent sense of the limits of Enlightenment rationality, postmodernism/poststructuralism has become the code name for the crisis of confidence in Western conceptual systems. A term coined in the United States but now a category that has migrated back and forth into many fields and countries, postmodernism "has an intertheoretical, international elasticity. . . . It is like the Toyota of thought: produced and assembled in several different places and then sold everywhere" (Rajchman, 1991, p. 125).

Current debates over the meaning and value of the postmodern flood the academic journals. There are those who warn that postmodernism fosters nihilism, relativism and political irresponsibility (Habermas, 1987; Anyon, 1994). Many fear that postmodernism is especially dangerous for the marginalized (Hartsock, 1987). Such positions on the part of those committed to emancipatory projects urge caution that, at best, the deconstructive dismantling of Enlightenment myths such as a self-correcting science, the transcendental, humanist subject and the assumption of teleological progress has both emancipatory and reactionary effects (McLaren, 1988). No few see postmodern theory as an outcome of capitalist decline and decadence, a new form of abstract, disengaged radical chic, of "nouveau smart" hyperintellectualism that marks the betrayal of the intellectuals. Habermas, for example, calls it "the new obscurity" (quoted in Rajchman, 1991, p. 46). Some few are unambiguously celebratory. Deleuze, for example, writes, "It is as if, finally, something new were emerging in the wake of Marx" (1988, p. 30). Many, myself included, remain ambivalent, attracted to some parts of postmodern thought and practice, worried about others.

Attempting to create a "postmodernism of resistance," such ambivalence can serve as a way "to interrogate the limits and powers of postmodern discourse" (Hutcheon, 1988, p. 8). Moving back and forth among the various contestatory discourses, those seeking to appropriate postmodernism in the name of liberatory politics urge that we use it to think more about how we think. A special focus is how oppositional work, while at war with the dominant system of knowledge production, is also inscribed in what it hopes to transform. Foucault writes, "My point is not that everything is bad, but that everything is dangerous" (quoted in Sawiki, 1988, p. 189). Within feminism, for example, recognition of the doubled movement of inscription and subversion presses one to acknowledge the ways in which feminism is both outside the discourse of the fathers and, simultaneously, inscribed in Western logocentricism, patriarchal rationality and imperialist practices.

It is this paradoxical "doubled movement" or "doubled consciousness"[4] regarding both the contestatory and reproductive dimensions of our efforts to make meaning that is the hallmark of postmodernism. Critical appropriations of postmodernism focus on the regulatory and transgressive functions of discourses that articulate and organize our everyday experiences of the world. To both confirm and complicate received codes is to see how language is inextricably bound to the social and the ideological. This moves social inquiry to new grounds, the grounds of "discourse," where the ways we talk and write are situated within social practices, the historical conditions of meaning, the positions from which texts are both produced and received. Premised on the recognition that all knowledge is "fallen" in some way, postmodernism argues that there is no unproblematic access to a foundation, no Archimedean standpoint outside of time and place. To paraphrase Foucault, it is not so much about emancipating truth from systems of power as it is about detaching the power of truth from the forms of dominance (1980).

The relationship between the postmodern and the postcolonial is also much contested. Concerns are raised about the prematurely celebratory nature of the "postcolonial," given both the recent academic marketability of the term and its obfuscation of a (neo)colonialism that continues to very much dominate the geopolitical space of transnational capitalism (McClintock, 1992). Shohat (1992) notes how it can serve to displace ethnic studies in United States academic contexts in ways that work against forging more effective counterhegemonic alliances across university disciplinary borders. As elaborated by both third world intellectuals in first world contexts (e.g., Spivak, 1990; Bhabha, 1990) and white critical intellectuals (e.g., Giroux, 1992), the term is used to signal both the ambivalence and the productivities of working within and against disciplinary and institutional matrices across cultural differences and positionalities. McGee (1992) considers the postmodern unthinkable outside the events we call postcolonialism, where the "native" emerges as a subject, subjected to the very discourse she or he uses to resist her inscription as Other to the West (p. 139). By this argument, postcolonialism is not so much about the end of colonialism as it is about the beginning of the historical possibility of situating Western thought as a local phenomenon alongside other localized systems of discourse; a profoundly skewed specificity, held to account for its need to unlearn privilege, including what Spivak notes as the hegemonizing effects of discourses and subject positions of critical intellectuals (1988).

In my survey of the contested ground around the concept of the "postcolonial," I find Emberley's definition particularly useful. Noting how the term maps onto the theoretic concerns of poststructuralism, she argues that the postcolonial signals "a contemporary configuration which implies a new direction" in the analysis of relations, given neocolonialism's recent dominance as a form of colonialism not reducible to the economic (1993, pp. 5–6). This new direction has to do with recognition of how "neocolonial" is as much a so-called first world projection of its own desire for a new world order as it is also a site of indigenous resistance to the imperial division of geopolitical space. Hence postcolonialism shifts the critique of colonialization from political/economic to cultural/subject formation, with a focus on how we live out hybrid and contradictory positions and symbolic debts and exchanges (S. Ahmed, 1996).

In this usage, postcolonialism is about "a world structured by both broad macrodependencies and localized distinctions" (Gordon, 1938, p. 1).

It constructs "a space of subversion of self-privileging positions" in the exploration of "a new democracy of theory" for a world full of "creative collision" of incommensurable voices that do not map easily onto one another (Rose, 1990, pp. 50–53). Within this context, feminist postcolonialism is positioned as "an open discourse, one of different participants with different stakes in its construction" (Gordon, 1988, p. 4). Here, the lesson to be learned from colonialism is that the question is not really how to better represent people, but rather, can critical academics "be accountable to people's struggles for self-representation and self-determination?" (Visweswaran, 1988, p. 39). This is in contrast to the "transformative intellectual" (Aronowitz and Giroux, 1985) who acts as the master of truth and justice. Postcolonialism argues the need for intellectuals with liberatory intentions to take responsibility for transforming academic practices. Positioned less as masters of truth and justice and more as creators of a space to decolonize the space of academic discourse that is accessed by privilege entails opening that space up in a way that contributes to the production of a politics of difference. Such a politics, as advocated in Beverley's chapter, recognizes the paradox, complexity and complicity at work in efforts to change as well as understand the world.

At the risk of what Derrida warns of as "peremptory diagnoses upon returning from a quick trip to a faraway land" (1994, p. 71), what follows is a sketch from six months in New Zealand as viewed through my continued wrestling with the meaning of a postcolonialism that charges my "white feminist imaginary" (Blunt and Rose, 1994, p. 4) with rethinking hegemonic maps of representation. My particular interest is in what such learnings might have to do with my efforts toward a research methodology that shifts from mapping the researched to interrogating mapping.

Postcolonial Tensions: Toward a Less Comfortable Social Science — Three Stories

Through the course of my New Zealand experience, I was asked an array of questions for which I had no ready answers and over which I have continued to brood. Wishing to probe the postcolonial tensions attendant upon my experiences as a visiting first world academic, with all of the dangers of "colonizing, self-aggrandizing feminism" (Jacobs, 1994, p. 187) implicit in that position, my desire is to begin to map a methodology attuned to postcolonial openings. Hence I will draw on Jennifer Robinson's essay, "White Women Researching/Representing 'Others'" (1994), in which she uses Spivak to move toward a "speaking with" model of engagement that results in "more experimental polyvocal texts" that displace the privileged fixed position from which the researcher interrogates and writes the researched. Disrupting "insider-outsider" positionalities, "speaking with" people across cultural differences is about negotiated and partial meanings arrived at through processes that create messy "spaces in between." In such spaces, centers and margins are both situated and yet constantly changing intersections of interpretation, contradiction and mutuality, produced via research practices that work toward a less comfortable social science. How does my experience in New Zealand help me move toward such spaces in inquiry?

Story 1

An early postcolonial tension grew out of what I learned about increased Maori resistance to "being studied." I met a Maori archaeologist who was contemplating a doctoral study of whites studying Maori. I reviewed an article for the *New Zealand Journal of Educational Research* written by white educational psychologists wrestling with "research as invasion." All of this came together for me at a lecture when an indigenous person asked, "Is this just a crisis of the West?"

In response, I remember using Spivak (1983) to argue that the white male's problem with decentering is not necessarily "ours" and yet marginalizing those already on the margins in this highly invested academic discourse was problematic. Unarticulated was the uneasy interface between the postcolonial subject and poststructural theory. Nativist discourse that positions poststructuralism as a "Western" neocolonialism, and employs a realist epistemology and a return to some noncontaminated space, is discredited by the antirealist project of postmodernism, with its claims of no innocent space. This is in tension with the efforts of Western counterhegemonic intellectuals to "do difference" differently via an internal critique of a deconstruction of Western thought (Niranjana, 1992, pp. 170–71). On the other hand, to dismiss poststructuralism as "the latest in white bourgeois thought," purveyed in a language so opaque it appears to mimic the thought of communication itself, occludes Lubiano's (1991) point: anything that gets the white academy asking itself hard questions is something that can be worked to the benefit of people of color. Noting that neither the Enlightenment nor modernity has worked to the advantage of people of color (p. 156), she urges African-American academics to both deploy postmodernism and mobilize a "constantly reinvigorated caution" (p. 153) in their "elbowing in" (p. 160) "as a way to negotiate particular material circumstances in order to attempt some constructions of justice" (p. 157).

Spivak (1992) circles around all of these issues yet again in "French Feminism Revisited: Ethics and Politics," in which she writes of her first visit to Algeria to explore "indigenous global feminism" (p. 54). Recognizing what she has "not yet learned to speak" (p. 55), Spivak's effort is to work the violence of intellectual configurations of Western thought against the dangers of a "too admiring ethnography" (p. 71) in order to delineate the possibilities of exchange between metropolitan and decolonized feminisms toward the undoing of imperialism. Tracing her own movements as an Indian diasporic feminist, Spivak argues that part of the "historical burden" of both colonizer and colonized is a complexity of hybridity that is lived across various registers, depending on subject position. Across such differences, "trying to think the international" (p. 80) on the basis of discontinuous and contradictory "oblique relationship[s] with the diversified Algerian womanspace" (p. 73), she urges postcolonial intellectuals to negotiate with "the larger critique of humanism" (p. 80) in "the task of decolonizing the mind through negotiating with structures of violence" (p. 81).

While I, like Spivak, have not learned how to speak such things, I do note that in delineating these issues in the context of this chapter, the question has shifted from "Is this a crisis of the West?" to "What is the desire to be the subject of one's own knowledge in the face of the rhizomatic dispersion of knowledges across unequal cultural transfers?"

Story 2

A second and related tension emerged around indigenous practices of self-determination versus postmodern deconstruction of identity politics. This emerged in a discussion with an expatriate American feminist lawyer much involved in Maori land rights litigations who saw poststructuralism as undercutting the very ground of such litigation. I tried, awkwardly, to articulate my suspicions of taking a position too easily reduced to indexing self with the set of cultural codings by which it is marked, such as race, class or sex. Denise Riley (1988) both deconstructs feminist usage of its central term, "women," and advises "foxiness" and "versatility" in negotiating between awareness of the indeterminancy of the term "women" and a strategic willingness to speak as if they existed, "since the world behaves as if they unambiguously did. . . . Sometimes it will be a soundly explosive tactic to deny it. . . . But at other times [it may be useful]. So feminism must be agile enough to say, 'Now we will be "women"—but now we will be persons, not these "women."'" (pp. 112–14).

What this might mean within a specific context is evoked by Rey Chow in a 1988 paper on reading the "China crisis" in terms of gender, as discussed by Kumar (1990).[5] Chow writes, "We do not, because at the moment of shock Chinese people are degendered and become simply 'Chinese.' To ask how we can use gender to 'read' a political crisis like the present one is to insist on the universal and timeless sufficiency of an analytical category" (p. 154). Yet Chow goes on to insist that this does not disallow the "instructive" role that the concept of "Third World women" can play. For example, Chinese women

> get short-shrifted on both ends: whenever there is a political crisis, they stop being women; when the crisis is over and the culture rebuilds itself, they resume their more traditional roles as wives and mothers as part of the concerted effort to restore order . . . the very efficacy with which we can use gender and sexuality as categories for historical inquiry is itself historical. (p. 154)

Kumar concludes her discussion of Chow's work by arguing for a "both/and" move that both rejects some totalizing concept of "gender" as a useful category and, outside the logic of noncontradiction, uses gender as a "critical operation to radically dislocate the center" (p. 155).

This is the strategic essentialism argument as articulated by Spivak,[6] but with which I was not familiar at the time. This ironic sort of move, however, continues to miss what Slemon (1990) notes as the difference beween postmodern and postcolonialist readings: the postmodern is about intertextual parody, whereas the postcolonial wants to retain a referential purchase on oppositional truth-claims as "a crucial strategy for survival in marginalized social groups" (p. 5). "This referential assumption" both draws on the poststructural suspension of the referent in order to deconstruct colonial power, and simultaneously reinstalls the referent in the service of resistant struggles. This creates a radically fractured and contradictory text, a dual agenda that is both within a realist problematic and a deconstructive reading of (neo)colonial rhetoric. This is about living in hybrid space, outside the logic of noncontradiction.

Story 3

Finally, there was the story among Pakeha academic feminists of a British feminist who had recently visited with the attitude of "bringing feminism to the colonials." My work benefited greatly from the thoughtful, critical reception that it received from a variety of New Zealand academics. And my own tentative, early forays into poststructuralism were much challenged and deepened by exposure to the more established leftist climate of New Zealand educational work in the academy. But in light of the desire for an indigenous New Zealand body of intellectual thought and practice, how was I not the metropole? And what might this mean in my desire to move toward a position of "speaking with" in terms of research participants and "within/against" in terms of disciplinary apparatuses, a move that is, furthermore, toward a de-stabilizing methodology as I explore the micropolitical practices of representation of self and others in situated inquiries?[7]

Postcolonial Tensions: Toward a Less Comfortable Social Science — Three Mappings

I conclude the discussion of movement toward a less comfortable social science with three mappings, the first a comparison of interpretation and genealogy as modernist/postmodernist analytical movements, the second a comparison of terms that oppose modernity/postmodernity, and the final chart a delineation of post-Kuhnian and postcolonial research paradigms as a gesture toward further exploration of what postcolonial space opens up in terms of research methodologies.

I begin in Table 1 with my charting of Ferguson's 1991 essay on combining/colliding interpretation and genealogy in feminist theory. I find it a particularly useful enactment of the "doubled" move of both/and that so characterizes deconstruction. Rather than some move "beyond" interpretation, Ferguson argues for a "both/and" move that uses the modernist moves of interpretation and the postmodern moves of genealogy as interruptors of one another: to hold together needed incompatibilities, both to "stay honest" and to "keep moving," such as an ironical stance that recognizes the need for both a longing for and a wariness of an ontological and epistemological home.

Table 1. Interpretation/Genealogy

INTERPRETATION	GENEALOGY
Appearance/reality distinction: hermeneutics of suspicion	*Subversion of mixed meaning-claims: suspicious of hermenuetics of suspicion*
Ontology of discovery	*Nietzsche: We make up our claims to truth then we forget we made them up; then we forget that we forgot*
Search for totality	*Interrogation of the limit*
Traces contradictions	*Produces acts of transgression outside of logic of noncontradiction*
Constructs successor regime of truth via movement of margins to center; engineers consensus reading	*Politics of difference/nothing innocent, creative collision of incommensurable voices that do not map easily onto one another*
Narrative or progressive enlightenment and eventual unity	*Return to interpretation, but informed by genealogical fragility, humility, embodiment, partiality*

The oppositions that the next chart constructs in Table 2 are unstable, with differences shifting and collapsing both within and between columns. While some model like a Moebius strip[5] might better represent the slides of inside and outside that so characterize the contemporary hybridity of positionalities and consequent knowledge forms, a chart such as this can serve to delineate the differences between the modern and the postmodern moments, in all their simultaneity and discontinuity. The task is how to diagram the becoming of history against the limits of our conceptual frameworks that are so much about what we have already ceased to be. In such an effort, the diagram becomes an abstract machine, provisional and schematic, designed to move us to some place where oppositions dissolve through the very thinking that they have facilitated.

Table 2. Modernity/Postmodernity

MODERNITY	POSTMODERNITY
Metaphysics: Idealist/materialist	*Antimetaphysics: Rhetorical turn*
Incorporates other into the same	*Nonreducible differences*
Logic of noncontradiction	*Logic of paradox*
Optimism/pessimism	*Double affirmative*
Nihilism	*Nonstupid optimism*
Critical, confident	*Meta/reflective, ironic*
Teleological progress	*Deferral, nomadology*
Originality/originary	*Parody/intertextuality*
Whole/authenticity, unified subject	*Fractured subject*
Voice/presence	*Polyphony of fragments*
Persuasion via reason	*Seduction via desire*
Identity politics	*Strategic practices (e.g., essentialism)*
Salvation narrative	*No way out of indeterminacy*
DIALECTICS (IDEOLOGY CRITIQUE)	RHIZOMATICS
Metaphysical oppositions	*Decentered multiplicities*
Categories/concepts	*Fold/invagination*
	Planes, intensities, flows
Depth models	*Surfaces/becomings*
	Linkages/assemblages
	Production of unconsciousness
Philosophy of consciousness	*Contingent, discontinuous*
Continuous systems-found world	*Lines, movements, speeds*
	Made world

The preceding chart in Table 2 draws on ideas from many fields, literary theory, philosophy, anthropology, sociology and authors aligned with diverse movements, groups and views; even contemporary Broadway theater in the case of Tony Kushner's "non stupid optimism" from *Angels in America*.[9] Key references would include Deleuze (Boundas and Olkowski, 1994) and Derrida (1994). The next and final chart articulates the post-Kuhnian versus the post-colonial moments in terms of research methodologies.

Table 3. Post-Kuhnian/Postcolonial

POST-KUHNIAN	POSTCOLONIAL
"Paradigm Wars"	"Post-Paradigmatic Diaspora"
Naturalistic	*Marxist*
Qualitative	*Race-specific*
Hermeneutic	*Feminist*
Phenomenological	*Freirian*
Constructivist	*Deconstructive*
Interpretive	*Poststructural/postmodern*
Humanist	*Posthumanist*
	Queer

Kuhnian frameworks deemphasize the political content of theories and methodologies and the dissolving of the world as structured by referential notions of language. They also diminish the play of multiple emergent knowledges vying for legitimacy. Caught in a representational logic, they search for codifications and standards instead of asking if something more fundamental than a "paradigm shift" in the academy might be going on.

Postcolonial frameworks attempt to unlearn accepted ways of thinking and analysis that reinscribe dominance within the context of a politics of difference and a vision of social justice. The goal of such inquiry is to act with others, not on or for. Much of this is about unlearning privilege and working against assumptions of the unified subject, historical laws and Enlightenment rationality. Rather than merely a shift in academic knowledge, it is about how we talk, listen and live our lives in a world marked by inequities and resistance struggles.

The particular oversimplifications of this chart are many; for example, its columns are more a hodgepodge listing than they are sets of binaries, and much if not most Marxist, race-specific and feminist empirical work is conducted within a representational logic. More complicatedly, its binaries reduce the possibilities for a reaction such as Derrida's to Geoff Bennington's writing about his work: "what is written 'up' there, beside or above me, *on* me, but also *for* me, in my favor, toward me and in my place" (Bennington and Derrida, 1993, p. 26).[10] Key references are Spanos (1993) and Dickens and Fontana (1994).

Offered here as a means toward some conceptual distinctions, fruits of my experiences with what James Clifford has termed "traveling theory" (1989), in spite of their limits, these charts are offered as a gesture toward the opening up of other sites of inquiry that are attuned to the postcolonial moment on the world historical stage.

CONCLUSION

Travel is not a word that can be easily evoked to talk about the Middle Passage, the Trail of Tears, the landing of Chinese immigrants at Ellis Island, the forced relocations of Japanese-Americans, the plight of the homeless. Theorizing diverse journeyings is crucial to our understanding of any politics of location. . . . From certain standpoints, to travel is to encounter the terrorizing force of white supremacy.

bell hooks, "Representing Whiteness in the Black Imaginary"

Hooks's caution regarding the use of "traveling" to speak of theory underscores that the international debate over postmodernism is about "the recirculation of the European Enlightenment in a non-Eurocentric age" (Rajchman, 1991, p. 116). Displacing the ideal of a global, totalizing project of emancipation with a celebration of dispersion and differences, postmodernism is characterized by the intersection of various bodies of thought and practice that interrupt any articulation of a new master discourse.

Turning and turning in the widening gyre
The falcon cannot hear the falconer;
Things fall apart; the centre cannot hold;

W.B. Yeats*

So begins William Butler Yeats's poem, "The Second Coming," written in horrified response to the Russian Revolution. I often think of these words as I struggle with what postmodernism is and might mean for those of us contesting how the world is understood in ways that challenge hegemonic knowledges. Rather than the sense of loss and even fear at "And what rough beast, its hour come round at last/Slouches towards Bethlehem to be born?",

I conclude feeling poised at some opening that bodes well for those of us who want our intellectual engagement to matter in the struggle for social justice. The conjunction in critical social theory of the various feminisms, postcolonialisms, neo-Marxisms and poststructuralisms feels fruitful ground for shifting us into ways of thinking that can take us beyond ourselves. This essay is intended to both mark and propel that shift.[11]

NOTES

*Reprinted with the permission of Simon & Schuster from *The Poems of W.B. Yeats: A New Edition*, edited by Richard J. Finneran. Copyright ©1924 by Macmillan Publishing Company. Renewed 1952 by Bertha Georgie Yeats.

1. An earlier version of this paper was presented at a symposium on Gender and Education: International Issues at the annual conference of the Comparative and International Education Society, Pittsburgh, Pennsylvania, March 14–17, 1991.

2. "New Times" refers to the shift in late capitalism that characterizes postmodernism (e.g., Harvey, 1989). A New Zealand journal, *Sites*, had a special issue on the topic (Autumn, 1990, 20), well ahead of any sustained US attention, especially in journals to which educational writers regularly contribute.

3. Pakeha is the Maori word for white settlers. See Jones, 1991; Middleton, 1993.

4. W.E.B. DuBois used the term "doubled consciousness" in 1903, as pointed out by Gilroy (1989) in an essay that deals with how "the prized linguistic insights of post-structuralism" were deduced earlier by many black thinkers and artists out of the historical and cultural substance of black life in the West (p. 110).

5. Chow has since published this essay in Rey Chow, *Writing Diaspora: Tactics of Intervention in Contemporary Cultural Studies* (1993).

6. In her essay on subaltern studies, Spivak (1987) argues for "a strategic use of positivist essentialism in a scrupulously visible political interest," a deployment of essentialism as a provisional gesture toward a history of dispossessed subjects (p. 205). The Subaltern Studies Group, a Marxist historical collective committed to historiographical critique and reinscription of South Asian history, is both critiqued and endorsed by Spivak for its reliance on humanist notions of agency, totality and presence. To quote Diana Fuss (1989), "in other words, when put into practice by the dispossessed themselves, essentialism can be powerfully displacing and disruptive" (p. 32). Spivak has since urged caution about the ways "strategic essentialism" can become a slogan that elides the continual need to deconstruct precisely that which we think we cannot think without (1993, p. 9).

7. The situated inquiry I am presently involved in is a study of women living with HIV/AIDS, where pressing issues of "telling the other" (McGee, 1992) are being much played out (Lather, 1995; Lather and Smithies, 1995).

8. A Moebius topology blurs "inside-out, thinking itself ahead, leaving itself behind, encountering the 'savage singularities' it draws and binds and strategically diagrams, the thinking line, folding point 'line of the outside' . . . that 'terrible line that brews all the diagrams'. . . . Homeless, pointless, without identity; absolute position without a location . . . it is utter nonsense. . . . It is a jumping point, 'leaping over itself' into a line . . . sweeping across itself onto a surface, folding its self-surface into a volume . . . without ceasing to be a plane surface, the surface a line, the line a point, the point a fractalescent chaos" (Canning, 1994, quoting Deleuze, Prigogine and Paul Klee, p. 90).

9. Kushner calls on Walter Benjamin and his *Angel of History* to set his work against "the stupidly optimistic" in an interview in the October 5, 1992, *Los Angeles Times* (pp. 74–76).

10. The Bennington and Derrida book is a split text, with the top two-thirds

Bennington's intellectual history of Derrida and the bottom one-third Derrida's reaction to Bennington.

11. In regard to the use of full names in the reference list, I break form with the *Publication Manual of the American Psychological Association* in order to make it possible to read the gender politics of the sources I draw on for this chapter.

REFERENCES

Ahmed, Sara. (1996). Moving spaces: Black feminism and postcolonial theory. *Theory, Culture and Society, 13*(1), 139–146.

Anyon, Jean. (1994). The retreat of Marxism and socialist feminism: Postmodern and poststructural theories in education. *Curriculum Inquiry, 24*(2), 115–134.

Aronowitz, Stanley, and Henry Giroux. (1985). Radical education and transformative intellectuals. *Canadian Journal of Political and Social Theory, 9*(3), 48–63.

Bennington, Geoffrey, and Jacques Derrida. (1993). *Jacques Derrida.* Chicago: University of Chicago Press.

Bhabha, Homi, ed. (1990). *Nation and narration.* New York: Routledge.

Blunt, Alison, and Gillian Rose, eds. (1994). *Writing women and space: Colonial and postcolonial geographies.* New York: Guilford.

Boundas, Constantin, and Dorothea Olkowski, eds. (1994). *Gilles Deleuze and the theater of philosophy.* New York: Routledge.

Canning, Peter. (1994). The crack of time and the ideal game. In C. Boundas and D. Olkowski (eds.), *Gilles Deleuze and the theatre of philosophy* (73–98). New York: Routledge.

Chow, Rey. (1993). *Writing diaspora: Tactics of intervention in contemporary cultural studies.* Bloomington: Indiana University Press.

Clifford, James. (1989). Notes on theory and travel. *Inscriptions, 5,* 177–188.

Cocks, Joan. (1989). *The oppositional imagination: Feminism, critique and political theory.* New York: Routledge.

Deleuze, Gilles. (1988). *Foucault.* Translated and edited by Sean Hand. Minneapolis; University of Minnesota Press.

Derrida, Jacques. (1994). *Specters of Marx: The state of the debt, the work of mourning, and the new international.* New York: Routledge.

Dickens, David R., and Andrea Fontana, eds. (1994). *Postmodernism and social inquiry.* New York: Guilford.

Emberley, Julia V. (1993). *Thresholds of difference: Feminist critique, native women's writings, postcolonial theory.* Toronto: University of Toronto Press.

Ferguson, Kathy. (1991). Interpretation and genealogy in feminism. *Signs, 16*(2), 211–239.

Foucault, Michel. (1980). *Power knowledge.* Translated and edited by Colin Gordon. New York: Pantheon.

Fuss, Diana. (1989). *Essentially speaking: Feminism, nature and difference.* New York: Routledge.

Gilroy, Paul. (1989). Cruciality and the frog's perspective: An agenda of difficulties for the black arts movement in Britain. *Art and Text, 32,* 106–117.

Giroux, Henry. (1992). Paulo Freire and the politics of postcolonialism *Journal of Advanced Composition, 12*(1), 15–26.

Gordon, Deborah. (1988). Introduction: Feminism and the critique of colonial discourse. *Inscriptions, 3/4,* 1–5.

Habermas, Jurgen. (1987). *The philosophical discourse of modernity.* Cambridge, MA: MIT Press.

Harding, Sandra. (1986). *The science question in feminism.* Ithaca, NY: Cornell University Press.

Hartsock, Nancy. (1987). Rethinking modernism: Minority vs. majority theories. *Cultural Critique, 7,* 187–206.

Harvey, David. (1989). *The condition of postmodernity*. Oxford: Basic Blackwell.

hooks, bell. (1992). Representing whiteness in the Black imaginary. In L. Grossberg, C. Nelson, Paula Treichler (eds.), *Cultural Studies* (338–346). New York: Routledge.

Hutcheon, Linda. (1988). *A poetics of postmodernism: History, theory, fiction*. New York: Routledge.

Jacobs, Jane. (1994). Earth honoring: Western desires and indigenous knowledges. In A. Blunt and G. Rose (eds.), *Writing women and space* (169–196). New York: Guilford.

Jones, Alison. (1991). At school I've got a change. *Culture/privilege: Pacific Islands and Pakeha Girls at School*. Palmerston North, New Zealand: Dunmore.

Kearney, Richard. (1984). *Dialogues with contemporary continental thinkers: The phenomenological heritage*. Manchester, England: Manchester University Press.

Kumar, Amitava. (1990). Toward a postmodern Marxist theory: Ideology, state, and the politics of critique. *Rethinking Marxism, 3*(3–4), 149–155.

Lather, Patti. (1991). *Getting smart: Feminist research and pedagogy with/in the postmodern*. New York: Routledge.

Lather, Patti. (1995). The validity of angels: Interpretive and textual strategies in researching the lives of women with HIV/AIDS. *Qualitative Inquiry, 1*(1), 41–68.

Lather, Patti, and Chris Smithies. (1995). *Troubling angels: Women living with HIV/AIDS*. Columbus, OH: Greyden Press.

Lattas, Judy. (1989). Feminism as a proper name. *Australian Feminist Studies 9*, 85–96.

Lubiano, Wahneema. (1991). Shuckin' off the African-American native other: What's "po-mo" got to do with it? *Cultural Critique, 18*, 149–186.

McClintock, Anne. (1992). The angel of progress: Pitfalls of the term "post-colonialism." *Social Text, 10*(2–3), 84–98.

McGee, Patrick. (1992). *Telling the other: The question of value in modern and postcolonial writing*. Ithaca, NY: Cornell University Press.

McLaren, Peter. (1988). Schooling the postmodern body: Critical pedagogy and the politics of enfleshment. *Journal of Education 170*(3), 53–83.

Middleton, Sue. (1993). *Educating feminists: Life histories and pedagogy*. New York: Teachers College Press.

Niranjana, Tejaswini. (1992). *Sitting translation: History, post-structuralism, and the colonial context*. Berkeley: University of California Press.

Ong, Aihwa. (1988). Colonialism and modernity: Feminist re-presentations of women in non-Western societies. *Inscriptions, 3/4*, 79–93.

Rajchman, John. (1991). *Philosophical events: Essays of the 80's*. New York: Columbia University Press.

Riley, Denise. (1988). *Am I that name? Feminism and the category of "women" in history*. Minneapolis: University of Minnesota Press.

Robinson, Jennifer. (1994). White women researching/representing "others": From antiapartheid to postcolonialism? In A. Blunt and G. Rose (eds.), *Writing women and space* (197–226). New York: Guilford.

Rose, Dan. (1990). *Living the ethnographic life*. Newbury Park, CA: Sage.

Sawiki, Jana. (1988). Identity politics and sexual freedom: Foucault and feminism. In Irene Diamond and Lee Quinby (eds.), *Feminism and Foucault: Reflections on resistance* (177–191). Boston: Northeastern University Press.

Shohat, Ella. (1992). Notes on the "post-colonial." *Social Text 10*(2/3), 99–113.

Slemon, Stephen. (1990). Modernism's last post. In Ian Adam and Helen Tiffin (eds.), *Past the last post: Theorizing post-colonialism and post-modernism* (1–12). Calgary: University of Calgary Press.

Smith, Linda. (1992). Maori women: Discourses, projects and Mana Wahine. In Sue Middleton and Alison Jones (eds.), *Women and education in Aotearoa* (33–51). Wellington, New Zealand: Bridget Williams.

Spanos, William. (1993). *The end of education: Toward posthumanism*. Minneapolis: University of Minnesota Press.

Spivak, Gayatri. (1981). French feminism in an international frame. *Yale French Studies, 62,* 154–84.

Spivak, Gayatri. (1983). Displacement and the discourse of women. In Mark Krupnick (ed.), *Displacement: Derrida and after* (169–195). Madison: University of Wisconsin Press.

Spivak, Gayatri. (1987). Subaltern studies: Deconstructing historiography. In *In other worlds: Essays in cultural politics* (197–221). New York: Methuen.

Spivak, Gayatri. (1988). Can the subaltern speak? In Cary Nelson and Lawrence Grossberg (eds.), *Marxism and the interpretation of culture* (271–313). Urbana: University of Illinois Press.

Spivak, Gayatri. (1990). *The post-colonial critic: Interviews, strategies, dialogues.* Edited by Sarah Harasym. London: Routledge.

Spivak, Gayatri. (1992). French feminism revisited: Ethics and politics. In Judith Butler and Joan Scott (eds.), *Feminists theorize the political* (54–85). New York: Routledge.

Spivak, Gayatri. (1993). In a word: Interview, with Ellen Rooney. In *Outside in the Teaching Machine* (1–24). New York: Routledge.

Spivak, Gayatri. (1994). Responsibility. *Boundary 2, 21*(3), 19–64.

Visweswaran, Kamala. (1988). Defining feminist ethnography. *Inscriptions, 3/4,* 27–46.

Yeats, W. B. (1924). *The poems of W.B. Yeats: A new edition*. London: Macmillan.

Mapping the Spaces of Capital

Crystal Bartolovich

There are no easy ways to map the rugged and shifting terrain of the intellectual territory known as Western Marxism. Indeed, its very boundaries and most prominent features have themselves been the source of heated dispute.

Martin Jay, *Marxism and Totality*

Pedagogy for "Analysis of Situations"

The subtitle of this volume ("Mapping Ways of Seeing Social and Educational Change") invites a consideration of the issues I will raise in this essay: How are we to think the practice of "mapping"? What roles might "mapping" and pedagogy play in effecting social change? "Spaces of Capital," a Ph.D. seminar I taught in the spring of 1995, was my attempt to bring a consideration of the metaphor and practice of cartography into the graduate cultural studies curriculum at a time when the economic, political and social relations of the globe on which we live seem to be in a crisis often understood *spatially*. Fredric Jameson (1991), for example, describes the disorientation of the "postmodern" subject in specifically spatial terms when he calls for a "cognitive mapping"—that is, an understanding of the material relations and forces in which all subjects are situated as a whole. This "mapping," he claims, will make it possible for us to "regain a capacity to act and struggle which is at present neutralized by our spatial as well as our social confusion" (p. 54). Robert Reich (1992) offers a more mainstream formulation: "What is the role of a nation within the emerging global economy, in which borders are ceasing to exist?" (p. 301). The "global economy," or process of "globalization," to which Reich refers may mark our entry into a new regime of capital accumulation, as commentators such as David Harvey (1990) have (cautiously) suggested, with far-reaching effects on all of our lives. If Jameson is at all correct in his assessment of the importance of "mapping," it seems pertinent to ask, at the current moment of spatial crisis Reich describes, how we might best theorize forces such as "globalization" and track its effects not only on the business pages of newspapers, but also in the capillaries of everyday life. By "mapping," then, I mean both a literal attention to "geo-graphy" in the sense of "world writing," the continuous process of producing a world out of a set of differential and differentiating relations, and the activity of theorizing in general.

By keeping these two activities—"world-writing" and theorization—yoked together, I emphasize the importance of directing theory to the analysis of concrete situations as Antonio Gramsci (1971) reminds us is necessary to any practical politics. As he explains:

> The study of how "situations" should be analyzed, in other words how to establish the various levels of the relations of force, offers an opportunity for an elementary exposition of the science and art of politics—understood as a body of practical rules for research and of detailed observations useful for awakening an interest in effective reality and for stimulating more vigorous political insights. (p. 175)

Such a project violates the academic specialist injunction to "disinterested" knowledge production, to be sure, but it offers in return the possibility of unpacking the unacknowledged or unknown agendas that inhabit such ostensible disinterest. Mapping is a useful metaphor for such a project since—as Jameson's use of it emphasizes—it signals an attempt to describe not only where one (as a social subject) is situated, but also where (theoretically and practically) one has been and might now go. Although mapping, understood as a particular kind of "analysis of situation" in Gramsci's sense, is in itself not a substitute for practical politics, it is nonetheless a crucial condition of emergence for such politics. Thus, its place in the classroom is not a politically indifferent matter.

Correspondingly, how one imagines and engages in mapping are hotly contested issues. Because of our spatial concerns, my "Spaces of Capital" class was particularly attentive to Fredric Jameson's work, since he has repeatedly returned to the metaphor of "cognitive mapping." However, we were also careful to take into account poststructuralist critiques of his position, especially the concept of "hegemony" as it has been taken up by post-Marxist theorists attempting to break out of what they perceive to be the prison house of totality that Jameson posits as irreducible to any Marxist project. "Totality"—thinking in terms of systematic relations among parts of a social/historical "whole"[1]—is one of the aspects of Marxist analysis that has come under the most intensive fire with the emergence of poststructuralist critiques, with their emphasis, as Derrida (1978) put it, on "holes" rather than wholes (p. 178). Stanley Aronowitz (1990) elaborates: "a new post-Marxist discourse has emerged to explain the transformation, principally in spatial terms, which have rendered suspect many Marxist assumptions" (p. xiv). Cognitive mapping of the Jamesonian kind has, thus, come into conflict with the more mobile mapping demanded by poststructuralist critics. Hence, I wanted to explore the possibilities of the claim made for the continued pertinence of cognitive mapping by Gayatri Spivak (1993): "Careful cognitive mapping . . . with the deconstructive awareness of complicity can have its uses for the internationalist activist" (p. 257). This deployed/deconstructed mapping might be called "strategic" cognitive mapping.

In the pages that follow, I will consider the implications of Spivak's formulation and the encounter it demands between Marxist analysis and poststructuralism. I begin with a discussion of the debates surrounding some of the key terms (totality, locality, hegemony) via a discussion of my class's first set of readings; I then provide a narrative of the way in which we worked through these terms, focusing on the specifically spatial understanding (thinking in terms of positions and relations) such terms encourage. Although Jameson's attempt to preserve the traditional Marxist concept of totality and poststructuralist critiques of it both rely on spatial metaphors, the mapping of the spaces proposed is quite different in each case (even allowing for the wide variation in poststructuralist positions). I end with a brief description of the research projects undertaken by seminar participants during the semester in order to explain how the class encouraged "analysis of situations" in Gramsci's sense. Throughout, I hope to indicate the continued pertinence of strategic cognitive mapping along with a recognition of its problems. As Stanley Aronowitz (1990) explains: "Even when sharply criticized, historical materialism remains the referent of all theorizing whose object is human emancipation" (p. xv). When we are faced with growing global economic inequalities, and the continuing exploitation of the majority of the planet in the interests of the few, an analysis that makes it possible to understand the relationship between extremes of poverty and wealth has hardly lost its usefulness. And yet when we examine specific situations closely, it has become increasingly evident that a class analysis alone simply will not suffice to trace out the complex power relations in which we are all entangled. Thus, the "Spaces of Capital" course attempted to provide some theoretical tactics not only for mapping specific situations, but also for coming to terms with the controversies over how they are best mapped.

"Spaces of Capital" enrolled nine graduate students in the Literary and Cultural Theory Program at Carnegie Mellon University (CMU). This is a rather specialized group, with a better background in theory than can be anticipated in many graduate classrooms (and certainly in virtually all undergraduate classrooms); however, I hope that this essay presents the material covered in such a way that the course does not seem impossible to reproduce in some fashion elsewhere. Indeed, CMU is continually reworking its courses and program to better fit the changing understanding of cultural studies and the needs of students. The point of this rather exhausting commitment to revision is a recognition that, although cultural studies does not mean anything and everything, it also should not be reified into a stable body of work, issues or methods. As Stuart Hall (1992) has put it: "Here one registers the tension between a refusal to close the field, to police it and, at the same time, a determination to stake out some positions within it and argue for them" (p. 278). Hall's characterization of cultural studies is notable both for its refusal to codify, and for its invocation of spatial relations in order to explain the predicament of a "field" that refuses to "close" its borders, and yet continues to "stake out . . . positions." How to map such an indeterminate field is an ongoing concern, and not without significance to broader questions about how to map social "(w)holes."

Although Hall's claim cited above might seem to make it difficult to provide any kind of definition of cultural studies, he goes on to describe his experience of it in some detail; what emerges is a series of intellectual and practical-political encounters, ongoing tensions and struggles: What happens when structuralism and semiotics meet up with Marxism? feminism? antiracism? each other? Cultural studies has attempted to deal with the ways in which meanings are made and society ordered not just in dominant discourses and institutions or "high-cultural" texts, but also in popular forms and the realm of everyday life as a site of contested meanings and multiple productions, of resistance as well as consent. Such meaning production does not, of course, take place on an even field, but is everywhere inflected with power relations, "preferred meanings," hierarchical orderings. It has been the task of cultural studies to make explicit and understand this process of hierarchization, articulating cultural/symbolic production with economic and political domains without subordinating it to them. Often in cultural studies this practice is described as a "mapping" task, as when Peter Stallybrass and Allon White (1986) claim that their book on transgression is "an attempt to map some interlinked hierarchies on the terrain of literary and cultural history" (p. 2). This knowledge can then be put to use for action in (not just reflection on) the social order by exposing, denaturalizing—and thus providing a means of resisting—dominant ideologies that disavow their production. Because the dominant social order is itself often flexible and responsive to changing conditions, the corresponding counterforces must be flexible as well, as Hall's "open" cultural studies recognizes.

"Spaces of Capital" was designed to be a mapping project for the sort of cultural studies Hall describes. At the crossroads of Marxism and poststructuralism, as they have been taken up by cultural studies, it sought to understand how the "globe" and other "spaces" (topographical and theoretical) are produced today in specific situations. To get the basic concepts (i.e., totality, locality, hegemony) on the table, several sessions were devoted to setting out in more detail the thematics of the class by way of readings of essays by Fredric Jameson, Michel Foucault, Stuart Hall and Paul Smith. The goal of the readings was to help the class establish a theoretical frame for, as well as provide a few concrete examples of, the sort of analysis I wanted us to engage in; specifically, thinking of the practice of mapping in terms of "totality" and critiques of totality—how mapping needed to be understood differently in cognitive (Jameson) versus poststructuralist (e.g., Foucault) terms. From Jameson's extensive oeuvre, students read the "Cognitive Mapping" essay he presented at the "Marxism and the Interpretation of Culture" conference at the University of Illinois in 1983. We compared his remarks at that time with his later refinements of them in the conclusion to his 1992 book *Postmodernism*.

What the later and the earlier Jameson discussions of cognitive mapping share is a commitment to Marxist totalizing analysis, both for coming to terms with the ("symptomatically"—or, pathologically—fragmented) current moment and planning for a more liberatory future. Marxist theorizations of totality, threatened by the forces of poststructuralism, become for him the conceptual space in which the battle between these two modes of knowledge production (and practical politics) are waged. Cognitive mapping eludes the problem of "representation" (as critiqued by poststructuralists) for Jameson because it is self-consciously abstract rather than adequately referential. For Jameson, representation is properly understood as "figuration," which does not imply a faithful mimetic copying of a pre-existent reality but simply an attempt to come to terms with "the real" in all its complexity of relations, even though that "real" will always remain an "absent cause" for which any representation is necessarily inadequate.

In spite of these claims, however, Jameson's own examples are often spatial in a topographical sense and at least *seem* referential (he later points out in the conclusion to *Postmodernism* that he did not intend to reduce "mapping" to everyday cartography in this way). They are also always grounded in a socioeconomic mode of production—a point on which he remains firm. He posits, for example, a spatiality appropriate to the different "stages" of capital, and traces out an increasing disjuncture between individual subjects and their "lived experience" of structure, such that seeing the world as a whole becomes increasingly difficult (although ever more necessary). Cognitive mapping is a mode of analysis appropriate to addressing and overcoming these difficulties imposed by the logic of capital because, for Jameson, it makes possible seeing a world—that appears to be fragmented and incoherent—as whole. The alienation in the face of fragmentation Jameson describes, however, is apparently profoundly Western and indicates some of the reasons why postcolonial critics have found Foucauldian analysis (to which I will turn shortly) strategically useful in combating the modes of production narrative in theorizing change.

For example, Jameson notes of the stage of "imperial" capital:

> the phenomenological experience of the individual subject . . . becomes limited to a tiny corner of the world, a fixed camera view of a certain section of London or the countryside or whatever. But the truth of that experience no longer coincides with the place in which it takes place. The truth of that limited daily experience of London lies, rather, in India or Jamaica or Hong Kong; it is bound up with the whole colonial system of the British Empire that determines the very quality of the individual's subjective life. Yet those structural coordinates are no longer accessible to immediate lived experience and are often not even conceptualizable to most people. (p. 349)

One sees Jameson's point: The expansion of capital has had the effect of dispersing its subjects' conditions of existence beyond their own immediate locales, making it much more difficult to understand (and, thus, resist) them. There are, however, a number of troubling aspects of this passage, including its profound nostalgia—the reliance on a "fall" myth of an originary wholeness, accessible to lived experience, that has been lost to the alienating relations of capital (a set of assumptions that can also be located in Raymond Williams's work).[2]

Even more problematically, however, Jameson's totality evades the question of the rather different experience of the Indian or Jamaican subjects, for whom the imperial relations in which their lives were implicated with those unknowing masses in London might have been entirely more explicit and evident. For such subjects, a theory of totality, centered and described as Jameson centers and describes it, may not seem so liberatory. That his view (for all his political good will) is not only ethnocentric but elitist reveals itself in formulations such as: "Everyone knows how, toward the end of the nineteenth century, a wide range of writers began to invent forms to express what I will call monadic relativism" (1988, p. 350). Everyone? Who is this everyone (who clearly does not include everyone at all) exactly? Like the globe of capital that simply ignores and excludes anything not relevant to its own needs, Jameson's "everyone" (albeit with better intentions) performs a suspect gesture of ignoring and excluding as well. In this dynamic of exclusion, necessary to constituting any "whole," the problem of "everyone" and the problem of "totalization" converge.

Jameson's critique of localizing politics is, however, salutary and the class spent some time working it through. His basic argument is that without a sustaining totalizing vision, both in order to make connections among struggles and to work toward a more desirable future, no long-term political transformation is possible. He gives the example of the Black Revolutionary Workers' Party and its astonishing rise into political prominence in Detroit in the late 1960s, a prominence that was eventually lost, he claims, because their local successes were not effectively tied into a network of national/international vision of, and efforts for, change. Indeed, in their attempt to effect such ties, the leaders of the movement neglected their local site of operations, causing their movement to lose momentum. The question, then, of how to imagine an effective "practical politics" while also dealing with the dilemma of unavoidable exclusion and incompleteness, became the focus of our attempt to understand totality and its discontents. For Jameson—and the Marxist tradition with which he allies himself—one of the goals of a proper historical materialism is "comprehending theoretically the historical moment as a whole," as Marx and Engels (1948) put it (p. 19). Postmodern critics, on the other hand, avoid the near occasion of such totalities. Stuart Hall's refusal to close the field of cultural studies, for example, indicates a certain sympathy with poststructuralist mistrust of totality. His desire to encourage practical political activity is qualified by recognition of the potential oppressiveness of any totalizing project. To interrogate Marxism in this way is not to reject it, but rather, as Aronowitz puts it, to recognize that "historical materialism is incomplete, not surpassed" (p. xvii).

To consider the trouble with totality from a poststructuralist perspective in more detail, we next took up Foucault's *Discipline and Punish* (1979), to work with at least one influential poststructuralist text to set out why the battle between Marxism and poststructuralism has taken the particular form that it has. The "whole of society," Foucault (1977) once remarked in an interview, should never be imagined except as "something to be destroyed" (p. 233). Totality for Foucault is necessarily oppressive; it marginalizes, excludes, homogenizes. Localized resistance, not utopian attempts to imagine a new (whole) social order, provides his preferred model for social change. Power—as is most emphatically demonstrated in *Discipline and Punish*—is dispersed and virtually ubiquitous, productive rather than merely repressive. It is also the insidious (though inescapable) figure of the drive to totalize in Foucault. Both power and totalization, then, must be resisted by way of localized struggles, limited incursions. Not only is the overthrow of any regime of power impossible, since it is not centralized in any site upon which to launch an attack, but any attempt to imagine (and work to bring into existence) a wholly new society participates in power's logic and reproduces its hold. Jameson and Foucault can be seen, thus, as representing the extremes of the debate about totality. The promise of a better totality (utopia), for the one, is simply a repetition of oppression for the other.

In response to critiques of totality such as those we find in Foucault, poststructuralist critics turned to a consideration of "local" forces and situations. Commenting on the insistently spatial metaphors that have accompanied this shift, Steven Connor (1989) has remarked: "the distinguishing postmodern problem is one of reflexivity, or the involvement of the activity of theory in the very field which it is attempting to theorize. For a map, at least the kind of map that we are used to in the advanced West, presupposes a position outside or suspended placelessly above the field that is being surveyed. The problem for a postmodern cultural theory is to construct a map of the world from inside that world" (p. 227).[3] Such a map, Connor stresses, "is more responsive to the small scale intimacies and complexities of political life" (p. 228). De Certeau's (1984) urban walkers and Deleuze and Guattari's (1987) nomads are two prominent figurations of the sort of transient subjectivity and reflexive modes of knowledge production Connor invokes here. This turn to the local has been heralded as a corrective to the ostensibly homogenizing totalization of history in terms of "master narratives," of which the modes-of-production narrative is a prominent example.

The turn to "locality," however, gives rise to a consideration of *its* politics as well. James Clifford (1992), for example, asks: "Local in whose terms? How is significant difference politically articulated, and challenged? Who determined where (and when) a community draws its lines, names its insiders and outsiders?" (p. 97). Addressing these questions, and the reflexivity dilemma Connor outlines, Clifford proposes the "itinerary" ("a history of location and a location of histories. . . . a dream of mapping without going 'off earth.'" [p. 105]) as a more helpful metaphor for knowledge production than a traditional map. An itinerary is a mapping practice that implicates the mapping/mapped subjects in "lines" produced by their collective movements under specific conditions. The dilemma Clifford is exposing here is the tendency of ostensibly "local" concepts to begin to exert a homogenizing and difference-effacing tyranny (as "global" concepts do) if they are not continuously interrogated.

For Jameson (1991), however, the Marxist tradition itself provides an answer to this problem in the concept of "hegemony," which offers an alternative to understanding totality as the sort of enforced homogeneity that poststructuralists claim it to be: "a mode of production is not a total system in that forbidding sense; it includes a variety of counterforces and new tendencies within itself, of 'residual' as well as 'emergent' forces, which it must attempt to manage or control (Gramsci's concept of hegemony)" (p. 406). "Hegemony," although often described as if Gramsci coined the term, was actually taken up by him out of Internationalist discourse, where it signified "leadership" of a vanguardist variety (Anderson, 1977, pp. 17–18). In Gramsci's work, it took on a different emphasis that stressed the complex set of social relations under a given set of conditions to which a social order with a well-developed "civil society" (public sphere) could give rise. It is properly thought of as a process of winning consent rather than a practice of deluding or coercing in a crude fashion:

> previously germinated ideologies become "party," come into confrontation and conflict, until only one of them, or at least a single combination of them, tends to prevail . . . bringing about not only a unison of economic and political aims, but also intellectual and moral unity, posing all the questions around which the struggle rages . . . on a universal plane . . . thus creating the hegemony of a fundamental social group over a series of subordinate groups. (Gramsci, 1971, pp. 181–82)

In other words, dominant ideology is not in any simple sense "imposed" on a passive mass public; rather, hegemony necessarily entails a continuous "working out" or settlement among competing groups with different agendas—not just among, but within, themselves. Hegemony, then, less resembles a map (with its stable lines and settled relations) than the activity of traveling.

Not all theorists have been as sanguine as Jameson, however, that hegemony, understood in its Gramscian sense, provides an answer to the dilemma of totality. Marking one of the influential moves from Marxism to post-Marxism, Laclau and Mouffe (1985) pushed Gramsci's hegemony further in the direction of poststructuralism in their influential *Hegemony and Socialist Strategy*. One crucial break that Gramsci makes with the dominant Internationalist thinking of his time, according to Laclau and Mouffe, is replacing "the principle of representation [in the theorization of hegemony] with that of *articulation*" (p. 65, emphasis theirs). In other words, rather than relying on an invariable underlying narrative of which concrete situations are more or less faithful, if displaced, representations, Gramsci described a world in which the situation itself, not some pre-existing world-historical mission, gives rise to particular hegemonic relations. He was able to do this, they suggest, because his view of ideology was not a mere "system of ideas," but rather "an organic and relational whole, embodied in institutions and apparatuses, which welds together a historical bloc around a number of basic articulatory principles" (p. 67). Such "principles" are not reducible to the "expression" of stable and determinate "class interests" per se, or to a purely economic logic, since in actual conditions of struggle the "relation of forces [is] in continuous motion and shift of equilibrium" (p. 172). Where hegemony is at work, transactions are complex and ongoing, requiring constant analysis and redirection on the parts of participants; hegemony, thus, provides a much more flexible totalizing alternative to the base/superstructure model. In other words, Gramsci's (1971) mapping practice, to return to the guiding metaphor of this paper, distinguishes between "conjunctural" (local, ephemeral) and "organic" (broad, enduring) elements in such a way that his maps are provisional rather than static: "to believe that one particular conception of the world, and of life generally, in itself possesses a superior predictive capacity is a crudely fatuous and superficial error" (p. 171).

Nevertheless, in the end, Laclau and Mouffe see Gramsci, in spite of all his flexibility and complexity, as not having pursued the implications of his own theorization of ideology far enough, because "even though the diverse social elements have a merely relational identity . . . there must always be [for Gramsci] a single unifying principle in every hegemonic formation, and this can only be a fundamental class" (1985, p. 69). This insistence on the "fundamental class" returns us to the very problems, Laclau and Mouffe argue, that Gramsci's analysis promised to remove us from: "the logic of hegemony does not unfold all of its deconstructive effects on the theoretical terrain of classical Marxism" (p. 85). Hence they propose a different theorization of hegemony: "Hegemony is, quite simply, a political type of relation . . . but not a determinable location within a topography of the social. In a given social formation, there can be a variety of hegemonic nodal points" (p. 139). From this reworked version of "hegemony," they go on to theorize progressive politics as a "radical democracy" in which a plurality of agents (feminists, for example, or eco-activists)—not simply the proletariat of classical Marxian analysis—may further progressive aims that are not simply interpreted as the historical mission of the working class in another form.[4] Without grounding in the working class, or any other ultimately determining force, Laclau and Mouffe's "social formation" is a shifting set of localized articulations rather than a "totalizable" whole with identifiable unifying force. Having pulled the most distinctively Marxist rug (its insistence on the irreducible reliance of progressive change upon a "fundamental class") from underneath Gramsci's hegemony, Laclau and Mouffe clear the theoretical field for an even more provisional and mobile mapping than Gramsci proposes.

The class further considered "hegemony" (perhaps the most privileged—if unrigorously deployed—concept in contemporary cultural studies) by way of Stuart Hall's work on Thatcherism, especially "Toad in the Garden," which was delivered at the same conference in which Jameson delivered his "Cognitive Mapping" essay. Hall (1988) is far more tolerant of poststructural rewritings of hegemony than Jameson, who would not be likely, for example, to concede the following, which is a cornerstone of Hall's own thought: "Class interest, class position, and material factors are useful, even necessary starting points in the analysis of any ideological formation. But they are not sufficient—because not sufficiently determining to account for the actual empirical disposition and movement of ideas in real historical societies" (p. 45). From this observation, Hall can then make another move that would be anathema to Jameson. Commenting on Gramsci's theorization of how new ideological complexes emerge, Hall notes (approvingly): "the whole of Laclau's subsequent elaboration of articulation/disarticulation is contained in the nucleus of that thought" (p. 56). Where Laclau goes, Hall can follow, but Jameson cannot, dedicated as he is to both a class-centered analysis and a concept of totality.

Paul Smith's (1988) "Visiting the Banana Republic" enters this fray with an analysis of the articulations of culture, economics and politics at the current conjuncture in a concrete situation—the advertising and sales practices of the clothing retailer Banana Republic—without resorting to a Jamesonian cognitive map. As Smith puts it in his discussion of the deployment of "colonial" motifs in Banana Republic decor and promotional materials, the

> aim [of historicization] is not satisfied by the imposition of historical templates, cyclical forms, and abstract totalities onto material history. So, too, the project of historicizing cannot selectively and eclec-

tically pass through the array of contemporary critical discourse only to emerge with yet another totalizing system, another monolithic view of history and totality. (p. 140)

The reason Smith offers for this mistrust of the "monolithic," and totality of the kind he attributes to Jameson, is that

> the project of historicizing must begin outside the parameters of a master narrative (which ultimately make of history one history or a transhistorical—and thus very nearly ahistorical—construct) and insist instead on the provisionality of history, spoken through and by representations that alter, submerge, and erase both themselves and other representations. (p. 141)

In other words, Smith's article asserts, in contrast with Jameson's, that "history," to be a progressive mode of analysis, must be understood (to paraphrase Irigaray) as that narrative which is not one. History does not come to us ready-made, but is continuously being made and remade in cultural forms as well as in economic and political life.

Mapping "The Spaces of Capital"

In spite of numerous postmodern critiques, there have been many attempts on the left to produce a Jamesonian cognitive map of global capital. Leslie Sklair's *Sociology of the Global System* (1991) and David Harvey's *Condition of Postmodernity* (1990), for example, abhor what they call "fragmenting" analyses, whether "comparative" social science studies in which parts of the world are looked at together without reference to the "system" in which both are inserted (Sklair), or "postmodern" critiques of totality and narrative (Harvey). As an alternative to the isolating "state-centrist" analysis it critiques, *Global System* suggests a rather rigid, ostensibly boundary-transcending schema for "global" analysis in which the "transnational corporation" is cast in the role of primary "economic" force, "the transnational capitalist class" (which he defines, to elude a productivist critique, as simply "those people who see their own interests or the interests of their nation, as best served by an identification with the interests of the capitalist global system," [p. 8]) are the "political" forces and the unifying "ideological-cultural" force is "consumerism." The counterforce Sklair proposes to combat these dominant global forces is a "democratic feminist socialism." Harvey, alternatively, takes a far more culture-oriented approach than Sklair and emphasizes the role of film, art, architecture and other cultural artifacts and processes in bringing Fordist and post-Fordist practices into everyday life. While at first glance, Harvey's position seems less mechanistic than Sklair's because it is more cultural in its orientation, in the end, as Angela McRobbie (1990) has observed, Harvey, like Jameson, returns to an unmediated determinism: "what happens in the economy has a direct effect on what happens in culture" (p. 7). Thus, while their critiques of the problems of "fragmenting" analysis in the traditional disciplines is important, Jameson, Sklair and Harvey produce (from a poststructuralist perspective) an incomplete narrative in turn. Their commitment to totality is at the same time an insistence on an economically determined whole obeying the "logic of capital" (understood, to be sure, as a social as well as an economic force). Such a story subordinates or effaces other stories, other ways of seeing, in ways that render it incomplete.

With this incompleteness in mind, the question of how cultural texts might be seen other than as reflections of economic conditions became one of the class's preoccupations as a focus on the "current moment" was followed by three weeks of "historical" work, in which we read Eric Wolf's (1982) *Europe and the People without History* and Aphra Behn's (1973) *Oroonoko*. This pairing, of a current Marxist history of the emergence of capital and a late-seventeenth-century proto-novel that combines romance with travel narrative in its story of the slave trade and European colonization of the "new world," may seem odd. However, it provides a way of pointing out how a careful reading of "cultural" texts can unsettle a confidently "total" political-economic account of the history of capitalism, and open the way for a different kind of mapping than a privileging of the economic sphere permits. Such a corrective task is especially important in dealing with Wolf's book, which takes as its project two major problems in Western social science: first, the fragmentation of disciplines so that no one field attempts to understand the world as a whole, as a set of complex and irreducible relations (a critique also made by Sklair explicitly, as well as Harvey and Jameson implicitly), and, second, the erasure of the histories of Europe's "others," such that even now they seem to be examples of "primitive" changeless societies. Both of these problems are important ones, and Wolf's book certainly deserves careful reading by students of capitalism and colonialism, but it is nonetheless flawed in certain important respects: a residual Eurocentrism (necessarily) inhabiting its model of totality (which I will take up in my discussion of postcolonial theory below), and a tendency to discount culture as itself a force equivalent to the economy or politics in the world system he describes.

Oroonoko, like Wolf's history, provides a representation of the slave trade in the early colonial period. Immensely popular when it was first published in 1688 (the year of the "Glorious Revolution"), Behn's narrative of the grisly fate of the "royal slave" (Oroonoko) is a fervent argument in favor of "natural" hierarchy. Although emphatically critical of the capture and enslavement of its eponymous hero, Behn's text is not at all a critique of the slave trade; Oroonoko himself ultimately disdains his fellow slaves, pronouncing them (after they fail to persevere in a revolt): "by Nature slaves, poor wretched Rogues, fit to be used as Christians, Tools" (p. 66). The book emphasizes, rather, Oroonoko's royal birth and nobility; it is *his* slavery that defies nature, not the slavery of Africans in general. Both Wolf and Behn trace the route of the triangle trade, describe the ordering of sugar plantations in the West Indies and give accounts of the interactions between Europeans and Africans in negotiating trade. Behn's text, however, is no mere reflection of mercantile interests (or, to use Jameson's preferred term, their "figuration") as Wolf's mode of analysis would suggest; as a cultural form, it is participating along with political and economic forces to help shape the complex social formation in which it emerged by disseminating particular codings (for example) of "blackness" and "whiteness," "high" and "low," which cannot be explained in economic terms alone.

Wolf's unwillingness to see culture as an autonomous force bespeaks a larger problem with his evasion of heterogeneity: He cannot imagine any story other than Western capital's story (or even see capital as having more than one story). After it emerges, capital, as a unified logic, is the only game in town, and becomes the irreducible reference text of world history, swallowing all counternarratives. The resulting Eurocentrism of this position is all the more striking in *Europe and the People without History* because it is devoted to (and largely succeeds in) overturning the Eurocentrism of most European history that simply assumed that "primitive" peoples (as defined by Europe) had no history, and thus, were static. Wolf's book provides an excellent corrective to this pervasive notion; however, by relying on the modes-of-production narrative instead, he manages to replace one form of Eurocentrism with another, as postcolonialist critiques of his book suggest.[5]

To help develop some alternative ways of dealing with the issues raised in Wolf's text, we turned to Arjun Appadurai's (1990) "Disjuncture and Difference in the Global Economy" and Dipesh Chakrabarty's (1992) "Provincializing Europe: Postcoloniality and the Critique of History." Gayatri Spivak (1988) has described the project of the Subaltern Studies Group (to which Chakrabarty belongs, and which has influenced the work of Appadurai) as moving from theories of "transition," which rely on a monologic modes-of-production narrative, to theories of "change," which attempt to pluralize modernity and history (p. 197). Chakrabarty's essay indicates both deep respect for the Marxian tradition and a recognition that its "totality" cannot account for all aspects of postcolonial societies. Specifically, he argues that the "modern individual" of Western capitalism, with its emphasis on "interiority" and "the private sphere" as well as "citizenry" and the "public sphere," has not fully replicated as such in India, where the latter emerges without the former: "Public without private?" (p. 339). Chakrabarty observes, worrying that this leaves the postcolonial subject ever outside the map of the modern.

Wolf himself critiques the tendency of bourgeois ideologists to banish the non-West from full modernity. He comments, in a critique of "modernization theory," that "it used the term modern, but meant by that term the United States, or rather an ideal of a democratic, pluralistic, rational and secular United States" (1982, p. 13). For Chakrabarty, however, the modes-of-production narrative does not provide an adequate alternative to bourgeois/colonialist theories (as it does for Wolf), because into its narrative a whole multitude of local trajectories must be subordinated, rendering history "the site where the struggle goes on to appropriate, on behalf of the modern (my hyper-real Europe), . . . other collocations of memory" (1992, p. 343). A traditional Marxian narrative is not much more acceptable than a colonialist one in this respect. The critique that Chakrabarty puts forth here is similar to those of feminist historians (among others) who foreground what even the most careful Marxist accounts are obliged to exclude or subordinate to their own privileged narratives.[6]

Appadurai (1992), too, chafes at attempts to assimilate into one tidy narrative, however complex, threads that in certain senses remain discontinuous. He takes a different approach than Chakrabarty, who situates his counterhistory within the space of specific geographic parameters labeled "India." Rather, he looks at the world as a set of flows and relations that can only be discussed as a "totality" by reifying and distorting them. For this reason, he complains, "even the most complex and flexible theories of global development which have come out of the Marxist tradition are inadequately quirky and have failed to come to terms with what Lash and Urry have called disorganized capitalism" (p. 328). He proposes a model of analysis that imagines the global system in terms of flows of people (ethnoscapes), culture (mediascapes), technology (technoscapes), capital (finanscapes) and information (ideoscapes); these various "scapes" are autonomous, though interactive, and, therefore, not coordinated by any overarching (and ultimately determining) instance: "the relationship of these various flows to one another, as they constellate into particular events and social forms, will be radically context-dependent" (p. 337). Although different in approach and implications, both Appadurai and Chakrabarty produce critiques of totality for what it renders unthinkable, what it cannot account for, how it is itself a problem.

Gilroy's (1993) *Black Atlantic*, from which the class read an excerpt next, steps into this moment of crisis with the observation that cultural studies itself has colluded with an oppressive identity politics from its founding moments by taking the nation-state—which Gilroy sees as inseparable from the African diaspora and colonial exploitation—as its fundamental unit of analysis. He suggests that both "race" and "nation" are concepts that require problematization in similar terms. To do so, he argues that alternative intermediate sites of analysis—between "the nation" and "the globe"—are needed. Like Chakrabarty, Gilroy is determined to fragment the monoliths of modernity. Unlike Wolf however, Gilroy does not claim "culture" and "nation" to be coextensive categories; rather, Gilroy explodes the notion of "over-integrated culture" and examines the ways in which supposedly intact cultures flow out in all directions, and rely on inflows from "outside" their supposedly self-generating cores. Specifically, he claims that diasporic African experience is best thought of not as a static and overlocalized hybrid (e.g., African-Americanness), but rather in terms of what he calls the "Black Atlantic," a series of interconnections and flows, not unlike a "scape" in Appadurai's terms (although James Clifford is the theorist Gilroy invokes). Through the figure of the Black Atlantic, Gilroy attempts to direct our attention to "routes" of identity-formation—itineraries, movements and processes—rather than the static and stable fantasy of "roots." Such an analysis, he claims, "makes blackness a matter of politics rather than a common cultural condition" (p. 27). Emphasizing how the sets of relations in a given time and place are produced and reproduced, Gilroy urges a complex understanding of fragmented black identities. His book's critique of the "nation-state" as a unit of analysis also points to a crucial question at this moment of so-called "globalization": how to understand the relationships among "state," "nation," "capital" and "culture"—in other words, how to articulate culture, economics and politics while dealing with a world whose spatial relations seem to be changing radically.

Having determined the relationships among "nation," "state" and "capital" to be crucial to understanding "globalization," the class read several influential descriptions of these relations: Maseo Miyoshi's (1993) "A Borderless World?" Kinichi Ohmae's (1995) "Putting Global Logic First," an excerpt from Robert Reich's (1992) *Work of Nations*, a selection from the 1994 *Socialist Register* (a special issue on "Globalism and Nationalism"), and Stuart Hall's (1994) "The Local and the Global: Globalization and

Ethnicity." All of these essays are attempts to come to terms with new global-bal realities in which transnational capital has disentangled itself from a settled relation with any one nation-state. Miyoshi argues that the corporation is now itself a "colonial" power, operating without the cumbersome apparatus of states, but asserting its influence just as emphatically.[7] He urges strategies of resistance to be developed appropriate to its specific oppression, as strategies of resistance were developed to state colonialism. Ohmae, on the other hand, an economist writing for the *Harvard Business Review*, declares the nation-state obsolete and celebrates what he sees as an increasing corporate freedom of movement over boundaries that once marked "national" markets. Examining the same situation from his "political" perspective, Reich observes that globalization is a crisis for the United States, which he argues must prepare a work force of "symbolic analysts" for the demands of a global economy, or stand to lose even its current weakened place within it. He also asks if there is anything to bind a nation, when an economy no longer corresponds with its borders. While Reich declares uncertainty, Manfred Bienefeld, writing for the *Socialist Register*, answers "yes," and insists that the nation-state must become the organizing field for combat with transnationalizing capital (hardly a position to be taken by Reich, who has nothing against capital per se, but would like to see a kinder, gentler [to the US] capitalism). Hall is concerned specifically with the unsettling disruption of identity politics that has accompanied the changes in the relations of capital and the state: "global and local are the two faces of the same movement from one epoch of globalization, the one which has been dominated by the nation-state, the national economies, the national cultural identities, to something new" (p. 27). Such a moment, he notes, is properly seen as a crisis, and, as in all moments of crisis, the transformations underway may lead in many different directions; it is our task both to understand and help guide them in a more liberatory one.

The course ended with a discussion of Bruce Sterling's (1989) *Islands in the Net*, a novel preoccupied with asserting the local in the face of a global information network (resembling a Foucauldian power network) that captures all the characters of the novel in its grasp. We read this ominous book alongside an essay on "Transnational Corporate Networks" (Mulgan, 1991) that describes the privatized system of information flows that large corporations have established on which to conduct their own business outside the more familiar "public" systems, like the Internet.

What sort of work do texts such as Sterling's perform in the sort of world(s) we inhabit? The book cannot control the fact that in its critique of the all-encompassing computer network, it also helps make such a network familiar and imaginable to its readers. The value of the local—indeed the potential for it to endure—is the pressing concern of the novel, although it focuses more on the individual than on the costs exacted on larger (explicitly social) locals, such as nation-states excluded from the Net. Ultimately, the main critique of Sterling's novel seems to be the havoc wrought on interpersonal relations by the global Net, which infiltrates the most intimate aspects of everyday life. The power of the Net, by the novel's end, resembles power in its most extreme Foucauldian form: ubiquitous, indefatigably recuperative, capable of generating even its own opposition as a means of further reinforcing itself. The Net seems inevitably pervasive, monstrously invulnerable. It is here that Sterling seems to participate in corporate fantasies of globalization, though he presents their dark underbelly. Does such a vision seem to render any resistance useless (M. Castells, 1996)? With this depiction of a world in which a cognitive map to encounter and resist the global forces of the Net seems called for, and yet in which any such totalization seems itself ominous given the example of the Net, the course ended.

ANALYZING SITUATIONS

Over the term, we examined many cultural artifacts in order to develop our capacity for nuanced theorization of concrete situations. For example, we looked at an advertisement for one of the major relief organizations to consider the "cartography of the deserving poor" (i.e., the nations that the organization did—and did not—serve) and the ways in which the ad reinforced a sense of paternalism of first world benefactors for their third world beneficiaries. We read newspaper accounts of current events, watched film clips, and looked at many different kinds of corporate self-representation in our attempt to theorize globalization at work. At one point we worked with a Ford advertisement in *Business Week* that depicted a world map with its corporate logo superimposed over the land masses where Ford cars had won auto races of various kinds; the accompanying copy declares, ominously enough: "And you thought the Earth's surface was dominated by water." Between the ad and articles by Richard Barnet and John Cavanagh (1994) and Nigel Thrift (1989), we were able to work out collaboratively as a class a conjunctural analysis of the advertisement as assisting corporate efforts to represent itself as a world-historical agent along the lines of the empire-building European nations from the sixteenth century onward. Whereas Britannia once ruled the waves, Ford now envisions itself as ruling the shore. I was careful to point out the extent to which this cognitive map for Ford is utopian; the articles that we read indicated that Ford's globalism, although certainly considerable, is not quite of the magnitude Ford's map suggests.

Our task, I argued, is to consider with great seriousness the work that such maps perform, and try to imagine the world in other ways. The need for cognitive maps that indicate how various oppressions are related to the expansion of capital (rather than being mere local accidents) manifests itself when we are confronted with the hard realities of actual factories, actual exploitation and actual systematic oppression of women and other groups in the international division of labor. At the same time, we must remain vigilant about, or open to the possibility of, the exclusions and potential tyrannies of any cognitive maps (as the numerous critiques considered in the previous pages suggest), since good intentions are not sufficient to safeguard totality from totalitarianism.

With this predicament in mind, each member of the seminar worked on a project during the term. The project I pursued concerned the use of maps by transnational corporations, especially in annual reports, and from time to time I brought along samples (such as the Ford map described above) for discussion. Maps and globes, I noted, are among the most common types of cover design employed by contemporary corporations in their annual reports, which inspired me to a literal-minded attempt to analyze corporate cognitive mapping practices (how they imagine themselves as part of a larger whole). A small subset of these maps helped me establish why a sense of history might be important to understanding how the world is spatially imagined today. Several annual reports in the early 1990s directly invoked colonial themes by deploying cover illustrations of antique maps and globes, or direct references to the early explorers. Drawing on the work of Jose Rabasa (1993) and Evitar Zerubavel (1992), I pointed out that mapping was one of the practices that helped transform the unsettling European "discovery" of the "New World" into imaginable—and therefore masterable—space.[8] I suggested that the citations of early modern colonial maps in current annual reports indicated that the unsettling forces of globalization (which render traditional maps, organized according to nation-state boundaries, in certain ways obsolete) demanded a similar "New World" mode of thinking, which has manifested itself, as in the early modern period, with a frenetic pace of new mapping projects.

Hence, I argued, even maps that do not directly invoke colonial themes are helping to rewrite nation-state colonization in terms of corporate imperialism in the new maps. One way that corporations signal this spatial reconfiguration is by producing maps in which no nation-state boundaries are marked at all, effecting terrains of what Paul Carter (1988) has called "map made emptiness" in which to deploy themselves (p. xx). Some maps subordinate the world to their product graphically (as in the maps printed across toothpaste tubes on one of Colgate's reports, or California Microwave's depiction of a world map contained within the limits of one of its own satellite dishes). Other corporations fragment the world, as if it were pieces for them to take apart and recombine, as in a Whirlpool report, which uses puzzle pieces as its mapping motif, or other reports, which display segments cut from world maps arranged in still life with the company's products as if they, rather than any other organizing principle, provide the possibility of (a new) world unity and order. Having put forth these examples, I asked how we might—in a moment of unsettling de- and re-territorialization of the globe in this way by transnational forces—best theorize these movements in ways other than a celebration of market logic.

Putting this question to different concrete situations, students worked on projects that helped us formulate a variety of ways of addressing it. They had been alerted before the start of the semester to be prepared to discuss their prospective term paper topics at our first meeting—and throughout the term—since their research, as well as my own, was part of the material on which we all would focus. Their projects included an examination of the implications of the "virtual office" in the operations of the advertising firm Chiat-Day, a discussion of sugar and the slave trade in the seventeenth century, an exploration of the role of the Andy Warhol museum in the attempt to portray Pittsburgh as a "World City," an investigation of corporate attempts to colonize the Internet for profitable enterprise, a study of panoptic practices utilized by Gap Corporation to discipline its employees, as well as analyses of the UN population conference, the Chiapas uprising, "New Americanist" discourse and "Fishery Science."

These diverse projects gave students a chance to work out a position on the totality debate with which we had been preoccupied all term by way of the analysis of a concrete situation. The theories we had studied thus did not remain mere content to be mastered, but ways of seeing to be worked through for better understanding of political, economic, and cultural situations. At this time of what Jameson has described as "spatial as well as social confusion," our collective labors provided not a definitive monologic map of the world, but an approach to attempting to understand changing relationships. The "class-scape," of course, was far less tidy and homogeneous than the previous account suggests. Indeed, this essay indicates both the promise and the problem of mapping: I was able to indicate some relationships among the themes and essays precisely because I examined them from a particular unifying point of view, but any such account—indeed any account—is necessarily incomplete. I acknowledge the incompleteness, encourage supplementation, and yet stand firm on the tactical usefulness of such a map. A cognitive map, as Spivak (1993) reminds us in the passage cited at the beginning of this paper, with a deconstructive sense of complicity, has its uses. "Spaces of Capital" was one attempt to use the classroom as a space to pursue some of those uses, and track some of those complicities.

Notes

1. Jay (1984) teases out two main strands in the conceptualization of "totality" in the Western Marxist tradition. The first is future-directed—totality, understood as plenitude and the achievement of social justice, is the *goal* of Marxist struggle. The other sense of totality (the one I emphasize here) views totality as a tool of analysis: "it stems from a methodological insistence that adequate understanding of complex phenomena can follow only from an appreciation of their relational integrity" (p. 24).

2. Within the cultural studies tradition, it is in Raymond Williams's work that we find a marked interest in something like a totality that would be more acceptable to Jameson than the perpetual incompleteness Hall invokes in the passage from one of his essays I quote above in the body of my essay. Williams's utopian hope was to return human interactions and everyday life to the organic wholeness from which the fragmented and fragmenting conditions of capital had alienated them. What he called the "study of relationships in a whole way of life" (p. 46) in *The Long Revolution* (1961) became a methodological outline for emergent cultural studies, albeit amended by E.P. Thompson's (1961) trenchant observation that we were caught up rather in "a whole . . . way of struggle" (p. 33). In either case, whether the focus was on consensus or conflict, there was a moment in the early (prior to poststructuralist influence) period of cultural studies in which wholeness was certainly the privileged model. With the rising insistence of poststructuralist critiques, however, the advocates of totalizing analysis were increasingly put in defensive positions.

3. Postmodernism and poststructuralism are not always synonymous terms, but the attention to reflexivity and mobility Connor emphasizes here are attributes of the strain of poststructuralism with which this paper concerns itself.

4. It is important to recall that for Laclau and Mouffe (1985) that the "new social movements" are not necessarily progressive. What matters is their articulation with other elements and forces in the social formation; see p. 87 of *Hegemony*.

5. See, for example, Talal Asad's review of Wolf's book: Are there histories of people without Europe? *Comparative Studies in Society and History*, 29(3), 594–607.

6. See, for example, Joan Scott (1988), *Gender and the Politics of History*, New York: Columbia University Press.

7. His critique of postcolonial discourse on the grounds that it does not recognize that colonial forces still operate under other guises in an ostensibly decolonized world is largely unwarranted, but his article is nonetheless important as an examination of how the transnational corporation can be seen as a colonial (I would prefer "imperial") agent.

8. See also the essay by D. Turnbull, "Constructing Knowledge Spaces and Locating Sites of Resistance in the Modern Cartographic Transformation," in this volume.

References

Anderson, P. (1976–1977). The antinomies of Antonio Gramsci. *New Left Review,* 100, 5–78.

Appadurai, A. (1990). Disjuncture and difference in the global cultural economy. *Public Culture,* 2(2), 1–24.

Aronowitz, S. (1990). *The crisis in historical materialism* 2nd ed. Basingstoke, England: Macmillan.

Barnet, R., and J. Cavanagh. (1994). *Global dreams: Imperial corporations and the new world order*. New York: Touchstone.

Behn, A. (1973). *Oroonoko, or, the royal slave*. New York: W.W. Norton.

Bienefeld, M. (1994). Capitalism and the nation state in the dog days of the twentieth century. In R. Miliband and L. Panitch (eds.), *Socialist register* (94–129). London: Merlin.

Carter, P. (1988). *The road to Botany Bay*. New York: Knopf.

Castells, M. (1996). The net and the self: Working notes for a critical theory of the informational society. *Critique of Anthropology, 16*(1), 9–38.

Chakrabarty, D. (1992). Provincializing Europe: Postcoloniality and the critique of history. *Cultural Studies, 6*(3), 337–357.

Clifford, J. (1992). Traveling cultures. In L. Grossberg, C. Nelson, and P. Treichler (eds.), *Cultural studies* (96–112). New York: Routledge.

Connor, S. (1989). *Postmodernist culture*. Oxford: Blackwell.

de Certeau, M. (1984). *The practice of everyday life*. Translated by S. Rendall. Berkeley: University of California Press.

Deleuze, G., and F. Guattari. (1987). *A thousand plateaus*. Translated by B. Massumi. Minneapolis: University of Minnesota Press.

Derrida, J. (1978). La parole soufflé. In A. Bass (trans.), *Writing and difference* (169–195). Chicago: University of Chicago Press.

Foucault, M. (1977). Revolutionary action "until now." In D. F. Bouchard and S. Simon (trans.), *Language, counter memory, practice*. Ithaca, NY: Cornell University Press.

Foucault, M. (1979). *Discipline and punish: The birth of the prison*. Translated by A. Sheridan. New York: Vintage.

Gilroy, P. (1993). *The black Atlantic: Modernity and double consciousness*. Cambridge, MA: Harvard University Press.

Gramsci, A. (1971). *Selections from the prison notebooks*. Translated by Q. Hoare and G.N. Smith. New York: International.

Hall, S. (1988). The toad in the garden: Thatcherism among the theorists. In C. Nelson and L. Grossberg (eds.), *Marxism and the interpretation of culture* (35–57). Urbana: University of Illinois Press.

Hall, S. (1992). Cultural studies and its theoretical legacies. In L. Grossberg, C. Nelson, and P. Treichler (eds.), *Cultural studies* 277–286. New York: Routledge.

Hall, S. (1994). The local and the global: Globalization and ethnicity. In A. King (ed.), *Culture, globalization, and the world system* (19–39). Basingstoke, England: Macmillan, in association with the Department of Art and Art History, SUNY Binghamton.

Harvey, D. (1990). *The condition of postmodernity*. Cambridge, MA: Blackwell.

Irigaray, L. (1985). *This sex which is not one*. Translated by Catherine Porter. Ithaca, NY: Cornell University Press.

Jameson, F. (1988). Cognitive mapping. In C. Nelson and L. Grossberg (eds.), *Marxism and the interpretation of culture* (347–356). Urbana: University of Illinois Press.

Jameson, F. (1991). *Postmodernism*. Durham, NC: Duke University Press.

Jay, M. (1984). *Marxism and totality*. Berkeley: University of California Press.

Laclau, E., and C. Mouffe. (1985). *Hegemony and socialist strategy*. Translated by W. Moore and P. Cammack. London: Verso.

Marx, K., and F. Engels. (1948). *The communist manifesto*. Translated by S. Moore. New York: International.

McRobbie, A. (1991). *New times in cultural studies. Working Paper 5*. Milwaukee, WI: Center for Twentieth Century Studies.

Miyoshi, M. (1993). A borderless world? From colonialism to transnationalism and the decline of the nation-state. *Critical Inquiry, 19*, 726–751.

Mulgan, G. (1991). *Communication and control: Networks and the new economies of communication*. New York: Guilford Press.

Ohmae, K. (1995). Putting global logic first. *Harvard Business Review, 73*(1), 119–125.

Rabasa, J. (1993). *Inventing A-M-E-R-I-C-A: Spanish historiography and the formation of Eurocentrism.* Norman: University of Oklahoma Press.

Reich, R. (1992). *The work of nations.* New York: Vintage.

Sklair, L. (1991). *Sociology of the global system.* Baltimore: Johns Hopkins University Press.

Smith, P. (1988). Visiting the banana republic. In A. Ross (ed.), *Universal abandon? The politics of postmodernism* 128–148. Minneapolis: University of Minnesota Press.

Soja, E. (1989). *Postmodern geographies.* New York: Verso.

Spivak, G. (1988). Subaltern studies: Deconstructing historiography. In *In other worlds,* 197–221. New York : Routledge.

Spivak, G. (1993). Scattered speculations on the question of cultural studies. In *Outside in the teaching machine* 255–284. New York: Routledge.

Stallybrass, P., and A. White. (1986). *The politics and poetics of transgression.* Ithaca, NY: Cornell University Press.

Sterling, B. (1989). *Islands in the net.* New York: Ace.

Thompson, E.P. (1961). The long revolution I. *New Left Review, 9,* 24–33 (continued in *NLR, 10,* 34–39).

Thrift, N. (1989). The geography of international economic disorder. In R. J. Johnston and P.J. Taylor (eds.), *A World in crisis* (16–78). Oxford: Blackwell.

Williams, R. (1961). *The long revolution.* New York: Columbia University Press.

Wolf, E.R. (1982). *Europe and the people without history.* Berkeley: University of California Press.

Zarubavel, E. (1992). *Terra cognita.* New Brunswick, NJ: Rutgers University Press.

Jameson's Project of Cognitive Mapping

A Critical Engagement

Robert T. Tally, Jr.

An aesthetic of cognitive mapping—a pedagogical political culture which seeks to endow the individual subject with some new heightened sense of its place in the global system—will necessarily have to respect this now enormously complex representational dialectic [of the postmodern condition] and invent radically new forms in order to do it justice. . . . The political form of postmodernism, if there ever is any, will have as its vocation the invention and projection of a global cognitive mapping, on a social as well as a spatial scale.

Fredric Jameson, *Postmodernism*

The map is open and connectable in all of its dimensions; it is detachable, reversible, susceptible to constant modification. It can be torn, reversed, adapted to any kind of mounting, reworked by an individual, group, or social formation. It can be drawn on a wall, conceived of as a work of art, constructed as a political action or as a meditation.

Gilles Deleuze and Félix Guattari, *A Thousand Plateaus*

Introduction

Of the various terms or keywords proliferating in the fields of postmodern social and cultural theory, Fredric Jameson's *cognitive mapping* has been one of the most influential.[1] At the same time, however, the definition of this term has been rather sketchy, and cognitive mapping has been used to describe a number of quite different activities or concepts. In large part, this is due to Jameson's own broad usage of the term. For example, he sometimes employs it in reference to an individual's "subjective" attempt to locate her- or himself in a complex social milieu; at other times, Jameson points to a supraindividual ("objective") production of space in the multinational, late capitalist world system. The difficulties of pinning down cognitive mapping are compounded by Jameson's own system of thought, in which multiple and even opposed elements are brought together under the power of some unifying or totalizing force (such as the dialectic, or capital itself). That is, Jameson does not hesitate to embrace disparate critical theories or practices that diverge significantly from the thesis he puts forth—such as Gilles Deleuze's "schizoanalysis"—only to reduce and incorporate them into his larger philosophical system: Jameson's own version of the Hegelian *Aufhebung*, canceling his opponents' arguments, while preserving a kernel of their ideas, and dialectically advancing his own position.

A key question of cognitive mapping that continues to be posed and debated is, "Are cartographic practices inherently liberatory or repressive?" Jameson clearly sees cognitive mapping—"a code word for [a new kind of] class consciousness" (1991, p. 418)—as a politically progressive activity. But many, including John Beverley in this volume, have criticized what they see as the hegemonic effects directly connected to mapping. Another position, one that I take to be preferable, follows Michel Foucault's and Deleuze's sense of spatial practices as strategies of power that can be used in a number of ways, both progressively and otherwise. Foucault has delineated many aspects of the spatialization of modern societies, vis-a-vis Panopticism and control over bodies. But he has also suggested that power, like Deleuze's "map," is reversible. Deleuze's nomadology, moreover, suggests a sort of mapping practice that subsumes restrictive boundaries even as it "maps" territories.[2]

In this chapter, I will examine Jameson's concept of cognitive mapping in terms of its usefulness as a critical practice. In the first section I will take up, as it were, the thing in itself, focusing on Jameson's (1991) arguments in *Postmodernism, or, The Cultural Logic of Late Capitalism*, and attempt to arrive at a definition of cognitive mapping. In the second section, I will take up the question of mapping as a hegemonic force "bound up with domination," as John Beverley has suggested. Sometimes in direct dialogue with Jameson's cognitive mapping, at other times more generally, a number of critics have argued that the increasing emphasis on spatiality and mapping has had the effect of closing down certain, for example, subaltern perspectives, or even, as Crystal Bartolovich suggests in her discussion of multinational corporations in this volume, of establishing a novel sort of imperialism. In the third section, I will look at alternative ways of imagining a kind of cognitive mapping, especially as they are presented by Foucault and Deleuze. While Foucault never completed a work that focused on geography, as he once suggested he would do (Foucault, 1980, p. 77), one can nevertheless see the concern for spatial relations and mapping in such books as *The Birth of the Clinic* and *Discipline and Punish*. Deleuze, following directly from Foucault, has elaborated further a theory of mapping that can be viewed as an alternative to Jameson's notion. Then, finally, and based on these examinations, I will argue that a form of cartographic practice is necessary for any pedagogical, not to say political, project that attempts to deal with the present condition. I believe that a synthesis of Jameson's project and Deleuze's somewhat Foucauldian theories can provide a useful understanding of cognitive mapping as a critical practice for social and cultural theory, one that can take certain antimapping perspectives into account while maintaining the importance of mapping to the contemporary concerns of comparative education.

In some ways, then, this chapter stands as a companion piece to Bartolovich's contribution to this volume. But whereas Bartolovich focuses her attention on the debate over the concept of "totality"—in which Jameson stands as a leading proponent and Foucault figures as the poststructuralist antagonist—I will focus on cognitive mapping as a critical practice, and leave question of totality, to a certain extent, on the back burner. I will, however, examine Bartolovich's argument more closely in the section titled "Can the Subaltern Map?", since Jameson's view of the social (now global) totality and the critiques thereof are extremely important to the critical debate surrounding cognitive mapping.

Jameson introduces the term cognitive mapping in his 1984 essay, "Postmodernism, or, The Cultural Logic of Late Capitalism" (reprinted with only slight modifications as the first chapter of his book of the same name [1991]). Not surprisingly then, this view of cognitive mapping has as its particular place of reference the historical moment of postmodernism and/or of late capitalism. Jameson begins with many descriptions of the ways in which our situation—in the global, or multinational, world economic system, the cultural and intellectual indications of the "crisis of representation" in art and (poststructuralist) theory, the combining of base and superstructure in "pop art" and consumer capitalism—is radically different from, and yet also, in a sense, linked with previous sociohistorical situations. He then proposes that older (realist or modernist) aesthetic practices are no longer suitable or even feasible, and that we need to develop an "aesthetic of cognitive mapping."[3]

Jameson has two main sources for this conception, one practical and one theoretical, although these cannot be long separated in Jameson's system. The first is Kevin Lynch's study of urban space, *The Image of the City* (1960), and the second is Louis Althusser's famous essay on ideology (first published in 1970, translated in 1971). Jameson is able to bring the two texts together in the figure of cognitive mapping, and indeed, as he says later, "cognitive mapping . . . can now be characterized as something of a synthesis between Althusser and Kevin Lynch" (1991, p. 415).

Lynch's study focuses on the ways in which individuals in cities imagine their environments. His framework is fundamentally phenomenological, inasmuch as it presupposes a psychological subject who can "map" the landscape of empirical data. By way of illustration, Lynch compares (and contrasts) the ways in which subjects see and "map" Boston and Jersey City. Boston, with its familiar landmarks and borders (such as the John Hancock building and the Charles River), is a city of which one can form a mental diagram (and through which one can move) with relatively little conceptual difficulty (at least, according to Lynch). Jersey City, lacking traditional markers, presents a model of the alienated city, the space that one has trouble mapping while living in it—a *Lebenswelt* characterized, in some significant part, by confusion. From Lynch (although Lynch does not, to my knowledge, use the term), Jameson gets his first definition of cognitive mapping: It "involves the practical reconquest of a sense of place and the construction or reconstruction of an articulated ensemble which can be retained in memory and which the individual can map and remap along moments of mobile, alternative trajectories" (p. 51). As Jameson quickly points out, however, Lynch's work was intended to apply to the urban experience alone, and—notwithstanding the ever-growing amount of work on cities and postmodernism—Jameson's argument calls for a postmodernism characterized by global, or multinational, socioeconomic relations, which would include but not be limited to those of the city. While Jameson would not discount the impact of the urban on the existential conditions of a city's inhabitants, he does want to emphasize the global vis-a-vis postmodernism over and above the urban—which is perhaps too closely tied to modernism. Jameson nevertheless feels that Lynch's model can be extended to the national or multinational without much difficulty, at least, so long as this extension occurs on the conceptual horizon. Jameson later notes, in his discussion of *Detroit: I Do Mind Dying* by Dan Georakis and Marvin Surkin (1975), that a successful city-level political program organized along the lines of local, urban experience will almost necessarily fail when it attempts to extend itself to the national or international level (1991, pp. 413–15).

Althusser provides Jameson a way to make such a conceptual extension. Using Althusser's by now well-known redefinition of ideology as "the representation of the subject's imaginary relationship to his or her real conditions of existence," Jameson suggests that this is precisely what, in theory, the individual in the city is attempting to do in practice when engaged in cognitive mapping. Earlier, in his *The Political Unconscious* (1981), Jameson had implied the same Althusserian formula to refer to the vocation of the novel, or the literary text, as a way of symbolizing imaginary solutions to real contradictions. So, reading Jameson in reverse, one might say that literary writing is already a form of cognitive mapping.

Althusser provides a theoretical framework for Lynch's more empirical or experiential analysis of the ways in which individuals negotiate their surroundings. "Ideology" provides a bigger picture than the "image of the city," insofar as it allows for a sense of the "real" as mode of production. By "synthesizing" Althusser and Lynch, Jameson is able to expand Lynch's city-model to a more global terrain, while grounding Althusser's abstract thesis in the practical "art" (the "aesthetic") of cognitive mapping. Thus, Jameson manages to cancel and preserve the two, while presumably elevating both to a new plane.

While Jameson's text consistently refers to cognitive mapping in relation to postmodernism—and indeed, he introduces cognitive mapping as a concept and practice of and for postmodernism—he nevertheless provides historical examples of such a practice that antedate his periodization of postmodernism. In an important "digression on cartography," Jameson first mentions that Lynch's "cognitive mapping" is really precartographic referring more to itineraries than to maps, to "diagrams organized around the still subject-centered or existential journey of the traveler, along which various significant key features are marked" (1991, pp. 51–52). Jameson compares Lynch's results to those of the sea charts of antiquity, "where coastal features are noted for the use of Mediterranean navigators who rarely venture out into the open sea" (1991, p. 52). The technology of the compass and the sextant, then, opens up a higher stage of cognitive mapping—understood in these examples in terms of maritime navigation—insofar as it adds a sense of the universal, for example, in relation to the stars, to the isolated experiential knowledge of individual mariners: "At this point, cognitive mapping in the broader sense comes to require the coordination of existential data (the empirical position of the subject) with unlived, abstract conceptions of the geographic totality" (1991, p. 52). Finally, with the advent of the globe (in 1490) and the Mercator projection (a bit later), "yet a third dimension of cartography emerges," which calls forth a new crisis in representation—Jameson here refers to the "dilemma of the transfer of curved space to flat charts" (1991, p. 52). The "naively" mimetic maps no longer work, and it becomes apparent that there can be no "true maps." Jameson further asserts that this moment represents a watershed in the history of map making, going so far as to suggest that a sort of beginning of history of cartography now comes into view: "at the same time it also becomes clear that there can be scientific progress, or better still, a dialectical advance, in the various historical moments of mapmaking" (1991, p. 52).

In addition to this very brief history of cartography, Jameson draws on Henri Lefebvre and Ernest Mandel to provide a glimpse of a history of the production (as distinguished here from the history of the representations) of space: "The three historical stages of capital have each generated a type of space unique to it, even though these three stages of capitalist space are obviously far more profoundly inter-related than are the spaces of other modes of production" (1991, p. 410). As Jameson sees it, the market space is that of the grid, and is related to the broad Enlightenment project of secularizing the world; the aesthetic form correlating to this is realism, and Jameson seems to locate the birth of the novel within this space of market capital. With the emergence of monopoly capital (and imperialism), a new sort of national space, already becoming quite international, opens up; the lived experience of, say, London no longer coincides with its own place, since the "truth" of that experience lies elsewhere, in Jamaica or India for example—that most British of traditional, daily practices, teatime, might require Indian tea and Jamaican sugar, even if the tea drinkers have no idea of those places and their histories. Jameson links this stage to modernism, although Edward Said has certainly demonstrated (in *Culture and Imperialism* [1993]) that the "overlapping territories" of the metropolitan and colonized spaces are every bit as apparent in so-called classical realist novels—that is, not just Joyce or Camus, but Austen and Dickens as well.

It is harder to detect the "quantum leap" from this form of nationalist space to the (current) postmodern space of late capitalism. Jameson does not clearly define the difference, or he suggests that there is merely a difference in degree—"If this is so for the age of imperialism, how much more it must hold for . . . 'late capitalism'" (1996, p. 412). But with the dismantling of great colonial empires (at least in their most visible forms, i.e., those of direct political control), a kind of withering away of the state has made room for a global (or multinational) space. This has been accompanied by greater (and more ominous) technologies that allow for almost instantaneous border-crossings and recrossings—such that, for example, in the time that it would take to put a stamp on a letter to France, an e-mail message has already arrived in Paris. Hence, Jameson's postmodern space involves a much greater "suppression of distance" and "saturation of space" than earlier types had, although these terms could certainly be applied to the effects of market and monopoly capitalism as well. Jameson takes the project of cognitive mapping to be the most suitable form of epistemological, as well as political, activity for dealing with the hyper–hurly-burly of the postmodern situation. This is what Thomas Mouat, in this volume, indicates by the phrase, "the timely emergence of social cartography."[4]

Given Jameson's exhaustive treatment of how postmodernism is defined by a crisis in representability, it may seem that the call for a critical practice of cognitive mapping is romantic at best, or even simply naive. Jameson has admitted that cognitive mapping is a "modernist strategy," and thus he implies that it may not necessarily be best suited for the present historical situation. In the next sections I will look at other ways of imagining a sort of critical mapping, but before moving on, I want to make sure that I have at least put forth (or taken from Jameson) a provisional definition of cognitive mapping. Given its apparent scope, the breadth of its possible usages, it might be best to imagine cognitive mapping in terms of Jameson's thinking in general, rather than in terms of his analysis of postmodernism alone. For, as we have seen, one of the paradoxes of Jameson's conception is that he locates cognitive mapping firmly within postmodern sensibilities at times, and, at others, he seems to suggest that cognitive mapping is an almost perennial human activity (as with the ancient itineraries and sea charts).

Colin MacCabe, in his preface to *The Geopolitical Aesthetic* (1992), has offered one very good characterization, if not strict definition, of cognitive mapping in terms of Jameson's project over the last fifteen years or so. MacCabe sees it as providing the psychology that was lacking in the concept of the political unconscious: "What Jameson requires is an account of the mechanisms which articulate individual fantasy and social organization" (p. xii). The "political unconscious" provides Jameson a key theoretical term, and "postmodernism," a key historical category (p. xii); but, as MacCabe tells us, "cognitive mapping" is the most crucial element of Jameson's philosophical system: "Crucial because it is the missing psychology of the political unconscious, the political edge of the historical analysis of Postmodernism, and the methodological justification of the Jamesonian undertaking" (p. xiv).

What allows MacCabe to say this is the essence of Jameson's nearly lifelong project, the question central to nearly all of his work: How does the psychically enclosed, subjective individual relate to the socially dispersed, objective totality? This question is put again and again, in a variety of forms—How does the apparently asocial and ahistorical, but very psychological, modernism reflect, represent or just come to grips with the sociohistorical? How do the private fantasies of one discrete individual subject point to the broad, intersubjective political realities? How does one map a totality?—but it comes back (in equally various forms) to an answer that Jameson, drawing from not only Althusser but a host of thinkers and writers also, consistently submits: There are something like imaginary solutions to real contradictions that are forever sought (or produced) by individuals in order to deal with the ever more refined and nuanced disjunctions or fragmentations in their daily lives. Jameson has admitted, for instance, that "'cognitive mapping' was [although "is" would seem more appropriate] in reality nothing but a code word for 'class consciousness'" (1991, p. 418), albeit a class consciousness suitable for our global situation with its inherent spatiality. Further, Jameson has since defended his conception of cognitive mapping, by asserting that the "idea has, at least on my view, the advantage of involving concrete content (imperialism, the world system, subalterity, dependency and hegemony), while necessarily involving a program of formal analysis of a new kind (since it is centrally defined by the dilemma of representation itself)" (1992, pp. 188–89). While a revised understanding of the term "class consciousness" might have sufficed, "cognitive mapping" more explicitly brings these elements to the fore.

Hence the literary category that has preoccupied Jameson's thinking, at least since *Marxism and Form* (1971), of allegory: essentially, one story which tells another—for example, a favorite of Jameson's, the narrative of an individual psychological subject which tells the story of great sociohistorical conditions. A cognitive map is also such an allegory, for just as there can be no "true map" (one that purely represents the geographic space) neither can the cognitive map be the true (perfectly mimetic) representation of "the Real," but rather an allegorical structure that attempts to "tell" another spatial tale. Cognitive mapping augments the political unconscious by providing a clearer link between the psychological and the social, and Jameson has, I think, linked the two key terms together in a new term (which has not yet gained, if it will at all, the cachet of these others), the *geopolitical unconscious*, where the "geo-" prefix lends the earlier book-title term a sense of that spatiality so important for Jameson's more recent thought. "Space, representability, allegory," then, are not just the points of reference for *The Geopolitical Aesthetic* (as he says in his introduction [1992, p. 5]), but for Jameson's entire project over more than twenty-five years, a project that can now perhaps be called, simply, "cognitive mapping."

Can the Subaltern Map?

Jameson's concept, or project, of cognitive mapping has elicited a great deal of spirited criticism. For example, as Bartolovich's analysis (in the present volume) demonstrates, Jameson's insistence on comprehending, and mapping, the social totality has drawn fire from a generation of critics who, following poststructualist interventions, take any notion of totality to be both an impossible and, indeed, an undesirable dream. Jameson has said that cognitive mapping is centrally defined by a crisis in representability, and the question of representation remains central to Jameson's antagonists as well. Representation not only causes trouble in regard to totality—How can one represent a totality?—but also in terms of who is doing the representing and who is being represented.

John Beverley's contribution to this book begins by asking the question, "Are maps inevitably bound up with domination?" While Beverley's essay is not specifically on this question, it does contain an underlying polemic against mapping, cognitive or otherwise. Beverley's fundamental objection involves the representation of the subaltern—in this case, primarily, Latino and Latin American people—in the age of multinational capitalism. According to this argument, cognitive mapping becomes another, albeit well-intended, attempt to represent the subaltern in such a way as to take away any self-representational authority from them. Following Spivak—who paradoxically holds that subalterns cannot "speak," insofar as their speech cannot (as Beverley puts it in this volume) "carry any sort of authority or meaning for us [who are presumably non-subaltern] without altering the relations of power/knowledge that constitute it as subaltern in the first place." Beverley suggests that studying, or mapping, the subaltern is "self-contradictory," insofar as it involves the acknowledgment that the subject itself (the subaltern) resists representation. Thus cognitive mapping, which Beverley also sees as firmly within the scope of Western, privileged academic discourse, fails, since it involves representing the unrepresentable. Beverley sees as the goal of subaltern studies not representing the subaltern, but registering the ways that such representation is impossible, and in so doing, offering a critique of the very institutions of "academic knowledge," or, as Beverley concludes, exploring "the possibility of a new kind of politics."

Of course, however academic in approach, Jameson is thoroughly aware of the inability to faithfully represent the subaltern or anything else. As Bartolovich points out in this volume, Jameson uses the term *figuration*, "which does not imply a faithful mimetic copying of a pre-existent reality but simply an attempt to come to terms with 'the real' in all its complexity, even though that 'real' will always remain an 'absent cause' for which any representation is necessarily inadequate." Nevertheless, for Bartolovich, Jameson's examples seem to indicate something more like representational thinking. Her main complaint with Jameson's project lies in his adherence to a theory of totality that, Bartolovich argues, excludes or at least marginalizes certain (subaltern) groups. In the example mentioned above, regarding the space of monopoly capital, Jameson describes how the "structural coordinates" of a Londoner's daily lived experience no longer coincide with the space of that city, but encompass India or Jamaica as well. Bartolovich points out that those Indian or Jamaican subjects would perhaps not find the figuration of an integrated totality quite so liberatory.

Bartolovich, like me but with her different emphasis on the theory of totality, turns to Foucault for an alternative to Jameson's system. But Foucault "represents" (her word) the other extreme: Whereas Jameson calls for a somewhat utopian totality, Foucault's theory of power—which holds that power is always decentered, not located in any particular place—seems to allow no grounds for resistance to gain purchase, and takes any totalized vision (utopian or not) to be merely a substitution of one form of domination for another.

The figure for this totalizing vision is, for Bartolovich, the map itself. She suggests that the itinerary, rather than the map, provides the better figure, or metaphor, since the itinerary is grounded in the actual "mapping/ mapped subjects in 'lines' produced by their collective movements under specific conditions." In a sense, then, Bartolovich seems to favor a return to Kevin Lynch's type of mapping—which, as I noted above, Jameson concedes is really precartographic—but Bartolovich has extended this beyond urban space alone. She sees the map as something with "stable lines and settled relations," whereas the itinerary "represents" something more like the activity of traveling. I do not, however, take the map to be, in any fundamental way, stable or settled, and in the next section, I address this issue in relation to Deleuze's work.

There is now quite an impressive bibliography of work dealing with the relations, historical and theoretical, between cartography and imperialism. Undoubtedly, European cartographic techniques advanced dramatically after the discovery of the New World, and these continued to be refined in the centuries of imperial expansion that followed. Benedict Anderson, following Foucault's methods, has argued that the development of mapping Southeast Asia and Africa had an enormous and facilitating impact on the Western colonization of those regions, not just in terms of direct political control, but also with regard to the production of knowledge in and about those areas (1991, especially pp. 170–78). And Timothy Mitchell (1988), also somewhat Foucauldianly, has asserted that the surveying of Egypt in the eighteenth century helped to make possible new techniques of surveillance both in the colony and in the metropolitan center—in fact, Mitchell notes that Bentham's famous panopticon principle was initially derived from his brother's regulatory treatment of colonial workers.

The passing of what Lenin called "the age of imperialism," the world-historical moment that (according to Jameson) precedes our "postmodern condition," has not entailed any concomitant waning of cartographic practices. On the contrary, as Jameson insists repeatedly, and to which Mouat in this volume and others testify with ever greater urgency, the spatialization of societies and the need for a kind of spatial analysis have never been more intense. The tendency of capital towards globalization, which Marx described in the *Grundrisse*, is today being celebrated or attacked as a fait accompli by many, while others at least acknowledge that the tendency is still there. Some multinational corporations now figure as greater economic entities than many powerful so-called first world states. And Bartolovich, in this volume, has noticed the ominous enthusiasm that their advertising agents seem to have for maps. She notes that maps and globes are "among the most common types of cover design employed by contemporary corporations in their annual reports," and she describes a well-published Ford advertisement that depicts a world map with its corporate logo superimposed over many land masses. Furthermore, popular among the visual maps used in advertising and annual reports are those "antique" sixteenth century maps with their quite explicit relations to explorers, discovery and colonization. Bartolovich suggests that multinational corporations are involved in representing themselves as the new colonial powers, now that, for the most part, direct imperialist control by one nation over another has withered away.

David Turnbull, also in this volume, notes that while the increased spatialization in modern societies and the growing centrality of cartography have benefited dominant economic and political forces, those who are "subaltern" nevertheless can, and do, use maps and mapping as a counter-hegemonic practice (as in the case of the Inuit remapping of official government maps).[5] Whereas Beverley seems to deny any possibility of using maps for political resistance—since the representational form itself is suspect—Bartolovich and Turnbull see the need for cartographically based resistance precisely because maps have come to be the dominant form of representation. For Bartolovich and Turnbull, mapping has to be useful as a counterpractice, even as it must necessarily implicate itself in the representation problems of cartography.

CARTOGRAPHIES OF POWER: FOUCAULT AND DELEUZE

Because so many critics of the politics of mapping, and of Jameson's cognitive mapping, derive their methods, theories or just inspiration from

Foucault's work, I will now turn to Foucault and examine what I take to be his contribution to these debates. An interesting aspect of the Foucault phenomenon is that, while Foucault enables through his historical analyses a number of antimapping positions, Foucault is not himself opposed to mapping. In fact, I contend that Foucault—in addition to, Deleuze—actually puts forth a conception of social mapping as a useful tool for critical inquiry.

Foucault has shown, in his archaeological and genealogical works, how modern societies have become increasingly spatialized. This spatialization has not only to do with geographic analysis, but with issues of demography, medicine, urban and regional planning and education. Foucault's work has fascinating resonances with Jameson's history of spatial formations, especially if one takes into account Deleuze's discussion of Foucault's "diagram" (1988, pp. 24–44). Foucault's cartography of power is not absolutely inconsistent with the historical mapping of the production of space in Jameson, but their methods and goals are quite different. Foucault lacks Jameson's sense of a dialectical movement of capital, but he does not at all dismiss the power relations that are in turn connected with (and even encompass) relations to the mode of production (labor, wage, monetary, and so on). It seems that despite Jameson's polemics—for example, he has called Foucault's power a "shadowy and mythical entity" (1991, p. 410)— Foucault's work can be used in connection with a type of cognitive mapping, for, in addition to those "diagrams" discussed in *Discipline and Punish* (and especially in the chapter on Panopticism), Foucault has discussed the emergence of an increasingly spatialized society—in shorthand, the modern. This spatialization not only appears in the direct sort of geographic boundary-drawing (as in the "strict spatial partitioning" in order to combat the plague), but also in the general ordering of demographic, economic and medical data (elaborated in *Birth of the Clinic*), such that a key feature of Foucault's "modern" society is the scientific distribution and codification of individuals in space. The importance given to *le regard* (as an almost technical term) in both of these books can be attributed to the fact that "the gaze"—broadly conceived as not only direct observation, but also that collecting and ordering of information—is the practical model of this spatial distribution. Turnbull's analysis in this volume resonates with Foucault's inasmuch as the emergence of modern cartography, science and the state are shown to be deeply interrelated and roughly contemporaneous.

As early as in *Madness and Civilization* (originally published in 1961, translated into English in 1965), Foucault had described the emergence of a society administered in terms of the organization and registration of individuals in a spatial array. In this work, Foucault focuses on the birth of the asylum—from the premodern exile of madmen with its haunting images of a ship of fools to what Foucault calls the "great confinement"—and so he deals with a centralization of the power to classify and "place" individuals in a certain recognizable group (the insane) and in a particular location (in this case the Hôpital Général). Foucault points out that the great confinement is contemporaneous with developments in capitalist production, such that asylums housed not only the insane, but the poor and "idle" as well—in fact, the distinction was not foregrounded, since a sign of moral and mental infirmity was idleness. This is the point at which Foucault also locates the urbanization of society, such that, for the French,[6] urban planning, surveillance and mapping become the model of national social organization as a whole.

In what might be called a fascinating ruse of history, one today sees not the forced confinement of the poor and idle, but the voluntary confinement *sought* by many wealthy and middle-class people in "new walled communities." As Dennis Judd (1995) has noted, such communities often are promoted with a rhetoric of grassroots, participatory democracy—the homeowners are all involved, sometimes necessarily and contractually, in the "local" governance of the community—while at the same time the community is sustained by rather authoritarian means (including strict codes concerning lawn care, number of children and pets and so on). Also, the public space of the market, the enclosed shopping malls, have further isolated and confined even that most "free" space where Adam Smith had seen no authoritative force operating other than the "invisible hand." Ironically, the most "poor and idle," the homeless, are not confined, but, on the contrary, left to the spaces outside of these "asylums."

By his next book, *The Birth of the Clinic* (1963, translated into English in 1973), still long before he had developed his theory of power characterized by its capillary and decentralized nature, Foucault already began to move away from the model of intense centralization. To be sure, medical practices and the knowledge to be gained by them were undoubtedly being centralized in the form of the state, but here Foucault notes the degree to which the spatial organization of individuals in society has less to do with confinement than with distribution. Individuals will be located, monitored and subject to registration, but without necessarily sequestering them in a particular location. The gaze is no longer limited to a particular place in which it operates but is generalized to cover the whole social field. Foucault's "archaeology of the medical gaze" (the book's subtitle) already points to his later genealogy of disciplinary practices, *Discipline and Punish* (1975, translated into English in 1977), which Deleuze would characterize as a cartography of power.

The famous chapter entitled "Panopticism" opens with a description of the organization of the plague-stricken city, thus recalling the arguments on madness and medicine from Foucault's earlier studies. Strict spatial partitioning, constant surveillance, individualizing distributions and the intensification of power typify the social organization of the plague-stricken town, which, Foucault claims, "is the utopia of the perfectly governed city" (1977, p. 198). Foucault takes this model in its instrumentality to be identical to that of Jeremy Bentham's *Panopticon*, an architectural apparatus in which those inside it can be assured of the possibility of always being seen, that is, of being located within a well-regulated matrix. Even more than in his earlier books, Foucault here takes spatial relations to be fundamental to the organization of society through the functioning of power. The Panopticon

> is the diagram of a mechanism of power reduced to its ideal form . . . It is a type of localization of bodies in space, of distribution of individuals in relation to one another, of hierarchical organization, of disposition of centers and channels of power, of definition of the instruments and modes of intervention of power, which can be implemented in hospitals, workshops, schools, prisons. (1977, p. 205)

Deleuze takes this to be the most important achievement of Foucault's *Discipline and Punish*: isolating and describing the diagram. The generalization of the panoptic diagram beyond merely architectural applications constitutes a new form of the social organization of space. As Deleuze writes, "the diagram is no longer an auditory or visual archive but a map, a cartography that is coextensive with the whole social field" (1988, p. 34). Unlike Jameson's version of the production of space, Deleuze's Foucauldian analysis focuses not on capital as the organizing power, but on power itself.

> We have seen that the relations between forces, or power relations, were microphysical, strategic, multipunctual and diffuse, that they determined particular features and constituted pure functions. The diagram or abstract machine is the map of relations between forces, a map of density, or intensity, which proceeds by primary non-localizable relations and at every moment passes through every point, "or rather in every relation from one point to another." (1988, p. 36, translation modified)

While Deleuze's analysis remains rather abstract, his reading of Foucault establishes the strict spatiality of power-relations in modern, disciplinary societies.

Deleuze, in his collaborative work with Félix Guattari, has supplemented this idea of the diagram with his later "nomadology."[7] Deleuze draws a distinction between nomads—who are understood not only by their border crossings and recrossings, but by their conceptual demolition of boundary lines themselves—and the state and "State philosophy," which are defined in terms of sedentary ordering and segmenting of the rank and file. By their deconstruction of boundaries, the nomads do not necessarily oppose mapping; on the contrary, they continually map and remap. They are, in Deleuzian language, forces of "deterritorialization," upsetting to a greater or lesser degree the order of the state.[8]

Deleuze further proposes that nomads have a qualitatively different kind of space than that of the state: "It is the difference between a *smooth* (vectoral, projective, or topological) space and a *striated* (metric) space: in the first case 'space is occupied without being counted,' and in the second case 'space is counted in order to be occupied'" (1987, pp. 361–62). For Deleuze and Guattari, the maritime model (as in Jameson's digression on cartography) provides an example of this distinction, for "the sea is a smooth space par excellence, and yet it was the first to encounter the demands of increasingly strict striation" (p. 479). Developments in cartographic techniques are partly responsible for such striation. For example, the Mercator projection, which establishes a grid composed of parallels, is perhaps the most obvious striating strategy—the smooth surface of the sea becomes a grid in which navigations between points are charted. However, we must not think that, because of this striating cartographic technique, Deleuze is somehow opposed to mapping. While Deleuze, often abstractly, provides a number of binaries such as these (for example, molecular/molar, rhizome/tree, map/trace), Deleuze also insists upon heeding Spinoza's caveat, *non opposita sed diversa* ("not opposite, but different"). For, as Deleuze states, "smooth spaces are not in themselves liberatory. But the struggle is changed or displaced in them, and life reconstitutes its stakes, confronts new obstacles, invents new paces, switches adversaries. Never believe that smooth space will suffice to save us" (p. 500).

Foucault's seemingly total, Orwellian vision of the panoptic organization of society is in fact no more total than Deleuze's state philosophy or striated space. True, Foucault does not allow for an "outside" of power, but at least this does not mean that there is nothing outside of oppression. For Foucault, power is productive—undoubtedly producing unpleasant things, but not exclusively so—and capillary, flowing through the social body. Nevertheless, this power is not unilinearly directed, by a particular class or a state, against subaltern groups or individuals. Deleuze's nomad provides one figure of a sort of resistance, a resistance that is not exterior to relations of power, but that exerts a force within their web.

Conclusion: Cartographics

I have suggested the term "cartographics" as the name for a set of critical practices that would engage with issues of space and spatial relations in connection with cultural and social theory. Cartographics has the advantage of situating "mapping" and spatial analysis firmly within the framework of these other broad fields of study, while remaining pliable enough to fit situations that are not properly in the domain of geographic inquiry. For example, cartographics would have to take into account the cultural forms that serve to map the terrain of an increasingly spatialized world view, including but not limited to the maps themselves. I believe that Jameson's idea of cognitive mapping is already a significant contribution to this sort of critical work.

It should be clear that the concept or practice of cognitive mapping as understood by Jameson's first definition (or characterization), drawn mainly from Lynch—where it is limited to a phenomenological sense—is not entirely appropriate for the critical project of cartographics, since the limited viewpoint of the itinerant subject cannot encompass the immense and multifaceted spatialized society. But Jameson does say that he takes Lynch's spatial analysis as emblematic or allegorical, where the "mapping," while obviously indicating the spatial, must also map the social (including not only geography, but education, political economy, media theory and others). Jameson also concedes that the cognitive map probably fails for the same reason that the "physical" maps fail—there can be no true maps—but he quickly adds that the inability to map (whether the results are "true" or not) is politically crippling, and thus cognitive mapping is still an integral part of any (Jameson adds "socialist") political project (1991, pp. 415–16). Jameson takes heart in the fact that many artists, who—notwithstanding many other critiques including a number from so-called third world or formerly colonized countries—have employed something like cognitive mapping, making the vocation of art itself "that of inventing new geotopical cartographies" (1992, p. 189).

By *cartographics*, I mean to indicate a sort of synthesis between the Jamesonian and the Deleuzian/Foucauldian projects. A project of cartographics would have to take into account the ways in which spatial practices—including, of course, geographical mapping, but also knowledge production, ethnography, economics and so on—are employed, both to repressive ends (as with neo-imperialism) and as a means to a kind of liberation (as Jameson would have it). In other words, to speak Foucauldianly, one should analyze the effects of such practices. I believe that mapping, understood broadly in terms of these manifold power/knowledge relations, presents the educator and critic with an important tool for social and cultural critical theory.[9]

1. Cognitive mapping is not, of course, exclusively a Jamesonian term. For the purposes of this chapter, however, I will focus only on cognitive mapping as Jameson uses it, and on the critical problems with his usage.

2. While the "Nomadology" appears as a chapter or "plateau' of Deleuze and Guattari's *A Thousand Plateaus* (1987), for the most part I refer to Deleuze alone throughout. This is not to suggest that Guattari's contribution to Deleuze's project is negligible, but rather that the project is Deleuze's. Deleuze, prior to meeting Guattari, had already begun to write of nomads and "nomadic distribution" in *Difference and Repetition* (1968, English translation 1994) New York: Columbia University Press.

3. Aesthetic, in this case, means not only a theoretical practice in relation to art, but also a way of seeing.

4. While Mouat's thesis is, like Jameson's project, based in a kind of dialectical movement of history, Jameson would probably consider it bizarrely millenarian, particularly since Mouat does not, as Jameson feels one must, ground his historical argument in terms of mode of production. Furthermore, Mouat's appendix seems to call for a kind of Second Coming, insofar as Christ is depicted as the individual who exceeds the limits of his social milieu by "fourth-level thinking."

5. Turnbull lists a number of examples of "cartographic resistence." The Inuit, for instance, "have resisted by taking government survey maps," removing the grid and scale and naming all of the significant points with Inuit names and then publishing it as the official Inuit map. Of course, Beverley might argue that this is a rather naive way of imagining resistance, since it involves merely substituting one name for another, and thus does not even approach the heart of the matter—that is, the power of maps and mapping. Foucault, for example, might take this to be a false form of resistance, like the prisoners who would take to the observation tower and fix their gaze on the wardens; the relations of power, in these cases, would not be altered.

6. Foucault's findings are often all too easily generalized to cover the "modern world," well beyond the situatedness, in France, of most of Foucault's historical research. Foucault has for the most part limited his research to French institutions, and many scholars, even those who are sympathetic to Foucault's work, have criticized Foucault's Eurocentrism (see, for example, Mitchell, 1988).

7. The chapter on *Discipline and Punish* in *Foucault* (1988) was originally published as a review essay in 1975. Thus the "Nomadology" section of *A Thousand Plateaus* (1987) represents later work.

8. It should be clear that Deleuze is not really referring to actual nomadic peoples, but rather to a certain conceptual organization of forces in space. For example, Deleuze cites Lucretius's Epicurean physics as an example of nomad science, as opposed to the "royal," Aristotelian science. Earlier, Deleuze (1977) had named Nietzsche as the perhaps paradigmatic "nomad thinker."

9. I am grateful to Rolland Paulston for his meticulous editorial and pedagogical assistance in preparing this chapter. John Beverley and Crystal Bartolovich also generously provided me with copies of their drafts. Jonathan Arac and Paul Bové gave me helpful comments on a much earlier version of this piece.

REFERENCES

Althusser, L. (1971). Ideology and ideological state apparatuses. In *Lenin and philosophy*, 127–186. Translated by B. Brewster. London: New Left Books.

Anderson, B. (1991). *Imagined communities: Reflections on the origin and spread of nationalism*. London: Verso.

Deleuze, G. (1977). Nomad thought. In D. Allison (ed.), *The new Nietzsche*, 142–149. Cambridge, MA: MIT Press.

Deleuze, G. (1988). *Foucault*. Translated by S. Hand. Minneapolis: University of Minnesota Press.

Deleuze, G. (1994). *Difference and repetition*. Translated by P. Patton. New York: Columbia University Press.

Deleuze, G., and F. Guattari. (1987). *A thousand plateaus*. Translated by B. Massumi. Minneapolis: University of Minnesota Press.

Foucault, M. (1965). *Madness and civilization: A history of insanity in the age of reason*. Translated by R. Howard. New York: Vintage.

Foucault, M. (1973). *The birth of the clinic: An archaeology of medical perception*. Translated by A.M. Sheridan Smith. New York: Vintage.

Foucault, M. (1977). *Discipline and punish: The birth of the prison*. Translated by A. Sheridan. New York: Vintage.

Foucault, M. (1980). Questions on geography. In *Power/knowledge* (120–146). Translated by C. Gordon et al. New York: Pantheon.

Jameson, F. (1971). *Marxism and form*. Princeton: Princeton University Press.

Jameson, F. (1981). *The political unconscious: Narrative as a socially symbolic act*. Ithaca, NY: Cornell University Press.

Jameson, F. (1991). *Postmodernism, or, the cultural logic of late capitalism*. Durham, NC: Duke University Press.

Jameson, F. (1992). *The geopolitical aesthetic: Cinema and space in the world system*. Bloomington: Indiana University Press.

Judd, D. (1995). The rise of the new walled cities. In H. Liggett and D. Perry (eds.), *Spatial practices* (144–166). London: Sage.

Lynch, K. (1960). *The image of the city*. Cambridge, MA: MIT Press.

MacCabe, C. (1992). Preface. In F. Jameson, *The geopolitical aesthetic: Cinema and space in the world system*. Bloomington: Indiana University Press.

Marx, K. (1973). *Grundrisse*. Translated by M. Nicolaus. New York: Vintage.

Mitchell, T. (1988). *Colonizing Egypt*. Cambridge, MA: Cambridge University Press.

Said, E. (1993). *Culture and imperialism*. New York: Knopf.

Social Cartography, Comparative Education and Critical Modernism

Afterthought

Carlos Alberto Torres

Critical Modernism, Social Cartography and Mapping

Our advocacy of social cartography has as its purpose the breaking of the image breakers, the encouraging of comparative analysis to become image makers, and, in doing so, the including of a visual-spatial imagery of the human in comparative discourse.

Rolland Paulston and Martin Liebman, "An Invitation to Postmodern Social Cartography"

This book is an attempt to introduce the genre of postmodern social cartography in comparative education. However, it contains a set of arguments that sometimes agree with the premises of postmodern social cartography and sometimes depart quite drastically from them. This afterthought appraises some of the assumptions made in the book, engaging the work as a whole in a critical dialogue from a critical modernist approach.

Contemporary social cartography is a tributary stream of postmodernism and as such benefits from postmodern criticisms of modernity. Let us consider the following statement by Liebman and Paulston: "Social cartography, in short, helps comparative educators, along with all participants in the educational enterprise, to order and interpret the relativism and growing fragmentation of our time" (Liebman and Paulston, 1994, p. 239). A postmodern perspective will take as a principle—if not foundation, despite the irony of the term given the positionality of the postmodern narrator—the notions of growing cultural fragmentation and interpretative relativism.[1] However, what are the scientific criteria for deciding about a synthetic principle for the codification and decodification of social reality—in brief the ordering of such fragmented reality—and what are the parameters for the interpretation, or analysis, or comprehension of such reality?

Before I engage in a dialogue, I would like to define the notion of critical modernism informing my work and advanced elsewhere (Morrow and Torres, 1995). I am basically concerned with the implications of postmodernity as a sociocultural context assuming that the nature of social and cultural reproduction has been fundamentally transformed in late capitalism. I agree with the postmodernist insight—taken as a fundamental premise in this book—that there is a serious breakdown in the linear explanations connecting theory and data, or in the logical consequences implied in the conventional view that social scientists can move somewhat comfortably from their own lived experience of social reality (life world) to the process of abstract representation of the life world, and assuming that they can do so with methodological rigor based on objectivity and value-neutrality. However, I will side with critical theory in assuming that despite poststructuralist critiques of representation, it is still possible to theorize about social reality, albeit in more self-reflexive and less totalizing ways than in the past. My point, to which I shall return, is that I view theorizing as an analytical procedure different from, and indeed more complex than, mapping per se, although that is not to say that theory-building and theoretical analysis cannot take advantage of mapping in many different ways. Fortunately enough, most if not all writers in this book reject the all-embracing interpretations of postmodernist theorists such as Lyotard, or some readings of Baudrillard and Foucault that culminate in relativism and solipsism (Morrow and Torres, 1995). I also side with the critical theories of postmodernity that take seriously both the need to revise—as good historicists—our theoretical constructs to take into account emergent and novel features of social and cultural life.

Indeed, the notion of critical modernism relates intimately to critical theory in the specific German tradition of the Frankfurt School. But it also takes into account contributions from a handful of neo-Marxists, such as Antonio Gramsci, Karel Kosic, Georg Luckas or Claus Offe, the work of Max Weber in political theory, and the continuation of Frankfurt School in the work of Jürgen Habermas. Other strong influences relate to critical theories of democracy a la C.B. McPherson or the early John Dewey, and other critical contributions such as Paulo Freire's work. In this context, and as an attempt to differentiate critical theory from traditional or positivist theory, the early Frankfurt School proposed that

> an alternative conception of social science was required, one that could grasp the nature of society as a historical totality, rather than an aggregate of mechanical determinants or abstract foundations.

Further, it was argued that such analysis could not take the form of an indifferent, value-free contemplation of social reality, but should be engaged consciously with the process of its transformation. (Morrow and Brown, 1994, p. 14)

These premises inform my notion of critical modernism, and from these premises I would like to engage in this afterthought—within the limitations of space provided—in a critical dialogue with the cornucopia of positions represented in this book regarding social cartography in comparative and international education.

The book is built around four axes: a mapping of imagination, a mapping of perspectives, a mapping of pragmatics, and a mapping of debates. A fundamental premise informing this book is the idea that the importance of social maps derives from the increase of social incongruities. This argument advanced by Liebman, as well as Paulston and Liebman, is indeed derived from postmodernism. Mouat endorses postmodernist analysis, arguing that reality is composed of disconnected fragments. Huff concludes her analysis arguing that mapping is a useful tool for research because it offers the possibility to appraise the elusive clues of cognition in a rigorous, reliable, and replicable way.

Despite this postmodernist stance, many of the authors of this book consider that metanarratives still have an important place in social research and exert societal influence. The project of modernity, for these authors, continues to have a powerful presence in the lifeworld. This claim is further documented in several chapters of the book. Stromquist, for example, argues that a visual representation that borrows concepts and metaphors from social geography will help to show the distinct traditions of academic feminism and popular feminism and how these approaches impact the objectives, content and pedagogical strategies of their educational interventions. Her analysis demonstrates how a social mapping methodology can be used to enhance the understanding of feminism as a new social movement. In her feminist work, Stromquist invites others to present their own views in the physical and conceptual space that she has drawn.

John Beverley, looking at testimonial narrators like Nobel Prize recipient Rigoberta Menchú, sees her narrative emerging as an example of the only narrative with the capacity of being a witness of history; in her case, of a historical process of repression and assassination of Guatemalan indigenous people by a military dictatorship that has its peculiar national and local roots but may also reflect a larger international and historical process of discrimination and/or annihilation of native peoples. From this standpoint,

Beverley argues about the importance of the subaltern as a transformative project, and the need to incorporate a subalternist perspective in pedagogy. A subalternist perspective drastically differs from postmodernism cognitive mapping because for Beverley, following Spivak, postmodernists have not been able to unlearn their privileged position and hence may be complicitous with what is a new academic sublime—an aesthetic cognitive remapping that may become itself a functional element of late capitalist culture just as the Romantic sublime was in the nineteenth century. For Beverley, a subalternist perspective does not claim to represent, map, or "let speak" the subaltern or alterity. On the contrary, a subalternist perspective registers instead the way in which the knowledge social scientists construct and impart as academics is structured by the absence of, the difficulty of, or the impossibility of representation of the subaltern and alterity. More strongly indeed, Beverley, in this volume, claims that

> the very idea of "studying" the subaltern is catachrestic, in a way that points to a new register of knowledge where the power of the university to understand and represent the world breaks down or reaches a limit. The idea of representing or studying the subaltern must itself confront the dilemma of subaltern resistance to and insurgency against elite conceptions.

A related claim is made by Patti Latter, that mapping the transnationalization of contemporary theoretical discourse runs the risk of, even from progressive perspectives, practicing theoretical imperialism. Crystal Bartolovich defends the notion that cognitive mapping could be used as a counterhegemonical practice. Mapping the self-representation of corporate capital, Bartolovich shows how the symbolic spaces of capital acquire different meanings when it shifts from state to suprastate forms. For Bartolovich, much in the tradition of critical modernism, cognitive mapping appears as a form of resistance as long as social scientists can develop mapping practices that are flexible, always in process, and capable of maintaining the tensions between local and global perspectives.

While some authors in this book favor the postmodern and the study of multiple perspectives, and therefore attempt to develop cultural clusters, narratives and influences, Paulston and Liebman, in this book, envision a move towards encompassing the perspectives and methods we can find that serve to advance both knowledge and understanding. Hence they speak for some of the authors in this book who seek not to replace one totalizing perspective with another, a situation that would not improve social and comparative research. They don't want to create a new focus for argument, misunderstanding and exclusion, and yet there are fundamental tensions in this.

Despite the watershed tension between modernist and postmodernist allegiances in this book, a central premise and what in fact makes for the narrative thread of this book is to consider mapping as a heuristical term, as a new tool and a metaphor for comparative education. From this assumption two premises follow: That the structures of multiple education knowledge systems can be recreated in one or more maps, and that comparativists should consider representing that space through the creation of maps. For Rust, for instance, social cartography appears as a novel way of seeing the intellectual landscape of comparative and international education. The advantage of social cartography, according to Rust, is that visual mapping allows for multidimensionality, nonlinearity and turbulence, and hence facilitates the introduction of a number of new ways of seeing, including chaos theory transformation theory and self-organization theory into comparative international education.

There are two choices in this process of creating a social cartography: to consider the map as a metaphor, or to take it as a more or less literal representation of reality. The hope is that these two options disallow the promotion of a new orthodoxy, and remain as a flexible tool and a framework for comparative international studies. In this vein, Turnbull argues that space (and hence geography), rather than time (and hence history), is the focus of social cartography. In an attempt to expand the dominant positivist view of a linear, individual and causal style of explanation, Turnbull argues, following Soja, that social scientists today need to include spatial, relational, systemic and metaphorical explanations to expand the dominant linear, individual and causal style of explanations.

Because mapping appears not only as a set of metaphors but as a heuristical tool, it is possible to explore its use in different terrains as, for example, implemented by Esther Gottlieb in mapping what she terms the utopia of professionalism in the Carnegie Foundation International Survey of the Academic Profession; by mapping the mythopoetic images of Western humanism (Buttimer, in this book); intercultural perspectives in educational consultancies; the use of spatial analysis in social cartography; the alternatives of popular participation in rural development projects; the different ways of seeing environmental education; or the examination of innovative mapping methods.

The key epistemological question that I would like to pose is to what extent does social cartography as defined and utilized in this book advance the debate beyond a positivist understanding of educational planning and normal science? Moreover, to what extent can an alternative perspective to positivism, namely constructivism, take advantage of the heuristical tools of social cartography and mapping in comparative education? I turn now to discuss the paradigms of positivism—as normal science and the dominant feature of educational planning—and the alternative approach of constructivism, highlighting the challenges that they pose to social cartography. In closing, I discuss some critical issues related to social cartography, mapping and social theorizing.

EDUCATIONAL PLANNING AND THE SOCIAL SCIENCES: POSITIVISM AS NORMAL SCIENCE

The logic of educational planning is closely linked with the model of normal social science, dominated by the epistemological paradigm of positivism (Wallerstein, 1991; Morrow and Brown, 1994). Positivism responds to a set of suppositions about how scientific work should be undertaken. Knowledge, for positivism, exists at three levels of abstraction and hence generalization: at the level of particular observations, at the level of laws and empirical generalizations and at the level of theoretical statements and definitions (Bredo and Feinberg, 1982, p. 16). In general terms, explanations are based on the possibility of establishing regularities or patterns of uniformities that can be differentiated from accidental generalizations or laws.

However, since strong positivist explanations are drawn from the natural sciences rather than the social sciences, positivism faces a principle of ambiguity in the social sciences that is not present in natural sciences: an event that doesn't conform to a rule of universality invalidates the rule, which in social sciences is virtually the case of any event because of its open-ended and intrinsically potentially idiosyncratic nature. This fact invites positivists to consider regularities in a more flexible form, by resorting to notions of statistical probabilistic models rather than generalization of laws or lawlike explanations. Laws, and lawlike models of explanations, should be differentiated from merely empirical generalizations, which address the issue of how to move from empirical observations to definitions of causality.

The problem is how to identify causal explanations without relying on interpretations—a relapse into metaphysics that positivists set out to avoid at any cost. The simple approach to this problem is to think of theoretical statements as hypothetical deductive postulations; that is, logical constructions rather than real entities (Bredo and Feinberg, 1982, p. 21). The question then is whether it is possible to sharply differentiate, as a logical distinction, between theory and observation, which opens the avenue for a challenge from an alternative paradigm, constructivism, as noted below.

Another issue is whether positivism's attempt to generate a social scientific model, separated from its theoretical foundations and universally applicable, is a valid scientific undertaking. This social scientific method attempts to establish a sense of certainty and analytical precision in a world that is increasingly unpredictable and imprecise, showing a great degree of variability and volatility. Positivism begins with a linear and evolutionary concept of knowledge around which deterministic inferences and deductive conclusions are based and empirical foundations are organized.

These empirical foundations are not based on metaphysics but on a self-evident and yet normative distinction between value judgments and empirical judgments. It is also important to reiterate that positivism is based on a search for patterns of regularity and universalizable and reproducible results. There is a sense that social reality, as concrete totality, can be parceled out in distinct domains. Each domain can be studied as such, finding, through specific specialized methods and means, recurrent patterns and regularities which can be studied with grand theories (embedded, nevertheless, in metatheories), middle-range theories, and empirical research as a third level of scientific practice, all of them clearly differentiated. Hence, as Joel Samoff suggests, positivism represents a scientific tendency that emphasizes disciplinary rather than interdisciplinary or transdisciplinary work (Samoff, 1990).

Positivism does not recognize the important of nonlinear events or the profound discontinuities of real-life phenomena. At the same time, the subjectivity and singularity of the researcher is disdained in the function of a notion of social objectivity. The notions of science and ideology are defined not only as potentially antagonistic and irreconcilable practices, but also as practices that are clearly differentiable and discernible through the systematic application of the scientific method and certain ethical and epistemological precepts regarding the separation of value judgments and empirical judgments.

Following these epistemological presuppositions, planners schooled in positivistic social sciences argue that there is a fundamental social order underlying the dynamic of things themselves. This order is discernible through the methodical and rigorous application of a specific method of social science. This method must reflect the premises of all scientific methods according to the model of natural sciences; that is, a method based on foundationalist objectivity, the search for control and manipulation of variables, experimentalism (or quasi-experimentalism), universality and rationality (Silos, 1995).

This scientific method permits the discovery of regularities that can be measured and quantified, applied in experimental or quasi-experimental analyses used to study correlations and causalities or manipulated (controlled) in future analyses. The goal of social science is to develop a set of arguments that study causal relations; when possible, these detected patterns or regularities can be applied like laws or empirical guidelines. These laws can be summed up in brief, concise, simple phrases and they can even be presented mathematically and used (previous empirical exam and proof subject to the falsifiability of the thesis) to manipulate and indeed plan the process of development of social reality. More complicated analyses trying to understand the historical nuances of things, their interrelations and the theoretical multidisciplinary analysis of numerous observations that may make the analysis problematic, tentative or uncertain, are ignored, or rejected as unnecessary. Or, if they are considered pertinent in theoretical terms, they are considered to lack planning based on well-defined problems, with a sense of urgency and immediacy, and motivated not by theoretical reasons, but by actions that quickly and efficiently resolve specific problems.

How does postmodern social cartography relate to the positivist bedrock? Theory is a mode of discourse that goes beyond mapping. Morrow and Brown have shown that "the narratives of scientific methodology are characterized by stories obsessed with questions about empirical evidence, proof and validity" (1994, p. 40). The challenge of social cartography from positivism is whether social cartography relates (represents or can substitute for) empirical theory, involving the "descriptive and analytical (formal) languages through which social phenomena are interpreted and explained" (Morrow and Brown, 1994, p. 41).

An important question is about the connections between mapping and measurement, considering the contributions of positivism to understanding social reality. For Liebman and Paulston, social cartography reflects a fundamental premise of analysis: that the constructed social world cannot be measured but viewed, reported and compared. Hence social cartography is post-paradigmatic. The problem with this view however, is that without measurement we cannot compare and contrast the time/space evolution of a given phenomenon. Let me consider for a minute the situation of increasing poverty in Latin America.

The 1980s have been called the "lost decade" for development in Latin America. This is a fascinating metaphor that dominated the literature of economics and development of Latin America for quite some time. It has the power of the commonsensical appraisal of a situation. It is fascinating both for what it reveals and for what it hides. The metaphor reveals a fact: the lack of economic growth as measured by different economic indicators. Hence, social and economic development in the 1980s can be written off.

The region moved backward, not forward. It is also fascinating for what this metaphor hides. While the 1980s are considered "lost" in terms of economic growth, especially with the decrease in the overall gross national product (GNP) and growing disparities in the distribution of income, it was also during this time that certain sectors of the Latin American financial and industrial bourgeoisie, especially those associated with the businesses and investments of the Latin American states, took substantial advantage of investment contracts and state protection through fiscal incentives, market protection or both. In other terms, the metaphor of the "lost decade" points to the commonsensical version of a systematic fall in income, when in reality this fall was not uniform for all sectors of the Latin American population. Groups of the economic and political elite became much richer and diversified their businesses in a disproportionate way. Metaphors cannot replace theory and should be, themselves, analyzed from a theoretical perspective.

But let me go back to the metaphor and ask whether, facing this "lost decade," we can and indeed need to measure what happened with the key economic variables in the region? In the 1960s, the average annual rate of growth of the GNP in the Latin American economies was 5.7%, while during the 1970s, this figure was around 5.6%, despite massive regional difficulties stemming from the problems of the international markets caused by the oil crisis. In the 1980s, the growth rate descended to a modest 1.3%, which, given the rapid increase in the size of the population, made the actual decrease in per capita income in Latin America almost 9%. In Argentina, this negative turn was even more pronounced, with per capita income falling 22%. In Brazil, the decrease was around 5%.

In 1990, the per capita income in almost all the countries of Latin America was less than the level it had been in 1980, with the exception of the figures for Chile and Colombia, which were slightly higher (CEPAL, 1990, 1991; World Bank, 1988).

There is evidence of the worsening of poverty in the region. During the 1980s, the per capita purchasing power of Argentineans and Brazilians fell by almost 30% and 6% respectively. At the same time, there were growing levels of unemployment and massive reductions in the social expenditures of the state. More specifically, the "classism" and the social polarization of Latin American societies were considerably accentuated: In the 1980s in metropolitan Buenos Aires, 25% of the poorest households lost 15% of their income, while 5% of the richest households increased their income by almost 20%. In the metropolitan areas of Rio de Janeiro and Sao Paulo, 25% of the poorest households lost almost 13% of their income, while 5% of the richest gained approximately 25%. However, income losses were not only experienced by the poorest of the poor; 50% of the households located in the middle of the scale lost between 3% and 10% of their income (CEPAL, 1991).

One may cast doubt on the validity of this data, or may argue that all official data are "political prisoners" of governments, or that most data of this kind are unreliable. Perhaps so. However, it is important that even with the ostensible limitations of the social sciences, we achieve some empirical measurement of changes in the social order, in this case of disparities in income distribution. I have argued elsewhere that as a result of government policies, the role of international and bilateral organizations and the same dynamics of capital accumulation of the elites in Latin America, class structures have become more polarized and dual, with the rich and poor sectors separated by a gap that is increasingly wider. This is also true in Mexico and Chile, where their neoliberal models were considered as the most successful cases of structural adjustment and economic stabilization. Until December 1994, Mexico was a linchpin of trade liberalization. Despite the economic success of Chile, as in the rest of Latin America, the poor are becoming poorer and more numerous. This well-known dualization of the class structure is not exclusive to Latin America. Rather, it is a generalized phenomenon found in all the varieties of capitalism inspired by the recommendations of the "Washington Consensus." The same patterns, with slight variations, are found in the history of the United States under Reagan and Bush, and in the United Kingdom under Thatcher and Major (Boron and Torres, 1994).

Perhaps this reasoning may be deemed simplistic in the light of postmodern notions of social fragmentation and interpretative relativism. After all, this notion of a growing gap between the rich and the poor that has grown dramatically wider in the past ten years is class-based. However class analysis also integrates gender and race studies, since single female heads of household are becoming a visible face of the new poor in Latin America, and the indigenous people, as the Chiapas conflict exemplifies, continue to suffer disproportionately the effects of structural adjustment and stabilization policies and continue to fill the ranks of the unemployed and poor. Indeed, the model of dominance that Eisler (1988) referred to is still thriving in Latin American dependent capitalism.

If we need to measure, we also need to theorize and to argue across different disciplinary and paradigmatic borders. Liebman and Paulston argue that "maps created by social cartographers are not to be replicable by other social mappers" (1994, p. 238). Moreover, "social cartography, mapping the plane of multiple social realities, locates itself eclectically. It is indifferent to ideological and theoretical controversies" (Liebman and Paulston, 1994, p. 240). Interestingly enough, however, since the early maps were drawn, some people and some regions were located in the upper level of maps (that is, the North) and some people and some regions were located in the lower level of maps (that is, the South). To reverse this situation, Jesse Levine designed the "Turnabout Map": a map of the world upside-down, geographically correct but with peoples and regions of the South in the top, and peoples and regions of the North at the bottom. How can we argue that maps and social cartography are indifferent to ideological and theoretical controversies, considering that most maps have been drawn originally by Northerners rather than Southerners, by conquerors and settlers rather than by conquered and original people of faraway lands? Measurement is also a contested terrain.

Constructivism as an Epistemological and Ethical Alternative to Positivism

The polar opposite to positivism is a constructivist model of social science, which reflects a strong alternative vision in which reality appears as a product of discontinuities and unpredictable effects. Learners, in the view of constructivists, actively participate in learning, a notion that applies to the most elementary forms of learning and the most advanced forms of research. Viewing all knowledge and learning as a social activity does not necessarily mean, as some postmodernists argue, that we cannot potentially represent reality; but it does imply that we must acknowledge the diversity of perspectives involved in the formation of a community—a community of inquirers and teachers, in particular. Abandoning the "quest for certainty" does not require abandoning the search for knowledge.

Methodological pluralism follows from a constructivist conception of scientific knowledge. This does not mean so much that "anything goes" as that we must acknowledge that there are diverse logics-in-use that make up inquiry. Accordingly, the key issue for policy research is the development of coherent research designs that link theory and research techniques appropriate to the questions asked and problems to be solved.

Another premise, in stark contrast with positivism, is that knowledge cannot be separate from meaning and value, hence education is necessarily a moral enterprise. But in a culturally diverse society, this does not imply an absolute moral code, as opposed to procedural principles for guiding ethical thinking and action. In the context of education, caring, justice and individual responsibility are central principles of moral action that should complement each other.

As feminist theorists have reminded us, the nurturing principle of caring is at the heart of all learning as an interactive process that must respect the dignity of others. In the context of male-dominated culture, an ethics of caring can only emerge through a feminist approach providing the foundations for change. Feminist scholars have argued that the male experience is overwhelmingly reflected in education, and more generally in public policy. We need to take very seriously the culture of "otherness," specifically women's culture and the cultures of minorities in schools, broadening the perspectives of citizenship education to include responsible and mutually respectful behavior in the domestic unit, the family, and in interpersonal relationships in society at large. Further, principles of social justice provide rational grounds for justifying criticism of social relations that undermine caring and the equitable fulfillment of human needs. Lastly, a notion of individual responsibility is central for the constitution of morality and ethics in education.

Learning is also an interactive process that should be organized around dialogical principles. Though not necessarily undermining the importance of intellectual authority and leadership, dialogical principles as predicated by John Dewey and Paulo Freire, among others, do pose questions about the education of the educators and researchers, the need for reflexivity about what is to be taught, and the social uses of research.

Finally, constructivists recognize that research and education are socially and historically situated activities in institutions that are constrained and enabled by the power relations in the society around them. For this reason, an understanding of the role of expert knowledge, research and education should be considered from the political sociology of education, paying attention to the relationships of the ideals and values embodied in researchers and research practices that seek to inform and guide educational policies.

The challenge to social cartography, from a constructivist perspective, is whether it can include normative theorizing that involves "the modes of theorizing that legitimate different ethical, ideological, or policy positions with respect to what ought to be" (Morrow and Brown, 1994, p. 41). Stromquist, in this volume, argues that

> mapping does not have to be mutually exclusive with the use of a particular set of normative principles that are to be applied to entire societies and even the world at large. For instance, maps can present multiple views regarding democratic rule and we can also map the values of what we might consider democratic rule to encompass.

Trying to reconcile a key principle of positivism (universality) with a key principle of constructivism (empowerment), Stromquist argues that "the de-

nial of all universals is intellectually disarming and politically naive"
(Stromquist, in this book). For Stromquist, a normative and emancipatory
view will object to saying, in a postmodern fashion, that "everybody is right."
Hence her criticism to the strategy of mapping is threefold: that mapping
invites thinking in terms of discrete entities occupying specific spaces; that
maps cannot show multiple dimensions in a given space; and finally that
maps cannot reflect border-crossing and borderwork.

Mapping and Theorizing: Concluding Remarks

Interestingly enough, Rust, in his presidential address to the Comparative
International Education Society, was one of the first to discuss the pertinence
of postmodernism for comparative and international education (Rust, 1991).
In agreeing with Seppi's view that cartography is useful because of its metaphorical utility, Rust insists that "it should not prevent us from appreciating the possibility of expanding the metaphor into an analogy" (Rust, in this
book).

The concept of metaphor means, literally, the transferring to one word
the sense of another. An analogy, on the other hand, refers to the similarity
between things in some circumstances or effects, when the things are otherwise entirely different. In logical terms, an analogy implies that a certain
resemblance between things implies the probability of further similarities.
Here we have then two fascinating epistemological problems. First, is it possible to "expand" a metaphor to become an analogy? Secondly, if this first
operation is possible, will the analogy or new concept reflect analogically
another entity that has the probability of matching in some way the quality
resembled by the original analogy?

Since Aristotle's *Nicomachean Ethics*, we have known that analogical reasoning constitutes the basis of any metaphor. The risk, however, is to
replace with a metaphor (or an analogy) scientific reasoning and explanation. The effectiveness of metaphors is that by enriching the language they
can convey powerful messages. For instance, consider the title of the book
by Riane Eisler, *The Chalice and the Blade* (1987), involving two metaphors
symbolizing the antinomies of *life* (the symbol of the ancient chalice as life-
generating and nurturing) and *death* (the symbol of the lethal power of the
blade). Yet the chalice and the blade are analogical proxies in the book for
analytical concepts showing the logical, cultural and political alternatives of
a model of social organization based on a partnership model (the chalice)
or a dominator model (the blade).

Another example is the multiple metaphors utilized by Karl Marx in
his work; for instance, the metaphor of society as a base and a superstruc-

ture *(liberbau)*, or the metaphor of religion as the opiate of the people. Given the constraints of space, I will only discuss the superstructure metaphor, since I discuss at length Marx's views on religion elsewhere (Torres, 1992). Interestingly enough, although Marx used the metaphor only a handful of times in his writings, the notion of superstructure became a hallmark theoretical explanation for certain variants of Marxism. This metaphor suggests that society is composed of an economic base or civil society with production and exchange, and a superstructure encompassing the legal and political structures, the state and the realm of ideas *(idealistischeii superstruktur)* (Silva, 1978). Many of the mechanist concepts associated with Marxist economic explanations result from the removal by many Marxists of dialectical and historical analysis and its replacement by this static and simple-minded metaphor. Undoubtedly, a large number of Marxist commentators resorted to the analytical short-cut of the superstructure metaphor instead of more refreshing and sophisticated theoretical analyses that can be developed from the massive corpus of Marx's writings.

The Spanish philosopher and novelist Miguel de Unamuno himself argued for the power of metaphors; he said that a metaphor was more valuable than a thousand concepts. The problem, however, is that metaphors can only play a synthetic role in calling attention to a relevant problem but cannot, by themselves, explain (or comprehend or understand) the nature of the problem. Analogies or metaphors cannot constitute appropriate generalizations about the possibilities of separating the idiosyncratic from the nonidiosyncratic characteristics of the problem investigated nor explain the meaning and form of the problem. Analogies or metaphors per se cannot explain the historical unfolding of a problem as it relates to its present spatial formulation. Analogies and metaphors do not necessarily reflect a synchronic/diachronic dynamic. For all the power of allegories, metaphors and analogies, they tend to be static, impressionistic-oriented and descriptive more than analytical. We have seen the simplification of the most powerful analogy of the sociology of consensus: equating society with a human organism. This analogical reasoning is indeed quite different from analytical reasoning or normative reasoning. In short, metaphors as heuristic tools should be welcome in social theorizing, and they help to bridge the worlds of social sciences and humanities. Metaphors, however, should have no place in social sciences if they substitute for social theorizing, including metatheory (or epistemology), empirical theory and normative theory.

Turnbull, in his historiographic account of the evolution of mapping in this book, follows Polanyi's premise that theory is a kind of map extended over time and space. Indeed, without maps, he seems to argue, science is impossible; yet maps are heterogeneous assemblages that cannot escape the conditions of instability and messiness of local and global conditions. Indeed, he concludes that

> the construction of the microworlds of science and cartography are directly analogous: Science and cartography are intimately related, the social processes in their construction are the same and together they help open up the knowledge space within which we produce and represent knowledge. However, they do not provide the only mode of assembling knowledge or the only social order; messiness can always re-emerge. (Turnbull, in this book)

I have emphasized in this afterword that metaphors cannot replace theory and metatheory. They can only be serviceable when embedded into a theoretical framework. Beverley warns us in this book, while making a reference to a very powerful theoretical current, that "in its current process of academic institutionalization cultural studies runs, in my opinion, the risk of becoming a primarily descriptive register for the emerging 'scapes' of global and local cultures it seeks to map." Here we encounter yet again another typical criticism of mapping, that it is essentially a theoretical strategy that can be reduced to a descriptive register, a static register, a mere photograph that, in the words of Paulston, can constitute a micronarrative of a narrative, but risk the impossibility of deciphering the origins, dynamics and social interests embedded in the narrative:

> By using maps as a part of our comparative studies we may provide an inside view, a visual dialogue of cultural flow and changing influences appropriate for future work in comparative education, particularly in those instances where cultural values and differences are revealed by competing knowledge claims. (Liebman and Paulston, 1994, p. 244)

If knowledge can be separated from value, then we can rely on empirical judgments and not on value judgments. But this central tenet of positivism is unacceptable for Liebman and Paulston, and hence the notion of competing knowledge claims should be examined not only in light of alternative explanations or representations but in light of the intractability of arguments. Language exchanges can reach a point where the collision of interests can make some concepts impossible to translate from a given theoretical paradigm into another theoretical paradigm. I suspect that mapping and social cartography cannot solve the intractability of discourses. Nor is mapping able to solve another important dilemma; that is, whether dialogue is even feasible or desirable, as Enrique Dussell suggests in Beverley's chapter, given asymmetrical power relationships among the dialogical individuals.

Despite these critical considerations, mapping and social cartography offer interesting avenues for research. However, as a heuristical tool for comparative education, mapping should face the challenges of both positivism and constructivism to demonstrate its epistemological value, practical utility, and ethical-political worth.

NOTE

1. Morrow and Torres have defined postmodernism as a cultural and analytical movement, as an attempt to account for what is perceived as involving various processes of fragmentation: a) a decentering and fragmentation of power that calls into question theories of domination and hegemony; b) an uncoupling of material interests and subjective expressions in collective action, resulting in the shift of demands of social movements from distributional to cultural-ethical issues; c) the emergence of heterogeneity as opposed to homogenization that has been previously characteristic of the world system; and d) a growing distrust and disillusionment with democracy resulting from the fragmentation of political communities and identities (Morrow and Torres, 1995, pp. 415–45).

REFERENCES

Boron, A., and C.A. Torres. (1994). *Education, poverty and citizenship in Latin America*. Paper presented at the International Symposium on Education and Poverty: From social inequality to equality, El Colegio Mexiquense, Zincantepec, Mexico, 26–28 October.

Bredo, E., and W. Feinberg, eds. (1982). *Knowledge and values in social and educational research*. Philadelphia: Temple University Press.

CEPAL (Comisión Económica para América Latina). (1990). *Transformación productiva con equidad*. Santiago, Chile: CEPAL.

CEPAL (1991). *Panorama social de América Latina*. Santiago, Chile: CEPAL.

Eisler, R. (1988). *The chalice and the blade: Our history, our future*. San Francisco: Harper.

Levine, J. (1982). Turnabout map. Distributed by Laguna Sales, Inc. San Jose, Ca.

Liebman, M., and R. Paulston. (1994). Social cartography: A new methodology for comparative studies. *Compare, 24*(3), 233–245.

Morrow, R., and D.D. Brown. (1994). *Critical theory and methodology*. Thousand Oaks, CA: Sage.

Morrow, R., and C.A. Torres. (1995). *Social theory and education: A critique of theories of social and cultural reproduction*. Albany, NY: SUNY Press.

Paulston, R., and M. Liebman. (1994). An invitation to postmodern social cartography. *Comparative Education Review, 38*(2), 215–232.

Rust, V. (1991). Postmodernism and its comparative education implications. *Comparative Education Review, 35* (4), 610–626.

Samoff, J. (1990). *More, less, none? Human resource development: Responses to economic constraint*. Paper prepared for the Interagency Task Force on Austerity, Adjustment, and Human Resources of the International Labor Office and the United Nations Educational, Scientific and Cultural Organization.

Silos, M. (1995). *Economics education and the politics of knowledge in the Carib-*

bean. Ph.D. dissertation, University of California at Los Angeles.

Silva, L. (1978). *El estilo literario de Marx*. Mexico, Mexico City: Siglo XXI.

Soja, E.W. (1989). *Postmodern geographies. The reassertion of space in critical social theory*. London: Verso.

Torres, C.A. (1992). *The Church, Society, and Hegemony: A Critical Sociology of Religion in Latin America*. Translated by Richard A. Young. Westport, CT: Praeger.

Wallerstein, I. (1991). *Unthinking social science: The limits of nineteenth-century paradigms*. Cambridge, MA: Polity.

World Bank. (1988). *Brazil: Public spending on social programs, issues and options*. Report 7086–BR. Washington, DC: World Bank.

Envoi

Gabriel thought maps should be banned. They gave the world an order and a reasonableness which it did not possess.

William Boyd, An Ice-Cream War

Map me no map Sir!
My mind is a map,
a map of all the world!

Henry Fielding, Rape upon Rape

As Boyd, Fielding and several chapter authors make clear, maps are not everybody's cup of tea. Some free spirits see mapping only as a panopticon, a prison house of fixed vision. Others who expound a master narrative tend to view open and participatory mapping as dangerous and potentially subversive.[1] But for those who might wish to become comparative mappers and enter into the visual play of similarity and difference in the shared construction of meaning, this volume offers numerous examples of how one might proceed. In this envoi, or short concluding section, I would like to present several maps of the multiple perspectives that construct this book. These representations may serve to illustrate and "send away" our social cartography thesis.

In Figure 1, a Cartesian plane, as described by Turnbull and Seppi in their chapters, is composed by crossing the two axes of "Actor/Society" and "Equilibrium/Transformation." These binary coordinates create the four paradigms—or what Said (1991) would call "arrogant categories"—used to type and situate the nineteen chapters. The ensuing modernist map of *Social Cartography* is thus static, centered and reduces complexity into a reason-driven or logocentric model. Particularistic messiness is made invisible by the seeming appearance of objectivity, symmetry and categorical order. Each paradigm is seen to hold incommensurable mutually exclusive under-

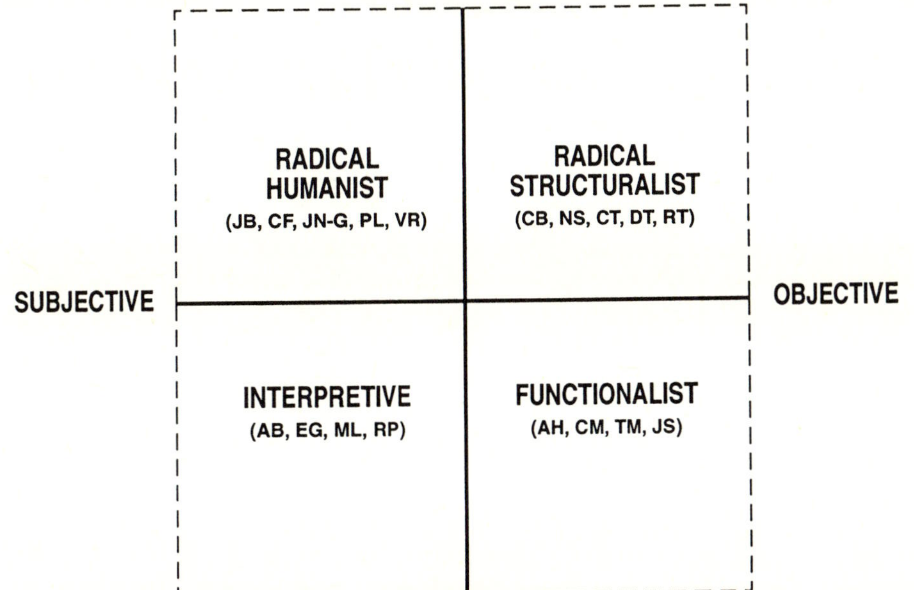

Figure 1. An intentionally modernist—i.e., "either/or"—mapping of Social Cartography in the style of Burrell and Morgan where the nineteen chapters are frozen into four "universal paradigms" located in a Cartesian coordinate plane: "The four paradigms are presented as mutually exclusive views of the social world. Each stands in its own right and generates its own distinctive analysis of social life." See G. Burrell and G. Morgan (1979), p. 22. (Initials refer to textual contributions by individual chapter authors in this volume.)

lying views and opposing metatheoretical assumptions. This figure represents the now rather conventional essentialist view of comparative education and the social sciences as they have come to be seen by many in "late modernity."

In marked contrast, Figure 2 is, in the words of several colleagues, "a mess." Here the Cartesian logic of linking the subject (that is, the author) with the object (that is, the work) is replaced with a poststructuralist preference for seeing practices (writing) as sites in an intertextual (field). Authors are, ironically, sent packing and the multiple perspectives I identified in context construct an acentered yet situated reality akin to Haraway's (1985) characterization of postmodern multiplicity as "a powerful infidel heteroglossia" (p. 100). As in Lefebvre's (1991) view, language now becomes our "instrument of veracity" with which free-form mapping seeks to "decode [to] bring forth from the depths not what is there but what is sayable, what is susceptible to figuration" (p. 139). In this elaboration of textual relations is the acknowledged presence of a "fiduciary subject," or embodied "mapper," who as a socially articulated self is the true site of agency. Here the overlapping of discursive and physical space reveals the body as the primary site of political authentication and political action (Stone, 1995, p. 91). From this perspective, social mapping escapes the violence of logocentric enclosure and instead elicits an embodied discourse system or set of reading that are frequently disrupted and in need of reordering. Social cartography provides a means to facilitate reordering and subject reconstruction within a physical field and a system of symbolic exchange. Identity is seen to be largely discursive and produced through the interaction of texts. This "legible social body" presents a set of cultural codes that "organize the way the body is apprehended and that determine the range of socially appropriate responses" (Stone, p. 41). Accordingly, Figure 2 represents my provisional and local structuring of "comparative education" as a set of contradictory yet complimentary cultural codes seen from what Griffin (1992) has paradoxically called a "constructive postmodern" perspective that takes pluralism seriously.

Lefebvre's resistance to categorization along with Deleuze and Guattari's call for "nomad" mapping (see Tally's chapter) provide inspiration for Figure 2, where six perspectives, or ways of seeing/knowing (that is, the ideological, structural, technical, ironic, textual and auratic[2]) I discover in the chapter texts and locate around the field's perimeter serve as nodes to link texts in a complex web of relations where no place/view is privileged over any other space or perspective. Binary oppositions are dissolved into a "virtual system—into an intertextual field of unresolved difference and creative tension" (Stone, 1995, p. 183). In this sense the free-form postmodern

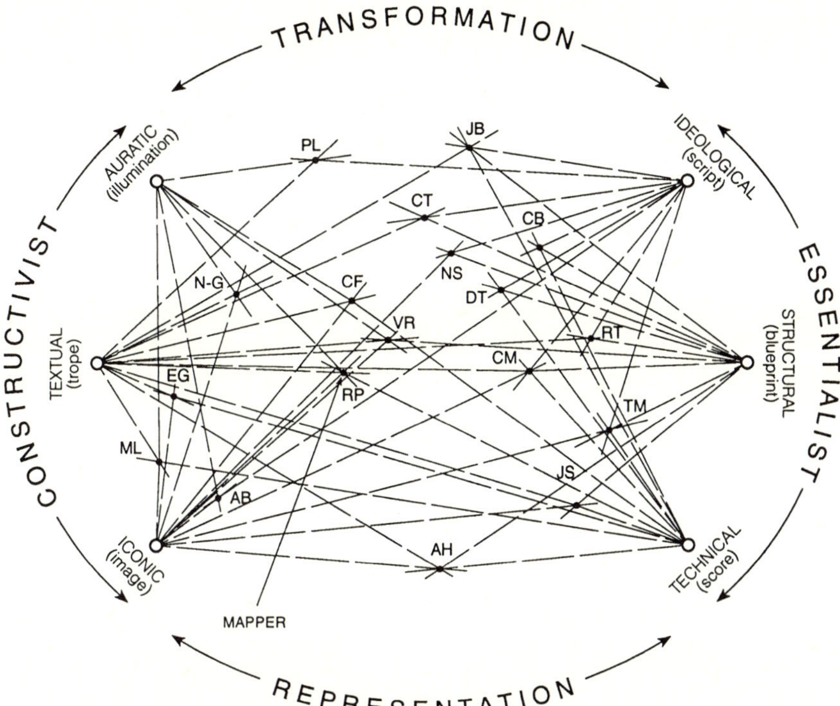

Figure 2. A perspectivist mapping of Social Cartography in the style of G. Deleuze and F. Guattari (1987), where the intertextual field is portrayed as a rhizome, or an acentered system. Here coordinates are not determined by theoretical analyses implying universals, as in Figure 1, but recovered "by a pragmatics composing multiplicities or aggregates of intensities" (p. 15). (Initials refer to textual contributions by individual authors in this volume.)

cartography attempted in Figure 2 may be seen as suggesting a "heterotopia" where the fragmentary and seemingly incongruous and incompatible spaces of postmodernity are presented as if they lacked definite centers, boundaries or common logic in a space in which everything is somehow out of place and seemingly without law or geometry (Foucault, 1986). Relph (1991, p. 104) claims that the notion of heterotopia bears the stamp of our age and our thought: "It is pluralistic, chaotic, designed in detail yet lacking universal foundations or principles, continually changing and [as in cyberspace] linked by centerless flows of information." The notion of heterotopias as countersites render doubtful most of our conventional ways of thinking about comparison and representation, and they provide fertile new terrain for a critical social cartography open to all views (Paulston and Liebman, 1994).

While comparative mapping offers no immediate resolution of our present heterotopic crisis of legitimacy, it does help both actors and communities provisionally represent and compare how they experience our world as ever changing perceptual fields of discourse and debate. In this way, social cartography—as a new form of paralogical cultural representation—addresses head-on today's challenge of difference and provides a valuable extension of earlier forms of critical social analysis into today's theoretical and ethical concerns (Calhoun, 1995).[3]

1. Exodus 20:4–6, the Bible's second commandment, presents an early illustration of this perceived danger: "Thou shalt not make any graven images, or likeness of anything . . . for I am . . . a jealous God. . . . " In the ensuing fight between art and religion (now science), music, poetry and even drama have largely been brought under realist authority. But the visual arts have remained suspect. For Plato, and also Seppi, "false images" are those of sensory appearance. True images—as with GIS (geographical information systems)—are seen to be found only in abstract, ideal forms of logic and mathematics. In the same way, Marx (and also Beverley and Torres), was deeply suspicious of anything but a "scientific" method whose goal was not the recovery of its own ideological origins. This demanded a so-called reproduction of the concrete situation in conformance with a universal truth system (Mitchell, 1986, p. 160). The contrast here to postmodern cartography suggests a profound perceptual shift. Increasingly we see our relationship to the world less as one of subject and object, as observer and observed, and more as mappers of a mutual inter-subjectivity that both accepts and problematizes all views and voices.

2. The auratic perspective draws on Walter Benjamin's (1968) notion of aura as immanent possibility, a moment of profane illumination when the spark of the dialectical has the power to intervene in the traditional bourgeois narrative—or song—of progress, empower emancipatory consciousness and open possibilities for counter mapping and remapping, perhaps radical remapping.

3. Paralogy, or postmodern "science," is represented in our book as a constant search for new ideas and concepts that introduce dissensus into consensus, that render the familiar strange, that open possibilities for new knowledge and a better understanding of our socially constructed world. Lyotard (1984, p. 10) explains paralogy as an agonistic tendency building upon Goedel's theorem and Kuhn's

theory, both of which clash with efficiency, production, and performance criteria favored by technocrats, bureaucrats and the larger system. Citing recent philosophical work in language and pragmatics, Lyotard suggests that language games (as proposed by Wittgenstein, Austin and Searle), with their always unstable and interactive exchange of messages, provide a credible model for mapping science and the social bond today.

REFERENCES

Benjamin, W. (1968). *Illuminations: Essays and reflections*. New York: Schocken.

Burrell, G., and G. Morgan. (1979). *Sociological paradigms and organizational analysis*. Portsmouth, NH: Heineman.

Calhoun, C. (1995). *Critical social theory: Culture history and the challenge of difference*. Cambridge, MA: Blackwell.

Deleuze, G., and F. Guattari. (1989). *A thousand plateaus*. Minneapolis: University of Minnesota Press.

Foucault, M. (1986). Of other spaces. *Diacritics, 16*, 22–27.

Griffin, D.R. (1992). Green spirituality: A postmodern convergence of science and religion. *Journal of Theology, 96*, 5–20.

Haraway, D. (1985). A manifesto for cyborgs. *Socialist Review, 80*, 65–107.

Lefebvre, H. (1991). *The production of space*. Oxford: Blackwell.

Lyotard, J.F. (1984). *The postmodern condition: A report on knowledge*. Minneapolis: University of Minnesota Press.

Mitchell, W.J.T. (1986). *Iconology: Image, text, ideology*. Chicago: The University of Chicago Press.

Paulston, R., and M. Liebman. (1994). The promise of critical social cartography. *La Educación, Inter-American Review of Educational Development, 38*(3), 491–508.

Relph, E. (1991). Post-modern geography. *The Canadian Geographer, 35*(1), 104–105.

Said, E. (1991). *Orientalism: Western conceptions of the Orient*. London: Penguin.

Stone, A.R. (1995). *The war of desire and technology at the close of the mechanical age*. Cambridge, MA: MIT Press.

CONTRIBUTORS

Crystal Bartolovich is an assistant professor of literary and cultural studies at Carnegie-Mellon University in Pittsburgh. Her research interests include the culture of early modern England, as well as Marxist and cultural theory, especially that involving spatial issues and historiography. She is currently completing a book on the role of land surveying in the transition from feudalism to capitalism in early modern England.

John Beverley is a professor in the Departments of Hispanic Languages and Literatures and Communications at the University of Pittsburgh. His recent books include *Literature and Politics in the Central American Revolutions* (1990) with Marc Zimmerman, *Against Literature* (1993), and a collection, *The Postmodernism Debate in Latin America* (1995), coedited with Jose Oviedo and Michael Aronna. He is a founding member of the Latin American Subaltern Studies Group.

Anne Buttimer is professor and head of the Department of Geography at University College, Dublin. She received her Ph.D. in geography at the University of Washington in Seattle and has held research and teaching positions in Belgium, Canada, France, Scotland, Sweden and the US. She is the author of numerous books and articles on subjects ranging from social space and urban planning to the history of ideas and environmental policy. She is chair of the International Geographical Union Commission on the History of Geographical Thought, member of Academia Europaea, and chair of the Irish National Committee for Geography. She has recently coordinated a European Union–sponsored research network on landscape, life and sustainable development with partner teams in Germany, the Netherlands and Sweden.

Christine Fox is president of the Australian and New Zealand Comparative and International Education Society and a vice president of the World

Council of Comparative Education Societies. She has worked extensively in educational planning and development in the South Pacific, and teaches curriculum theory and intercultural and international education at the University of Wollongong, Australia. Her Ph.D. and ongoing research and publications have been in the critical analysis of intercultural communication. Her latest publication, with Robyn Iredale, is *Immigration, Education and Training in New South Wales* (1994).

Esther E. Gottlieb is Director of Program Development, Humanities and Social Sciences in the office of Research and Economic Development at West Virginia University. She has experience in the areas of social development, international education, school and society, policy analysis and interpretive research methods. Her ongoing research has been in the analysis of pragmatic discourses, including the discourses of comparative education and educational reform movements. She has coauthored, with Don Adams, *Education and Social Change in Korea* (1993).

Anne Sigismund Huff is a professor of management studies in the College of Business at the University of Colorado, Boulder. She has an M.A. in sociology and a Ph.D. in management from Northwestern University. Her research interests focus on strategic change. She has carried out cognitive mapping projects at the level of individual decision makers, studied group political processes and considered organization- and industry-level causes and consequences of change. Her publications include *Mapping Strategic Thought* (1990), and the coedited series *Advances in Strategic Management*.

Patti Lather teaches qualitative research in education and feminist pedagogy at Ohio State University. She has a Ph.D. in curriculum, inquiry methodology and women's studies from Indiana University (1983). Her publications include *Getting Smart: Feminist Research and Pedagogy With/In the Postmodern* (1991), articles on the methodological implications of the intersection of the various neo-Marxisms, feminisms and poststructuralisms, and an in-process book, *Troubling Angels: Women Living with HIV/AIDS*.

Martin Liebman has explored social mapping from the time Rolland Paulston launched the form with his "Macro Mapping of Paradigms and Theories in Comparative and International Education Texts Seen as an Intellectual Field." His articles on social mapping, coauthored with Rolland Paulston, have appeared in journals in the US, England and Latin America, as well as been translated into Polish as a chapter for an education text. His work in this book furthers the development of social mapping theory and relates the project to the study of both metaphor and the fields of comparative literature and comparative education.

Christopher Mausolff is a doctoral candidate in public administration at the Graduate School of Public and International Affairs of the University of Pittsburgh. He is the author of "An Economic Analysis of Ecological Agricultural Technologies among Peasant Farmers in Honduras," which was published in the journal *Ecological Economics*. His research interests include organizational culture, organizational learning and program evaluation.

Thomas W. Mouat IV is a doctoral candidate in educational policy at the University of Calgary in Alberta, Canada. Utilizing a transdisciplinary approach, his research aspires to a unified theory of human psychosocial development. He gratefully acknowledges the ongoing financial support for his research provided by the Social Sciences and Humanities Research Council of Canada. He also wishes to acknowledge the inspired supervision of Professor Mathew Zachariah.

JoVictoria Nicholson-Goodman is a doctoral candidate in educational policy studies at the University of Pittsburgh. She has teaching and curriculum experience in social studies and science education. Her professional publications have focused on environmental problems.

Rolland G. Paulston received his master's degree in economic geography from the University of Stockholm and his doctorate in comparative education from Columbia University (1966). He is professor of educational policy studies at the University of Pittsburgh and a past president of the Comparative and International Education Society. He has published extensively on social theory, educational change efforts in Sweden, Peru and Cuba, and on education in social movements. His current research examines how the postmodern turn opens opportunities for remapping educational change theory and practice.

Val D. Rust is a professor of comparative and international education at the University of California, Los Angeles, and a past president of the Comparative and International Education Society. He has published extensively on educational change and reform in Europe. His most recent books include *Education and the Values Crisis in Central and Eastern Europe* (1994), *The Unification of German Education* (1995), and *Toward Education for the Twenty-first Century* (1995).

Joseph R. Seppi is a geologist and landscape planner. Mapping sciences have been the common thread in his research and consulting. In the areas of geographical information systems (GIS) and spatial analysis, he has brought innovative technology to numerous projects, i.e., visualizing data in contaminated groundwater, digital image analysis of satellite imagery for natural resource mapping, network allocation modeling for enhanced 911 dispatching, and visual resource analysis using digital terrain modeling. He is keenly interested in cross-fertilization between the natural and social sciences. Using GIS applications, Seppi helped further the technology transfer for the Social Cartography Project.

Nelly P. Stromquist is professor of international development education in the School of Education, University of Southern California, and a past president of the Comparative and International Education Society. She specializes in gender issues, particularly education and empowerment, adult literacy, and state policies and practices in education for girls and women. She recently authored *Gender and Basic Education in International Development Cooperation* (1994) and is the chief editor of *The Encyclopedia of Third World Women* (1996).

Robert T. Tally, Jr., teaches English at the University of Pittsburgh. His interests include nineteenth- and twentieth-century American literature and literacy theory. He has published reviews of Gilles Deleuze and Gianni Vattimo, and is currently working on a study of the cartographic imagination in Melville's writings.

Carlos Alberto Torres is professor in the Graduate School of Education and Information Studies, University of California, Los Angeles, and director of the Latin American Studies Center. Author or editor of numerous books and research articles in Spanish, Portuguese and English, his latest book is with Raymond Morrow, *Social Theory and Education: A Critique of Theories of Social and Cultural Reproduction* (1995). He is the vice president-elect of the Comparative and International Education Society.

David Turnbull is a lecturer in the social studies of science at Deakin University, Geelong, Australia. He has recently published *Maps Are Territories: Science Is an Atlas* (1993). His current research interests include indigenous mapping, comparative scientific traditions, and sociology of scientific knowledge with a special emphasis on the spatial and cartographic components of knowledge systems.

Subject Index

dialogue, 282, 295, 297–98, 362–63, 419,
        432
difference, xvi, xix, xx, 32–33, 297, 363
        crisis of, 358–59
        made visible, 243–44
        renegotiation of, 228
        and social cartography, 298
        universalized, 40
different similarities, 226
dimensions of humanness, as poetic univer-
        sals, 144
disadvantage, 297
disputatious communities, 322
diversity:
        of mapping methods, 163–64
        and postmodern sensibility, 268, 280,
        287
dominance, 297
domination, 103, 107
donor agencies, 234

Eastern Europe, 38
ecology, 308, 310, 318
ecological debt, 316–17
educational change, 292, 322–23
educational interventions, 224
educational rationalism, 292
emancipatory projects, 359–62
embodied identity, 293
empirical science, 92–98, 99–100
empowerment, 223–24
Enlightenment, 29, 39, 363, 404
environmental education concerns:
        aesthetic perspective, 317–19
        cultural ecology perspective, 315–16
        differential development perspective,
        316–17
        and directions of discourse, 324
        fictive narrative perspective, 317
        historical perspective, 310–11
        intertextual field of, 310–20
        radicalized science perspective, 314–15
        resacralized nature perspective, 319
        relationship aspect, 308
        risk aspect, 308
        science literacy perspective, 311–13
        Science-Technology-Society perspective,
        313–14
        social relations of science and, 311, 313
        sustainable development perspective, 315
environmental education discourse, 307
environmentalism, 107, 318
equilibrium, 47–49
ergon, 151
essentialism, 333, 364–65

Eurocentrism, 387–88
explanations, 422, 424

feminism, 40, 360, 363–64, 428
        post-colonial, 357
feminist academics, 241–42, 365–66, 419
feminist metaphors, 228
feminist theory, 357, 366
figuration, 407–8
fragmentation, 379–80, 387, 390, 406, 417
free-form mapping metaphors, 123
functionalism, 32

gender and development (GAD), 223
Gaia, 141, 152
gender, 364–65
gender issues:
        and economic growth, 230–31
        and educational interventions, 224
        and ideological norms, 232
        limited dialogue on, 235
        spatial approach to studying, 223
gender work:
        and cyberspace, 240
        and NGOs, 240, 244
Geographical Information Systems:
        and spatial organization of knowledge,
        55, 136
        and universal representation, 69
geography, and facets of humanness, 142
global analysis, 386
global and local, 391
global consensus, 315
global environmental problems, 118, 152
global perspectives, representation of, 251
global social structure, 97, 111
        and global education, 107
globalization, 375–76, 390, 391, 392, 393,
        409

habitus, 323
hegemony, 350, 383, 384
heuristic mapping, 14, 421
history, 386
history of cartography, 63, 70, 404
history of social development, 92–101
human subjectivity, 148
human social development, 90–92
humanism, 93, 141, 157
        and cultural self-reflection, 142
        and cyclical drama of social change, 141
        emancipatory role of, 141, 142
        mythopoeic characterization of, 141
        strands of inquiry within, 149–51
humanist metaphors, 142, 144, 145, 153–54

conceptual, 324
and constructed similarities, 192
and excavation of knowledge, 195–96
and framing of certainty, 207
human mapper in, 196–97, 303–4
importance of, 191–92
and incongruent beliefs, 191–92, 419
as interactive metaphor, 192–93
and multiple spaces, 195
phenomenographic, 324
and social centrality, 196–97
and synthesis of social incongruities,
192–93
as visual dialogue, 298, 309, 322
social network theory, 44–47
social spheres, 227
sociality and ecology, 155–57
spatial analysis, 409
spatial practices, 400, 414–15
spatial relations, 390, 401, 412, 414
spatial understanding, 153
state/science/cartography relationship, 53–75,
410
strategies of resistance, 391
structural adjustment programs (SAPs), 231
impact of, on women, 239
structures of violence, 364
subaltern, 347–54, 400, 407, 413, 420
and elite status, 349
representation of, 353–54
silence of, 349–50
studies, 347, 351–52, 354, 407
Subaltern Studies Group, 388–89
subject, the, 380, 399–400, 403, 405–6
sustaining qualifiers, 251
symbolic spaces of capital, 420
synthesis of incongruities, 208–10
systematic cartography, 66, 67
systems of explanation, 38–39, 42–44
systems theory, 42–44
systems types, 43–44

terrain mapping metaphor, 134
textual interpretation, 34, 252
thematic map inventory, 127
theoretical orientations, relations between, 49
theory models, 335–39
totalitarian theories, 31–32, 41
totality, 381, 389, 394, 401, 407–8
totalizing categories, role of, 226
transformation of discourse, 260–61
transnationalization of discourse, 420
transnationalizing capital, 391
truth:
aletheia and, 9

constructed by language, 30
dichtung and, 9–10

universalizing difference, 40
universals, 226

visual dialogue, 309, 322

Weltgrund, 152
Western legacy, 148–49
women as Other, 238
women in development (WID), 223, 235
women NGOs, 236–40
women's advocacy, 236
women's education, 235
women's problems, 235–36
world professoriate, 251
as Leibniz's God, 263

zero-degree writing, 259

Name Index

EDUCATION AND DISABILITY IN
CROSS-CULTURAL PERSPECTIVE
edited by Susan J. Peters

RUSSIAN EDUCATION
Tradition and Transition
by Brian Holmes, Gerald H. Read,
and Natalya Voskresenskaya

LEARNING TO TEACH
IN TWO CULTURES
Japan and the United States
by Nobuo K. Shimahara
and Akira Sakai

EDUCATING IMMIGRANT
CHILDREN
*Schools and Language
Minorities in Twelve Nations*
by Charles L. Glenn
with Ester J. de Jong

TEACHER EDUCATION
IN INDUSTRIALIZED NATIONS
Issues in Changing Social Contexts
edited by Nobuo K. Shimahara
and Ivan Z. Holowinsky

EDUCATION AND
DEVELOPMENT IN EAST ASIA
edited by Paul Morris
and Anthony Sweeting

THE UNIFICATION OF
GERMAN EDUCATION
by Val D. Rust and Diane Rust

QUALITATIVE EDUCATIONAL
RESEARCH IN DEVELOPING
COUNTRIES
Current Perspectives
edited by Michael Crossley
and Graham Vulliamy

SOCIAL CARTOGRAPHY
*Mapping Ways of Seeing
Social and Educational Change*
edited by Rolland G. Paulston

SOCIAL JUSTICE AND
THIRD WORLD EDUCATION
edited by Timothy J. Scrase